Developments in Mineral Processing, 1

MINERAL CRUSHING AND GRINDING CIRCUITS

Their Simulation, Optimisation, Design and Control

Developments in Mineral Processing, 1

MINERAL CRUSHING AND GRINDING CIRCUITS

Their Simulation, Optimisation, Design and Control

A.J. LYNCH

Julius Kruttschnitt Mineral Research Centre, Department of Mining and Metallurgical Engineering, University of Queensland, Brisbane, Qld., Australia

with contributions by:
P.D. Bush, G.A. Gault, M.J. Lees, R.D. Morrison, T.C. Rao, G.G. Stanley, G.W. Walter, W.J. Whiten, R.L. Wiegel

and case studies by staff members of:
ASARCO Inc., Silver Bell Unit, Arizona; Bougainville Copper Ltd., Papua New Guinea; Mount Isa Mines Ltd., Queensland; New Broken Hill Cons. Ltd., New South Wales; Cyprus Pima Mining Comp., Arizona; Renison Ltd., Tasmania.

ELSEVIER SCIENTIFIC PUBLISHING COMPANY

Amsterdam — Oxford — New York 1977

ELSEVIER SCIENTIFIC PUBLISHING COMPANY
335 Jan van Galenstraat
P.O. Box 211, Amsterdam, The Netherlands

Distributors for the United States and Canada:

ELSEVIER NORTH-HOLLAND INC.
52, Vanderbilt Avenue
New York, N.Y. 10017

Library of Congress Cataloging in Publication Data

Lynch, A J
Mineral crushing and grinding circuits.

(Developments in mineral processing ; 1)
Bibliography: p.
Includes index.
1. Ore-dressing plants--Equipment and supplies.
2. Crushing machinery. 3. Milling machinery.
I. Bush, P. D. II. Asarco Inc. Silver Bell Unit.
III. Title. IV. Series.
TN504.L93 622'.73 77-2830
ISBN 0-444-41528-9

Printed in Great Britain

PREFACE

Mathematical modelling and automatic control of mineral crushing and grinding circuits have been studied at the Queensland Julius Kruttschnitt Mineral Research Centre for several years. While much is still to be done, it is considered that knowledge of the subject has reached the point at which a useful monograph may be published.

The emphasis in this monograph is on the simulation and control of industrial machines and circuits. It was with some misgiving that I decided to restrict the discussions on the theoretical aspects of size reduction, size separation and material balances to include only the information which is necessary to understand the simulation and control sections. A complete discussion of these subjects would have required a much longer monograph. However, references for further reading in these cases are cited in the bibliography.

The order of presentation follows an order developed during many series of lectures which have been given to graduate and undergraduate students, plant metallurgists and research scientists during the past ten years. Considerable use has been made of numerical examples to illustrate models which have been discussed. The reason for this is that models are not generally understood until calculations are carried out using these models.

A major objective of comminution is to liberate minerals for concentration processes and a model of mineral liberation is discussed. This discussion is also restricted for the reason mentioned above.

Much of the work on which this monograph is based was carried out as thesis projects by candidates for higher degrees at the University of Queensland while attached to the Julius Kruttschnitt Mineral Research Centre. Their work has been of a particularly high quality and I have been fortunate to have had the opportunity of working with them. The theses are listed in the bibliography.

Case studies are most important in a monograph of this type to illustrate the practical application of the concepts which have been introduced. The case studies which are described cover several different types of grinding circuits, and the Julius Kruttschnitt Mineral Research Centre has been associated at some stage or another with each case study except that in which the development of a control system for the semiautogeneous mill circuit is discussed. This work was carried out jointly by Cyprus Pima Mining Company and Industrial Nucleonics Corporation, and I wish to thank these companies for their ready response to my request for the study.

I am indebted also to the contributors to all other case studies; each different circuit had its own problems and the discussions of why these occurred and how they have been overcome are most valuable.

I would like to thank in particular the management and staff of Mount Isa Mines Limited and MIM Holdings Limited for their continuing encouragement and extensive financial and technical support over the years. Many other companies have also supported the research work through the Australian Mineral Industries Research Association Limited, and I would like to acknowledge this with gratitude. These companies are : ASARCO Inc., BH South Limited, Bougainville Copper Limited, Broken Hill Proprietary Company Limited, Consolidated Gold Fields Australia Limited and subsidiaries, Conzinc Riotinto of Australia Limited and subsidiaries, Electrolytic Zinc Company of Australasia Limited, Foxboro Pty. Ltd., Mount Newman Mining Company Limited, North Broken Hill Limited, Peko Wallsend Pty. Ltd. and subsidiaries, Savage River Mines, and Western Mining Corporation Limited.

My compliments and gratitude are due to Mrs. Barbara Kirkcaldie who typed this manuscript several times, and to Mr. Vince Dooley who prepared all the diagrams, most of them from rough sketches.

A.J. LYNCH

SYMBOL NOMENCLATURE

The meanings of symbols used frequently in the text are given. The meanings of other symbols are given in the text.

Size reduction:

f, **p** size distributions of the feed to and product from a size reduction process expressed as $n \times 1$ matrices

X $n \times n$ matrix describing the breakage process within a size reduction unit

S selection or probability of breakage function, $n \times n$ diagonal matrix

C classification of size-dependent diffusion function, $n \times n$ diagonal matrix

B breakage function, $n \times n$ lower triangular matrix, frequently a step matrix

I unit matrix

v number of breakage events or stages of breakage in the matrix model

F feed rate to the mill

A appearance function in the perfect mixing model; almost synonymous with **B** in the matrix model

R breakage rate function in the perfect mixing model; almost synonymous with **S** in the matrix model

D discharge rate function in the perfect mixing model; almost synonymous with **C** in the matrix model

s size distribution of the mill contents

Size separation:

Q volumetric throughput of a hydrocyclone; units: litres per minute or l/m

P operating pressure of a hydrocyclone; units: kilopascal or kpa

Inlet, *VF*, *Spig* diameters of the hydrocyclone inlet and outlets (vortex finder and spigot); units: centimetres or cm.

FPS, *FPW* percent solids and percent water (by weight) in the hydrocyclone feed slurry

WOF, *WF* mass flow rates of water in the hydrocyclone overflow (fine product) and feed; units: tonnes per hour or t/h

CONTENTS

CHAPTER 1

Introduction

Size reduction is an important step in many of the processes by which raw materials are converted into final products. The quarrying, metal and cement production industries make extensive use of size-reduction processes on a large scale, and many secondary industries such as the paint, food and pharmaceutical industries also involve size reduction at some stage. In the mineral field alone the average annual tonnage of ores which were processed by crushing and grinding in the early 1970's was approximately 2500 by 10^6, but even this is small compared with future requirements.

There will be major increases in metal and mineral production in future years and this may be illustrated with reference to copper. The average annual growth rate for copper over many years has been 4.45%, and 6.3 by 10^6 tonnes of copper were produced from ores in 1971. If this growth rate continues, the requirement for copper in the year 2000 will be approximately 24 by 10^6 tonnes. The rate may not be sustained due to more efficient use of copper and to substitution by other materials, and some of the requirement may be met by production from other sources, such as recycling of wastes and in-situ leaching, but it is probable that at least 12 by 10^6 tonnes will be required from ores which require fine grinding and flotation. Approximately 300 by 10^6 tonnes of copper ore were processed in 1971 and since the average grade of ore to be processed in 2000 will be less than the average grade in 1971, it may be estimated conservatively that the tonnes processed in 2000 will be of the order of 750 by 10^6 tonnes. There are now several plants in various areas of the world with capacities in the range of 50—100 by 10^3 tonnes per day and by the year 2000 it can be expected that plants with capacities exceeding 250 by 10^3 tonnes per day will be common. The development and proving of an entirely new method of extracting metal from ore may alter this picture but this is unlikely, particularly in the short span of 25 years.

This argument concerning plant capacities also applies to other minerals; for instance, the growth rate for iron and steel has been quoted as 6% and aluminium as 8%, and although the grades of iron ore and bauxite from which these metals are extracted may remain fairly constant, the required increases in size-reduction capacity will still be large.

Thus the growing importance of size reduction as a unit operation must be recognised.

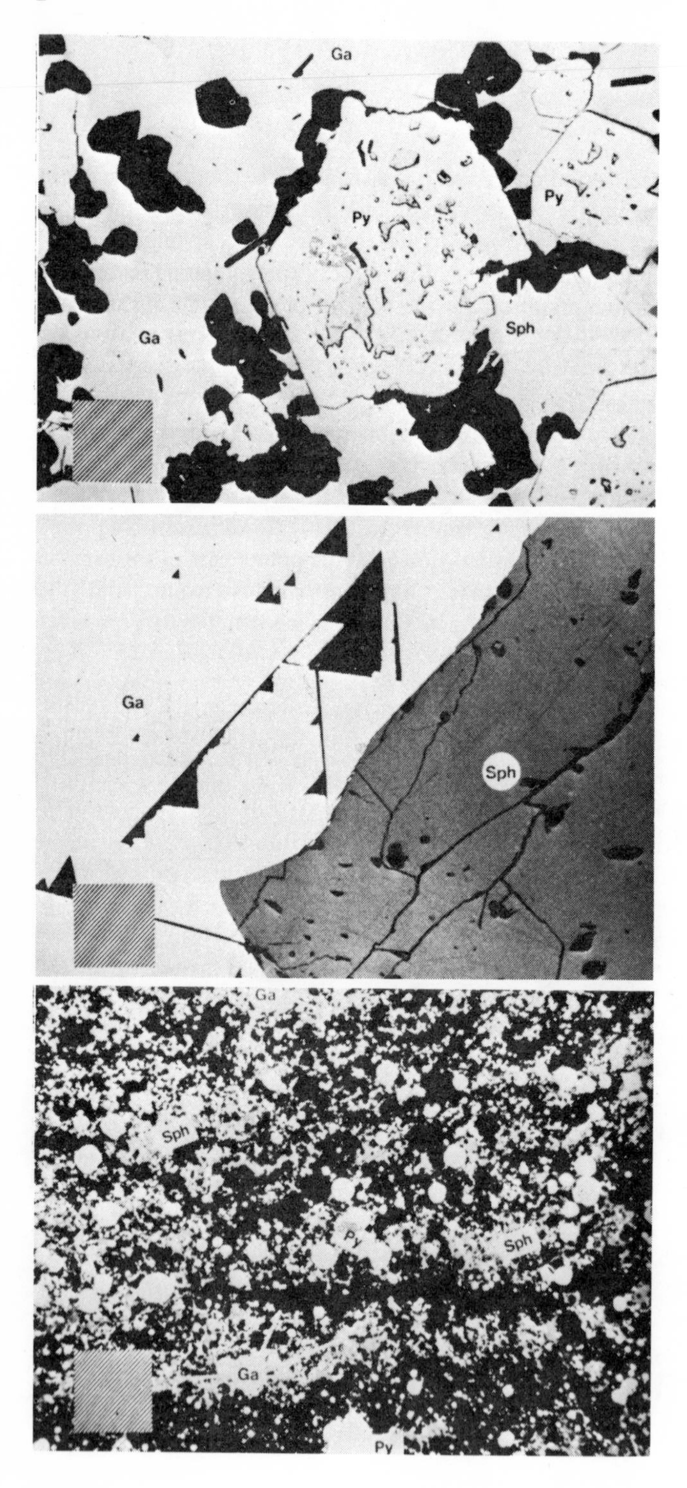
Ga
Py
Py
Sph
Ga
Ga
Sph
Ga
Sph
Py
Sph
Ga
Py

1.1 SIZE-REDUCTION CIRCUITS

The purpose of reducing the size of solid particles may be:

(1) to liberate valuable minerals from waste prior to concentration, such as in the treatment of metalliferous ores by flotation;

(2) to increase the surface area available for chemical reaction, such as in the reaction of limestone with silica and other minerals in a rotary kiln; or

(3) to produce mineral particles or dimension stone of required size and shape.

In the case of size reduction prior to concentration, the extent to which breakage must proceed depends on the fineness of intergrowth, or the "natural grain size" of the valuable particles. The natural grain size may vary widely for the same mineral in different ores and this is shown for galena in ores from Broken Hill, Mount Isa and MacArthur River in Australia in Fig. 1-1. The Mount Isa ore must be ground much more intensively for adequate liberation than the Broken Hill ore while adequate liberation of galena in the McArthur River ore is difficult to obtain.

Size-reduction processes generally involve several stages in series as shown in Fig. 1-2. It is common to include concentration machines within comminution circuits either to prevent "over-grinding" of valuable particles or to reject waste particles from the circuit as soon as they are produced in liberated form. Prevention of over-grinding is important because the efficiency of concentration processes may vary significantly with particle size, as shown in Fig. 1-3, while rejection of particles from the circuit at as coarse a size as possible reduces grinding costs. The flow diagram shown in Fig. 1-2 illustrates how particles may be extracted by flotation at a coarse size and only the partially liberated, slow floating particles broken further thus reducing over-grinding of other particles. The flow diagram of a taconite (magnetite-silica) circuit, from which particles are rejected as coarse as possible, is shown in Fig. 1-4.

Another objective of size-reduction circuits is to produce particles of a required size, for instance, aggregate for cement or road building. Crushing of iron ore for blast furnace feed may be included in this category because the crushing plants are usually a considerable distance from the blast furnaces and operate independently of them. A simplified flow diagram of an iron-ore crushing circuit is shown in Fig. 1-5.

The machines used in size-reduction circuits are described briefly below.

Fig. 1-1. Range of variation in natural grain size which can occur with metalliferous ores. Lead—zinc ores from Mount Isa (top), Broken Hill (centre) and McArthur River (bottom) are shown. The side of the hatched square in each case is 44 μm. *Ga* represents galena, *Sph* sphalerite and *Py* pyrite.

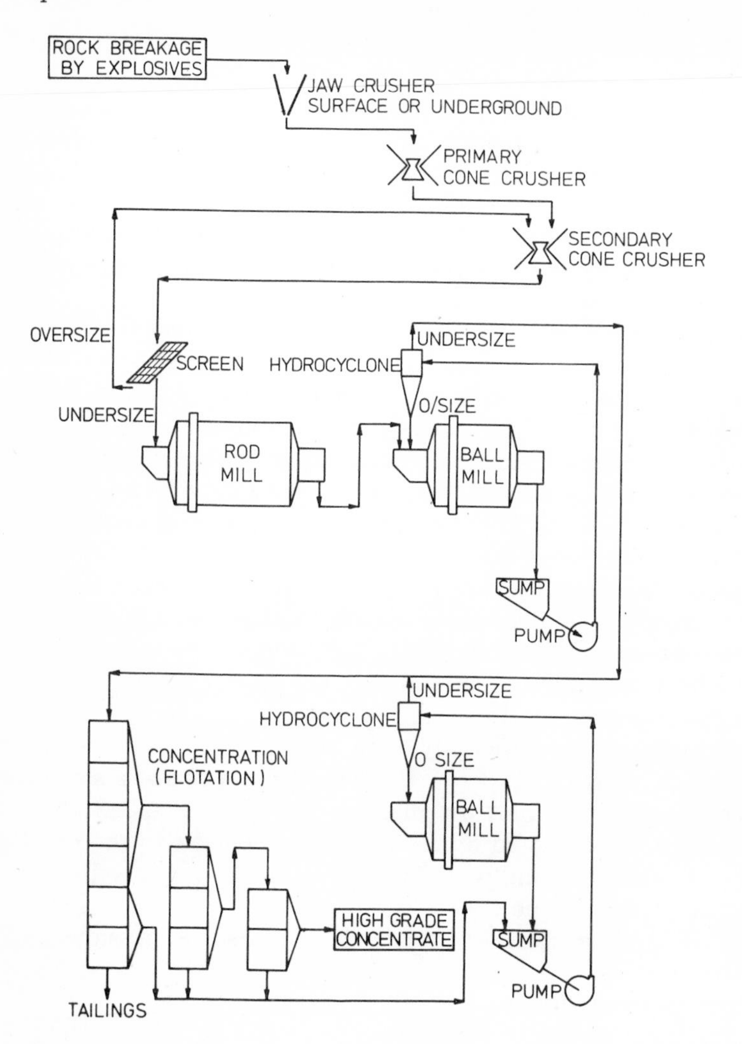

Fig. 1-2. Flowsheet of a typical size-reduction and concentration circuit.

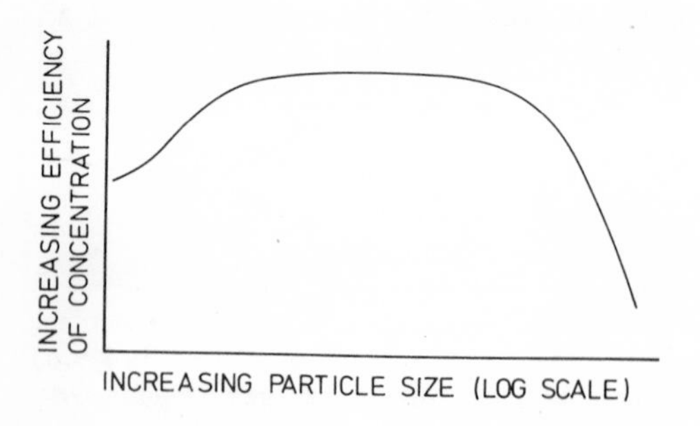

Fig. 1-3. Efficiency of concentration processes as a function of particle size.

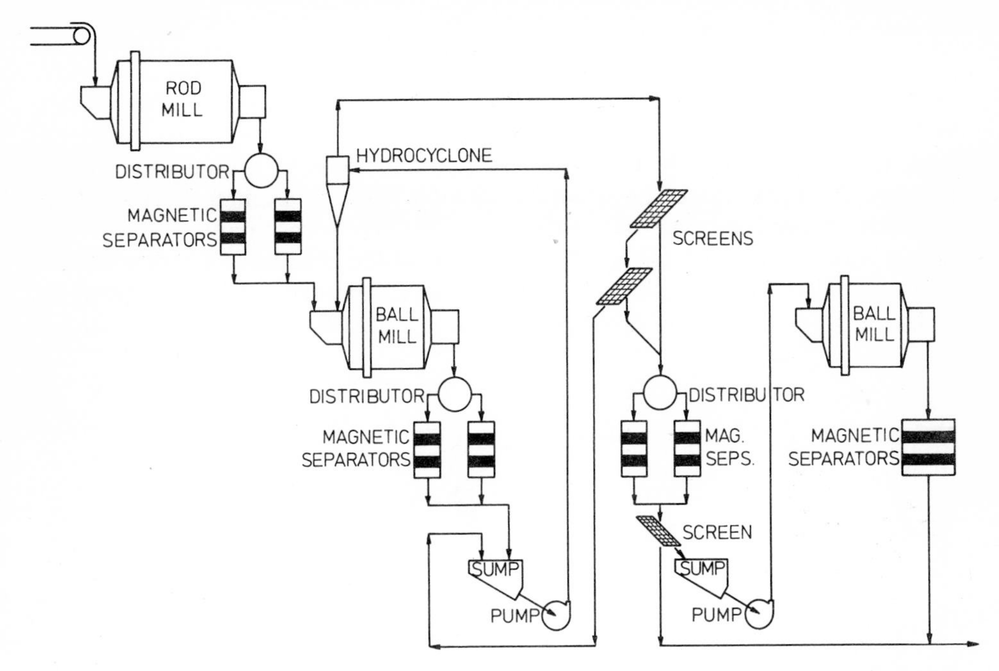

Fig. 1-4. Flowsheet of a taconite concentration circuit. Only the concentrate from each magnetic separator proceeds to further processing, the tailing is rejected from the circuit.

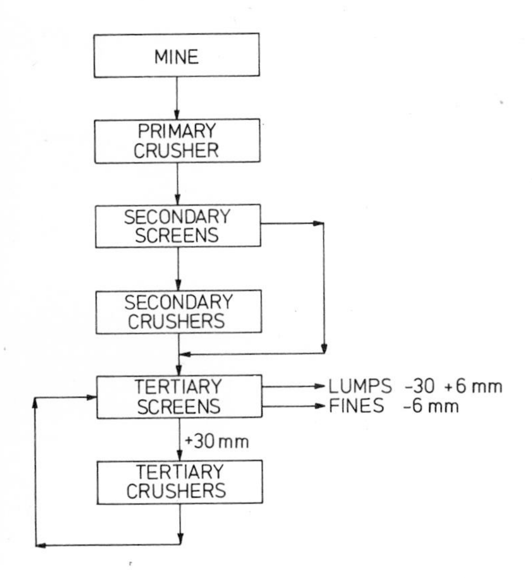

Fig. 1-5. Flowsheet of a typical iron-ore crushing circuit.

1.2 SIZE-REDUCTION MACHINES

Machines must be designed according to the task they are required to perform, and this changes as the average particle size decreases. When this size is large there are relatively few particles per unit mass of ore and the energy required for breakage of each particle is high. When it is small there are large numbers of particles per unit mass of ore but the energy required for breakage of each particle is low. Thus, a machine which is suitable for breakage of large particles will not generally be suitable for breakage of small particles.

Jaw crushers and cone crushers, shown in Fig. 1-6, are used for the break-

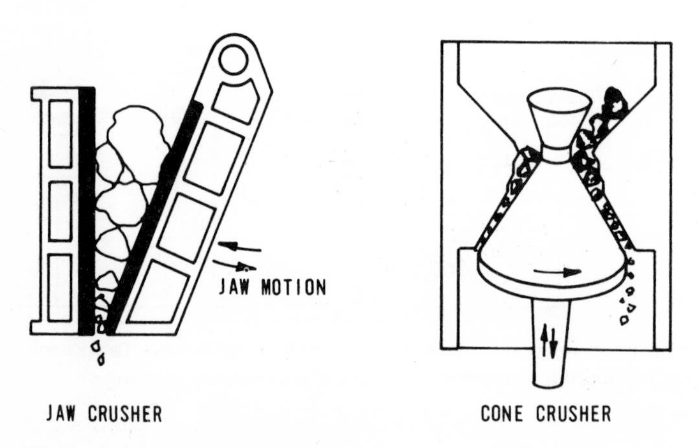

Fig. 1-6. Jaw and cone crushers.

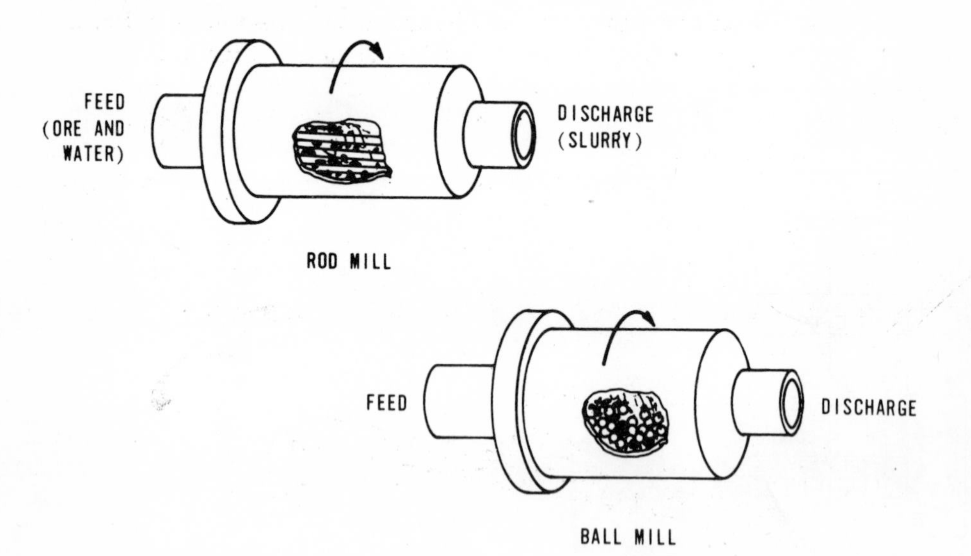

Fig. 1-7. Rod and ball mills.

age of rocks and coarse particles larger than 3 cm in diameter, while rod mills and balls mills, shown in Fig. 1-7, are used for the breakage of smaller particles. The energy consumption in size-reduction machines is related to the hardness of the particle, the initial size and the reduction achieved.

1.3 SIZE-SEPARATION MACHINES

Different types of machines are used depending on the size of the particles to be separated. Vibrating screens, shown in Fig. 1-8, can be used to separate

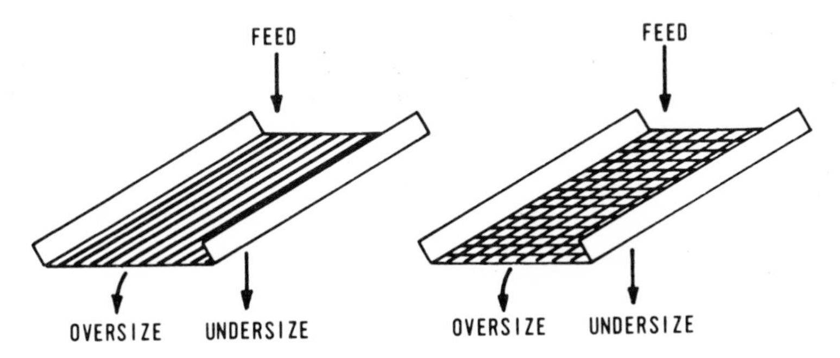

Fig. 1-8. Bar and wire-mesh screens.

particles down to about 0.25 cm in diameter but below this, screening is inefficient and size-separation processes are based on the differences in the settling rates of particles in fluids. This is generally called classification.

A particle settling in a fluid under the influence of gravity experiences a resistance and attains a maximum velocity — the terminal settling velocity — when the resistance becomes equal to the effective weight of the particle. The terminal velocity is related to particle size and density. If the fluid has an upward velocity intermediate between the terminal velocities of the particles to be separated, it is possible to divide the particles into those which move with the fluid and those which move against it. This is the principle which is used in particle separation by hydraulic classification.

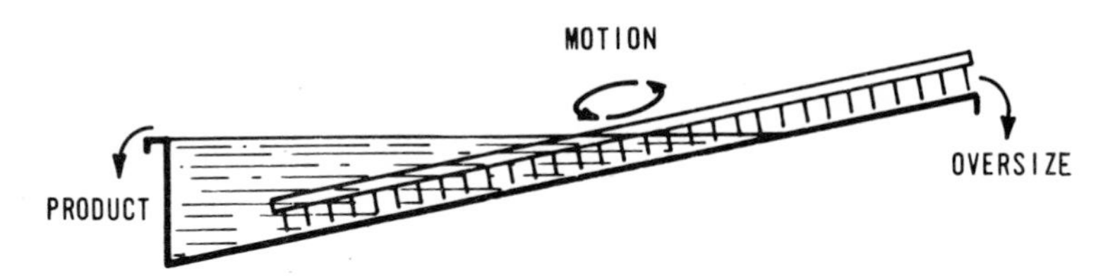

Fig. 1-9. Rake classifier.

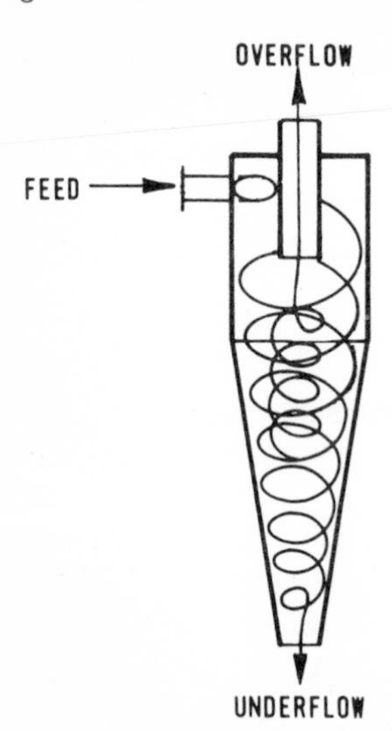

Fig. 1-10. Hydrocyclone.

Fig. 1-11. Mount Newman Mining Company Limited iron-ore crushing installation at Newman, Western Australia. View shows the parallel crushing and screening lines and the train loading facility. The rail line to the port, 426 km away, may be seen in the background.

Fig. 1-12. Mount Isa Mines Limited chalcopyrite ore processing line. Buildings include, from top, the coarse-ore bins, crushing plant and concentrator. The three bays in the concentrator include the fine-ore bins, grinding circuits and flotation circuit.

The operation of classifiers which use gravity as the separating force is shown in Fig. 1-9 and of hydrocyclones, in which centrifugal force is used as the separating force, is shown in Fig. 1-10.

Photographs illustrating different aspects of crushing and grinding installations are given in Fig. 1-11 to 1-15.

1.4 CIRCUIT DESIGN AND OPERATION

Problems in achieving maximum efficiency in a size-reduction circuit may occur at three stages:

Fig. 1-13. Crushing plant at Mount Isa Mines Limited.

(1) at the design stage when the optimum size-reduction flowsheet must be selected with respect to the types, numbers and sizes of processing units;

(2) at the operating stage when the correct values must be selected for those variables which may be altered while the circuit is off-line, but which must be constant while the circuit is on-line, for example, the inlet- and vortex-finder diameters of the cyclones; and

(3) under conditions of continuous operation when changes in the circuit feed must be compensated for, otherwise undesirable changes in the circuit product will occur.

The first two problems are concerned with optimisation, that is, with making the best choice from many possibilities, and the last with control.

Fig. 1-14. Grinding circuits at Mount Isa Mines Limited. Rod mills on left, ball mills on right, regrind ball mills in background.

Some typical questions which arise when the optimisation problem is considered are:

(1) for the crusher station, what is the relationship between the feed and product sizes for changes in screen aperture, crusher setting, feed rate or feed size?

(2) for the rod or ball mill circuits, what is the feed-size—product-size relationship for changes in feed rate or feed size, or classification conditions?

(3) what is the optimum flow chart for a circuit which includes several mills and classifiers? and

(4) what is the effect of using one large, rather than two small units for the same duty?

Perhaps the best approach to the solution of optimisation problems is to use simulation techniques to investigate the results of the alternatives which may be available. To "simulate" means "to feign or resemble", and in this context simulation will be regarded as meaning representation by mathematical models. The reason for using simulation techniques for the optimisation of grinding-circuit performance is that they are faster, less expensive, and

Fig. 1-15. The grinding line at Bougainville Copper Limited. The ball mills are the largest (5.49 m diameter by 6.41 m long) currently in use (1976).

more accurate than experimental techniques provided that appropriate mathematical models are used. Ideally, the models should include parameters which represent the unique physical characteristics of the ores and the machines, but the physical complexities of the processes are such that it has not yet been possible to do this. Only in the case of the simplest processes, such as single impact mills operating on a single layer of identical particles under conditions which are devised and controlled to preclude rebreakage of fragments from primary breakage, is it possible to be explicit about the physical sequence of events which transform the feed into the product. In the case of crushers, the sequence of events is complex, while in the case of tumbling and vibrating mills the events which cause the transformation of the feed into the product are understood only qualitatively. Consequently, it is not yet possible to write models of these processes which take all the physical factors into account in a manner which is completely correct, and the development of models completely from first principles is not possible.

However, less exact models of the processes are sufficient for purposes of process simulation provided that they allow prediction of the product characteristics for known feed characteristics and values of the operating and design variables. Such models have been developed and are discussed in this book.

The third problem mentioned above is concerned with the control of operating circuits on a minute-to-minute basis so that no matter how the ore entering the system varies, or the physical conditions of the processing machines change, the product characteristics remain at or close to the required characteristics. Efficient control of operating circuits by manual methods only, is difficult and some means of automatic control is necessary if good circuit performance is to be obtained. The dynamic simulation of circuit operation by the use of mathematical models is useful in developing and evaluating possible control schemes.

1.5 SCOPE OF THIS MONOGRAPH

The first part of this book is concerned with the development of process models, the second part with their use in the simulation of industrial size-reduction circuits on a digital computer, and the third part with the automatic control of these circuits. Some additional chapters are included to provide useful background information. For instance, energy—size-reduction relationships are discussed briefly in Chapter 2, and the movement of particles in fluids — which is important in hydraulic classification — is discussed in Chapter 5.

CHAPTER 2

Fundamentals of comminution

2.1 INTRODUCTION

For almost a century, the process of size reduction was studied in terms of the energy consumed during the operation of a grinding mill. This was a logical starting point because size reduction is responsible for a large proportion of the costs of ore treatment and the energy consumed is the major cost in size reduction. This basis of investigation was influenced more by the economics of the operation than by any other factor.

While metallurgists are still interested in reducing costs, they have now approached the problem in a different way, firstly by studying the process of size reduction itself, and secondly by obtaining mathematical relationships linking the operating variables.

Various aspects of these two approaches are discussed in this chapter.

2.2 ENERGY—SIZE-REDUCTION RELATIONSHIPS

2.2.1 *The general form of the relationship*

Early investigations aimed at the better understanding of the comminution process were concerned with the relationship between the energy consumed by a grinding mill and the amount of size reduction that the consumption of this energy brought about. Size reduction was studied as a function of: (a) the amount of new surface area of particles produced; (b) the volume of material broken; and (c) the diameter of the product particles.

It was observed experimentally that, in a size-reduction process, the small size change produced was proportional to the energy expended per unit weight of particles, and that the energy required to bring about the same relative size change was inversely proportional to some function of the initial particle size. The relationship between energy and breakage may be expressed in the equation:

$$\mathrm{d}E = -K \cdot \mathrm{d}x/x^n \tag{2-1}$$

Various workers have given different interpretations of this relationship. Thus, Rittinger (1867) suggested that the new surface area produced is proportional to the energy consumed and it may be noted that for spherical particles of a given diameter, the surface area per unit weight is inversely proportional to the diameter. Kick (1885) suggested that the same relative reduction in volume is obtained for constant energy input per unit mass irrespective of the original size. Both relationships may be derived from eq. 2-1 by substituting 1.0 and 2.0 for the value of n and integrating the differential equation. The resulting equations are as follows:

(1) Kick's equation:

$$E = K \cdot \ln(x_1/x_2) \tag{2-2}$$

(2) Rittinger's equation:

$$E = K \cdot (1/x_2 - 1/x_1) \tag{2-3}$$

A third relationship proposed by Bond (1952) is as follows:

$$E = 2 \cdot K \cdot (1/\sqrt{x_2} - 1/\sqrt{x_1}) \tag{2-4}$$

and this may be derived by substituting 1.5 for the exponent of x. The proportionality constant in this equation was defined by Bond as the "Work Index" and is now considered to be a function of the particle size. The relationship given by Hukki (1961) for the general form of the energy—size-reduction relationship:

$$\mathrm{d}E = -K \cdot \mathrm{d}x/x^{f(x)} \tag{2-5}$$

is a better description of the dependence of required energy on particle size than is eq. 2-1. This equation indicates that the constants of proportionality for the equations of Kick and Rittinger will also vary with particle size.

Much controversy arose about the hypotheses of Kick and Rittinger years after these were published as other workers produced results to satisfy either one or the other, and discussion increased when Bond published his "third theory" in 1952. The energy—particle-size theory of Bond may be considered as an empirical method of grouping a mass of industrial and laboratory results that are obtained so that some degree of extrapolation and interpolation is possible for known materials and machines, using the "Work Index" value.

Much has been published on studies of energy—size-reduction relationships (Charles, 1957; Schuhmann, 1960), and the validity of various methods for defining these relationships has been compared and reviewed (Austin and Klimpel, 1964; Harris, 1966). In general, the proposed relationships are only valid over limited ranges of variables in specific cases. It is true, however, that the energy required for equal size reduction of particles does increase as the size of particle being broken decreases and the various hypotheses outlined above may be compared with observed energy—size relationships as shown in Fig. 2-1 (Hukki, 1961).

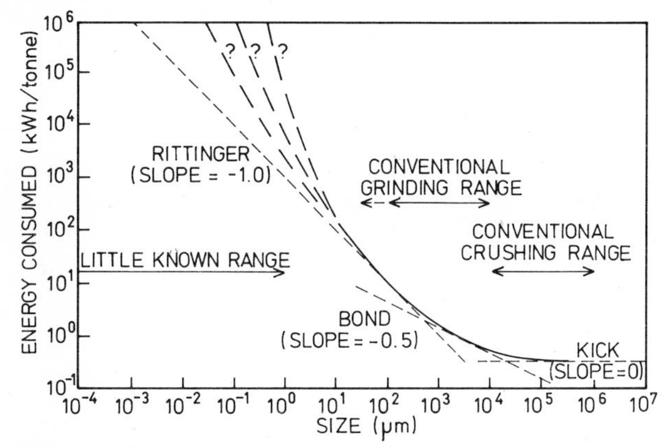

Fig. 2-1. Relationship between energy input and particle size in comminution. (After Hukki, 1961).

2.2.2 *Limitation of the energy approach*

At first sight, the consideration of energy input as a function of the grinding system was very attractive but it was much more complicated than originally realised. Not all of the energy supplied to a tumbling mill is dissipated in the breakage of particles and the power consumption of a mill may be almost independent of the incidence of fracture. A rotating ball mill is best considered as an unbalanced flywheel. It approaches its greatest degree of unbalance when it is merely rotating the dry tumbling media and is doing no useful work. In this condition, maximum power is required to keep it rotating. When pulp is being fed into the mill, it is more balanced and the power required to drive it decreases. The power required to drive a ball mill does not change greatly with a change in feed rate of material to the mill, although the amount of breakage taking place may change considerably.

Large energy losses occur in the transmission driving the mill in the form of friction and sound. While these losses may be determined, there are other losses within the mill itself that are impossible to assess. These include energy losses as a result of particle friction without breakage, kinetic and potential energy losses, elastic and plastic deformation of particles and the production of sound. The elastic and plastic deformation of particles may or may not influence the amount of energy required for breakage when favourable breakage conditions do occur. Much of the loss dissipated in the form of heat does not play any part in particle breakage. It is quite possible, however, that input energy is converted to another form as a necessary by-product of the breakage process. The production of sound and heat may be necessary during any breakage operation as shown by the relationship:

large particle + energy → smaller particles + sound + heat

If this is true, it would not be correct to subtract all sound and heat energy produced from the total energy to a breakage process in order to obtain a net energy value. Beke (1964) quoted a figure of 0.6% as being the amount of input energy used in "theoretical" size reduction. Austin (1964) gave a value of less than 3% as the proportion of the total energy used for this purpose. It is generally agreed that the energy consumed in the actual breakage operation is low compared with the total energy consumed. Since insufficient is known about the internal energy balance of a tumbling mill, it is impossible to determine this figure with any pretension to accuracy whatsoever for such a mill.

2.2.3 *Conclusion*

Comminution is best considered as a result of the mechanical operation of the mill. This mechanical operation consumes the energy and the size reduction is an indirect result of energy consumption. It is considered that in general the energy—size-reduction relationships do not suitably define the process of size reduction. The relationship between feed to and product from a mill is the necessary approach and can be determined more directly. Effects of the operating variables, such as solids feed rate, mill size and mill speed, on this relationship are the factors that require understanding.

In any quantitative discussion of size reduction, the variable of major importance is the size distribution. Consequently, the methods of measuring and expressing size distributions are discussed briefly in this chapter in order to provide background information to a detailed discussion of feed-size—product-size relationship in Chapter 3.

2.3 MEASUREMENT OF THE SIZE DISTRIBUTION OF BROKEN MATERIAL

An irregularly shaped particle has no unique dimension and its size is usually expressed in terms of the diameter of a sphere that is equivalent to the particle with regard to some stated property. Several definitions of particle size based on an equivalent sphere are given below (Allen, 1968).

1. *Surface diameter*, d_s, is the diameter of a sphere having the same surface area as the particle.

2. *Volume diameter*, d_v, is the diameter of a sphere having the same volume as the particle.

3. *Projected area diameter*, d_a, is the diameter of a sphere having the same projected area as the particle when viewed in a direction perpendicular to the plane of stability.

4. *Projected perimeter diameter*, d_p, is the diameter of a sphere having the same projected perimeter as the particle when viewed in a direction perpendicular to the plane of stability.

5. *Drag diameter*, d_d, is the diameter of a sphere having the same resistance to motion as the particle in a fluid with the same viscosity and at the same velocity.

6. *Free-falling diameter*, d_f, is the diameter of a sphere having the same density and the same free-falling speed as the particle in a fluid of the same density and viscosity.

7. *Stokes' diameter*, d_{St}, is the free-falling diameter in a laminar flow region (Re = 0.2), $d_{St} = (d_v^3/d_d)^{1/2}$.

8. *Sieve diameter*, d_A, is the width of the minimum square aperture through which the particle will pass.

9. *Volume specific surface diameter*, d_{vs}, is the diameter of a sphere having the same ratio of surface area to volume as the particle, $d_{vs} = d_v^3/d_s^2$.

In addition there are two statistical diameters, namely Martin's M and Feret's F. These are statistical equivalents of the projected area diameter, d_a, and the projected perimeter diameter, d_p, respectively.

For irregular particles, the measured size usually depends on the method of measurement. For example, sedimentation and elutriation analyses are expressed in terms of Stokes' diameter, d_{St}, sieves provide the sieve diameter, d_A, while microscopes may yield the surface volume, projected area and projected perimeter diameters as well as Martin's and Feret's diameters. The particle-sizing technique chosen should measure some dimension of the particle which is important in the process in which the particle is involved. This may provide problems where different units in the process are affected by different measurements such as a ball mill and a hydrocyclone in a closed grinding circuit.

However, as Allen (1968) pointed out, the equivalent sphere diameters can be related to each other. The ratio of any pair of the above diameters is found to be fairly constant over quite wide size ranges for any material derived from the same source or produced in the same way. This enables the correlation of analyses using more than one method, and the comparison of the results from such instruments.

In general, conventional sieve sizes will be used in this monograph when specifying a size distribution. The common sieve sizes are given in Table 2-1.

2.4 REPRESENTATION OF THE SIZE DISTRIBUTION OF BROKEN MATERIAL

Feed to and product from comminution can consist of particles of a large number of sizes and the range of possible particle sizes varies from submicron dimensions to an upper dimension dependent on the previous history of the material.

The first step necessary to understand the factors affecting the comminution process is to devise a method for the accurate mathematical representation of the distribution of particle sizes in the feed and product. Various methods for this are discussed below.

TABLE 2-1

Aperture sizes for some sets of testing screens

Tyler screens				British Standard screens			
mesh	aperture		mesh	mesh	aperture		mesh
	(inches)	(mm)			(inches)	(mm)	
	2.968						
	2.496						
	2.100						
	1.766						
	1.484						
	1.248						
	1.050						
	0.883						
	0.742						
	0.624						
	0.525						
	0.441						
	0.371						
	0.312	7.925	$2\frac{1}{2}$				
3	0.263	6.680					
	0.221	5.613	$3\frac{1}{2}$				
4	0.185	4.699					
	0.156	3.962	5				
6	0.131	3.327		5	0.1320	3.34	
	0.110	2.794	7		0.1107	2.81	6
8	0.093	2.362		7	0.0949	2.41	
	0.078	1.981	9		0.0810	2.05	8
10	0.065	1.651		10	0.0660	1.67	
	0.055	1.397	12		0.0553	1.40	12
14	0.046	1.168		14	0.0474	1.20	
	0.039	0.991	16		0.0395	1.00	16
20	0.0328	0.833		18	0.0336	0.85	
	0.0276	0.701	24		0.0275	0.70	22
28	0.0232	0.589		25	0.0236	0.60	
	0.0195	0.495	32		0.0197	0.50	30
35	0.0164	0.417		36	0.0166	0.421	
	0.0138	0.351	42		0.0139	0.353	44
48	0.0116	0.295		52	0.0116	0.295	
	0.0097	0.246	60		0.0099	0.252	60
65	0.0082	0.208		72	0.0083	0.211	
	0.0069	0.175	80		0.0070	0.177	85
100	0.0058	0.147		100	0.0060	0.152	
	0.0049	0.124	115		0.0049	0.125	120
150	0.0041	0.104		150	0.0041	0.105	
	0.0035	0.088	170		0.0035	0.088	170
200	0.0029	0.074		200	0.0030	0.076	
	0.0024	0.061	250		0.0026	0.065	240
270	0.0021	0.053		300	0.0021	0.053	
	0.0017	0.043	325				
400	0.0015	0.037					
	0.0012	0.030					
	0.00106	0.026					

2.4.1 *Continuous functions*

The sizes of the mineral particles can be represented by a continuous function defining the frequency with which they are present in any infinitesimal interval of particle-size range. Numerous continuous functions have been postulated to describe the size distribution of products of comminution. These have been dealt with fully by Fagerholt (1945). Each of the functions may be regarded as a special case of the more general function:

$$W(x) \cdot \mathrm{d}x = a \cdot x^m \cdot \mathrm{e}^{-b \cdot x^n} \cdot \mathrm{d}x \tag{2-6}$$

where $W(x)$ is the weight of particles of size x and a, b, m and n are parameters.

A theoretical treatment of the breakage of single particles was carried out by Gilvarry (1961) and Gaudin and Meloy (1962). Gilvarry has shown theoretically that:

$$y = 1 - \exp\,[-(\gamma \cdot x) - (\gamma_s \cdot x)^2 - (\gamma_v \cdot x)^3] \tag{2-7}$$

where γ, γ_s and γ_v are measures of the activated edge, area, and volume flaw densities, respectively, and y is the weight or volume fraction passing size x. This equation may be reduced to the form:

$$y = 1 - \exp\,(-b \cdot x^m) \tag{2-8}$$

when the assumption is made that edge flaws are predominant in the fracture process. This is the Rosin-Rammler (1933) equation. The Gaudin-Schuhmann distribution:

$$y = (x/k)^\alpha \tag{2-9}$$

is derived from the Gilvarry equation when x is small. Gaudin and Meloy derived the following form for the impact-fracture distribution equation:

$$y = 1 - (1 - x/x_o)^r \tag{2-10}$$

where r is a measure of the number of cracks in the crystal and x_o is the size of the original specimen. Gilvarry's law requires a knowledge of edge, facial, and volume flaws while that of Gaudin and Meloy requires that the number of micro-cracks cutting unit length of line segment be evaluated. A similar knowledge of micro-cracks was required for the theoretical equation derived by Bennett (1936).

A full discussion of the methods of expressing size distributions is beyond the scope of this book and it is sufficient to say that the use of these types of distribution functions is limited. They either refer to a portion only of the total size distribution or they require information that is difficult to obtain. However, the general method of representing the size distributions of particles by an equation is much more desirable than that used by Bond which

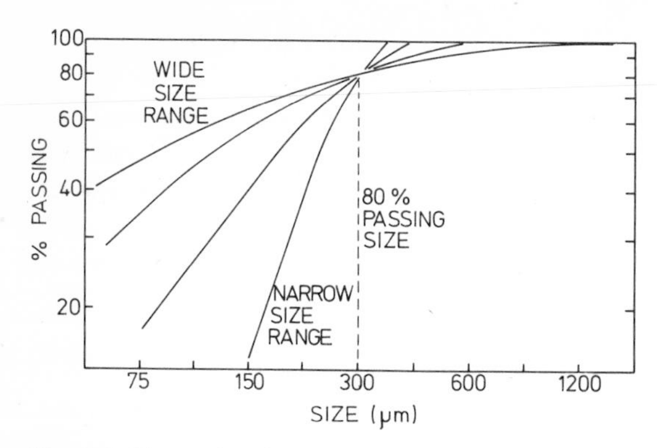

Fig. 2-2. Example of a single size representing several size distributions.

involved representing a size distribution by a mean defined by the sieve size at which 80% by weight of the material passes and 20% is retained. By this latter method, it is possible to represent many different size analyses by one value and, as shown in Fig. 2-2, this is inadequate.

For the complete description of a size distribution, it is necessary to use either the full size analysis or one of the continuous functions mentioned above. It was this problem of representing feed and product size analyses in a manner suitable for use in mathematical relationships that led to the use of matrix methods for such representations.

2.4.2 *Discontinuous functions*

Broadbent and Callcott (1956) used a convenient method of representing the size distribution of broken materials. They defined the continuous size distribution $W(x)$ at a set of points x_0, x_1, x_2, . . ., x_l, and represented the difference between consecutive points by a column vector, which was an $l \times 1$ matrix. The size-distribution curve was approximated to a series of straight lines between points of definition. The greater the number of such points considered, the closer was this approximation to the original curve. If more than four intervals were considered over the curve, the method was found in general to be of sufficient accuracy. The size distribution of broken material was conveniently represented since, if the points of definition correspond to screen sizes, the result is a size analysis of the particles, with the fraction retained between consecutive screens forming the matrix.

This method of representation has the following advantages: (1) the whole size distribution is described; (2) no approximations are made in order to fit a continuous function; (3) the values can be read directly without further graphical or mathematical manipulation; (4) all size distributions may be represented in this manner; (5) the use of this type of representation is very suitable for manipulation on a digital computer.

It is desirable to form a constant ratio between the points of definition such that $x_0/x_1 = x_1/x_2 = \ldots = x_{(l-1)}/x_l$. This is equivalent to the use of screens whose apertures are in a geometric progression. If screens of this nature are not used then adjustment can be made to the $W(x)$ values. This is done by plotting the function $W(x)$ against x in a convenient manner and reading off the $W(x)$ values at the desired intercepts. The resulting column vector has the form:

$$\begin{array}{c} 1 - W(x_1) \\ W(x_1) - W(x_2) \\ W(x_2) - W(x_3) \\ \cdots \\ \cdots \\ \cdots \\ W(x_{l-1}) - W(x_l) \end{array}$$

2.4.3 *Statistical functions*

Statistical functions have been found to be suitable to describe only a limited number of size distributions and are inadequate for general use. They are mentioned below for purposes of completeness and are discussed in textbooks on statistics, for instance, Weatherburn (1961).

Normal distribution. This distribution occurs when the measured value of some property of a system is determined by a large number of small effects, each of which may or may not operate.

The equation representing the normal distribution is:

$$N = [(\Sigma N) \cdot \exp\{-(d - \bar{d})^2/2s^2\}]/s \cdot \sqrt{2\pi} \qquad (2\text{-}11)$$

where $\bar{d}$ = mean diameter = $\Sigma(N \cdot d)/\Sigma N$; s = standard deviation = $[\Sigma N \cdot (d - \bar{d})^2/\Sigma N]^{1/2}$; N = number of measurements (in the present case N is the number of particles falling between sizes d and $\bar{d}$).

This function is applicable to biologically occurring systems though many actual distributions are skewed towards the larger sizes.

Log-normal distribution. This distribution applies to asymmetric distributions of the type that are produced by crushing. A useful property of this function is that if the number distribution is log-normal so are the surface and volume distributions.

The equation for the log-normal distribution is given as:

$$N = [(\Sigma N) \cdot \exp\{-(\ln d - \ln \bar{d})^2/2 \cdot (\ln s)^2\}]/(\ln s \cdot \sqrt{2\pi}) \qquad (2\text{-}12)$$

where $\ln \bar{d}$ = geometric mean diameter = $\Sigma(N \cdot \ln\ d)/\Sigma N$; $\ln s$ = geometric standard deviation = $[\Sigma N \cdot (\ln d - \ln \bar{d})^2/\Sigma N]^{1/2}$.

2.5 PARTICLE BREAKAGE

2.5.1 *Breakage as a single event*

The theoretical derivation of the size-distribution equations depends on the statistical analysis of the breakage of single particles. The operation of size-reduction machines is such that particles are not only subjected to simple primary breakage but also undergo a considerable amount of subsequent breakage. Because of this, it is not possible to explain industrial comminution processes as single-event operations. They must be considered as repetitive in form and composed of a summation of numerous single events, influenced by the operating mechanics of the size-reduction machine. The equations of Gilvarry (1961), Gaudin and Meloy (1962) and others may be used to explain the breakage portion of the process but the description of the complete process requires a more complex treatment.

2.5.2 *Breakage as a summation of events*

The study of energy consumption during grinding of mineral particles does not constitute a theory of grinding; nor is it sufficient to describe this grinding process purely in terms of the ideas of crack propagation in solid material. Breakage of particles in a machine is not as simple to analyse as the primary breakage of solid pieces of material. It is necessary to divide the process of comminution into its various sections and to define the operations that take place. The process may be defined as consisting of repetitive steps, each step consisting of two basic operations. These two basic operations, first defined by Epstein (1948), are those of selection of material for breakage and the subsequent breakage of this selected material by the machine. Epstein defined the two operations as:

(1) $p_n(y)$, the probability of breakage of a particle, size y, in the nth step of the breakage process; and

(2) $F(x, y)$, the cumulative distribution by weight of particles of size $x < y$ arising from the breakage of a unit mass of size y.

The introduction of these basic functions, coupled with the underlying assumption that a breakage process can be broken up into steps, gave a framework within which the changes in the particle-size distribution could be studied as a function of the number of steps in the process. The first function dealt with the probability of breakage of pieces of solid material during one step of the breakage process, while the second function dealt with the distribution by weight of particles in different size ranges arising from the breakage of unit amount of the parent material.

Each pair of the processes of selection and breakage constitutes a single step in degradation of the original material in the operation. After each breakage process, the function $F_n(x)$, denoting the original weight distribution of material equal to or less than size x, will have changed to $F_{n+1}(x)$, which is the new weight distribution after breakage. The form of the function $F(x, y)$ which operates on particles of initial size y may take any one of the many forms that have been derived by numerous authors as being descriptive of the fundamental breakage of solid particles. These basic ideas of breakage and selection have been used by Broadbent and Callcott (1956) in their matrix analysis of the grinding process.

During the past ten years, several other workers have regarded comminution as a continuous process and have proposed mathematical models of grinding in the form of continuous equations. Both the discrete and the continuous forms of the grinding model will be discussed in the next chapter.

CHAPTER 3

Mathematical models of size-reduction processes

3.1 INTRODUCTION

The mechanistic approach to the modelling of size-reduction processes is based on recognition of the physical events which occur and has been found to give models which are satisfactory for simulation. The basic idea underlying mechanistic models, proposed by Epstein (1948), was discussed in subsection 2.5.2. Epstein showed that the distribution function after n steps in a repetitive breakage process which can be described by a probability function and a distribution function is asymptotically logarithmico-normal, and this conforms to a frequently observed characteristic of sizing distributions of comminuted products. This concept has been used in what have come to be known as the matrix and kinetic models.

In the matrix model, comminution is considered as a succession of breakage events, the feed to each event being the product from the preceding event. The longer the period of grinding, the greater is the number of events and the size reduction attained. In the kinetic model, comminution is considered as a continuous process and the longer the period of grinding, the greater is the size reduction attained.

Both models are based on the concepts of:

(1) probability of breakage, and this has been called a selection or a breakage-rate function;

(2) characteristic size distribution after breakage, and this has been called a breakage, distribution, or appearance function; and

(3) differential movement of particles through or out of a continuous mill.

This is generally size-dependent and has been called a classification or discharge-rate function, or size-dependent diffusion coefficient. The occurrence of back-mixing in continuous mills may be recognised in any model by adding flow and mixing terms to the basic probability-breakage matrix. The models are described in the following sections although more attention will be given to the matrix model because this model will be used in the simulation work. A third model, called the perfect mixing model, which combines the better features of the matrix and kinetic models and which is applicable under some conditions, is also described.

3.2 MATRIX MODEL

3.2.1 *Description of the model*

This concept of probability of breakage of each size range and size distribution of each broken product was embodied in a matrix model of comminution by Broadbent and Callcott (1956), although they used the terms breakage and selection functions instead of distribution and probability functions. In this model, the feed to and the product from a size-reduction process may be expressed as sizing distributions in terms of n size ranges as shown in Table 3-I.

TABLE 3-I

Sizing distributions of feed to and product from a size-reduction process

Size range	Feed	Product
1	f_1	p_1
2	f_2	p_2
3	f_3	p_3
.	.	.
.	.	.
n	f_n	p_n
$n+1$	f_{n+1}	p_{n+1}

Size range 1 in Table 3-1 is the maximum size and the $(n + 1)$th size range refers to the residue, that is, to the particles which pass the finest screen in a nest of screens.

During a grinding process, particles in all size ranges have some probability of being broken and the products of breakage may fall in that size interval and in any smaller size interval. It will be noted that a particle may undergo minor breakage or chipping which is insufficient to ensure that all fragments are smaller than the original lower size of the size range. Consequently, a mass balance of a grinding process may be written as shown in Table 3-II.

Column 1 in the product refers to the products of breakage of the top size range in the feed, column 2 to the second size range, etc.. The elements in the product have been written in the form $p_{i,j}$ where i refers to the size range in which the element occurs and j refers to the size of the feed particle from which it came.

The following points will be noted about this array of elements which represent the product.

(1) The product size may be determined by summing the elements in successive rows of the array.

TABLE 3-II

Mass balance for a size-reduction process

Size range	Feed	Product						
1	f_1	$p_{1,1}$	0	0	.	.	0	0
2	f_2	$p_{2,1}$	$p_{2,2}$	0	.	.	0	0
3	f_3	$p_{3,1}$	$p_{3,2}$	$p_{3,3}$	.	.	0	0
.	.	.	.	.	.	.	.	.
.	.	.	.	.	.	.	.	.
n	f_n	$p_{n,1}$	$p_{n,2}$	$p_{n,3}$	.	.	$p_{n,n}$	0
$n+1$	f_{n+1}	$p_{n+1,1}$	$p_{n+1,2}$	$p_{n+1,3}$	.	.	$p_{n+1,n}$	$p_{n+1,n+1}$

(2) $\sum_{1}^{n+1} f_i$ represents the total feed mass F. The particles falling in the residue, that is, size range $n + 1$, may always be calculated for both feed and product by subtracting the cumulative weight retained on size n from F. The feed and product may be represented by arrays containing n by 1 elements.

The element $p_{i,j}$ in the array representing products of breakage may also be written: $p_{i,j} = X_{i,j} \cdot f_j$ where $X_{i,j}$ represents the mass fraction of the particles in the jth size range in the feed which fall in the ith size range in the product. The product array may be re-written as shown in Table 3-III.

TABLE 3-III

The product from a size-reduction process expressed in terms of the feed

$X_{1,1} \cdot f_1$	0	0	. . . . .	0
$X_{2,1} \cdot f_1$	$X_{2,2} \cdot f_2$	0	. . . . .	0
$X_{3,1} \cdot f_1$	$X_{3,2} \cdot f_2$	$X_{3,3} \cdot f_3$	. . . . .	0
.	.	.	. . . . .	.
.	.	.	. . . . .	.
$X_{n,1} \cdot f_1$	$X_{n,2} \cdot f_2$	$X_{n,3} \cdot f_3$	. . . . .	$X_{n,n} \cdot f_n$

If the feed and product size distributions are now written as n by 1 matrices and the X array as an n by n matrix, the size-reduction process may be represented by the matrix equation shown in Table 3-IV. Consequently, the simple matrix equation:

$$\mathbf{p} = \mathbf{X} \cdot \mathbf{f} \tag{3-1}$$

completely defines a size-reduction process. It will be noted that corresponding elements in **f** and **p** refer to the same size intervals and it is convenient for purposes of calculation if a constant geometric ratio exists between successive intervals.

Eq. 3-1 is a true statement about a breakage process but it is only useful if **X** is known. **X** cannot be deduced directly without further information. Consequently, it is necessary to consider how **X** may be divided into its components.

TABLE 3-IV

The development of the general matrix equation of breakage

$$
\begin{bmatrix}
X_{1,1} & 0 & 0 & \cdots & 0 \\
X_{2,1} & X_{2,2} & 0 & \cdots & 0 \\
X_{3,1} & X_{3,2} & X_{3,3} & \cdots & 0 \\
X_{4,1} & X_{4,2} & X_{4,3} & \cdots & 0 \\
X_{5,1} & X_{5,2} & X_{5,3} & \cdots & 0 \\
\cdot & \cdot & \cdot & \cdots & 0 \\
\cdot & \cdot & \cdot & \cdots & 0 \\
X_{n,1} & X_{n,2} & X_{n,3} & \cdots & x_{n,n}
\end{bmatrix}
\cdot
\begin{bmatrix} f_1 \\ f_2 \\ f_3 \\ f_4 \\ f_5 \\ \cdot \\ \cdot \\ f_n \end{bmatrix}
=
\begin{bmatrix}
X_{1,1} \cdot f_1 + 0 \quad \ldots + 0 \\
X_{2,1} \cdot f_1 + X_{2,2} \cdot f_2 + 0 \quad \ldots + 0 \\
X_{3,1} \cdot f_1 + X_{3,2} \cdot f_2 + X_{3,3} \cdot f_3 + \ldots + 0 \\
X_{4,1} \cdot f_1 + X_{4,2} \cdot f_2 + X_{4,3} \cdot f_3 + \ldots + 0 \\
X_{5,1} \cdot f_1 + X_{5,2} \cdot f_2 + X_{5,3} \cdot f_3 + \ldots + 0 \\
\cdot \quad \cdot \quad \cdot \quad \ldots \quad \cdot \\
\cdot \quad \cdot \quad \cdot \quad \ldots \quad \cdot \\
X_{n,1} \cdot f_1 + X_{n,2} \cdot f_2 + X_{n,3} \cdot f_3 + \ldots + X_{n,n} \cdot f_n
\end{bmatrix}
=
\begin{bmatrix} p_1 \\ p_2 \\ p_3 \\ p_4 \\ p_5 \\ \cdot \\ \cdot \\ p_n \end{bmatrix}
$$

3.2.2 *The selection function*

Particles of all sizes which enter a grinding process have some probability of being broken and this probability may change as the size of the particle changes. During the process a certain proportion of the particles in each size range is selected for breakage and the remainder pass through the process unbroken.

If S_1 is the proportion of particles in the largest size range which is selected for breakage, then the mass of particles in that size range which is broken is $S_1 \cdot f_1$. Similarly, the mass of particles broken in the nth size range is $S_n \cdot f_n$ and the following matrix equation shown in Table 3-V may be written.

TABLE 3-V

Selection in a size-reduction process

$$\begin{bmatrix} S_1 & 0 & 0 & 0 & . & . & 0 \\ 0 & S_2 & 0 & 0 & . & . & 0 \\ 0 & 0 & S_3 & 0 & . & . & 0 \\ \vdots & \vdots & \vdots & \vdots & \vdots & \vdots & \vdots \\ 0 & 0 & 0 & 0 & . & . & S_n \end{bmatrix} \cdot \begin{bmatrix} f_1 \\ f_2 \\ f_3 \\ \vdots \\ f_n \end{bmatrix} = \begin{bmatrix} S_1 \cdot f_1 \\ S_2 \cdot f_2 \\ S_3 \cdot f_3 \\ \vdots \\ S_n \cdot f_n \end{bmatrix} \qquad (3\text{-}2)$$

If the selection function is represented by **S**, the particles which are broken may be represented by the function **S** · **f**. The remainder of particles will pass through the process unbroken and for the nth size range the unbroken fraction will be $(1 - S_n) \cdot f_n$. The total mass of particles which pass through the process unbroken may be represented by the product $(\mathbf{I} - \mathbf{S}) \cdot \mathbf{f}$.

When **X** refers to those particles in the feed which are actually broken and not to be entire feed mass it may be replaced by the symbol **B** and the equation for a breakage process may be written:

$$\mathbf{p} = \mathbf{B} \cdot \mathbf{S} \cdot \mathbf{f} + (\mathbf{I} - \mathbf{S}) \cdot \mathbf{f} \quad \text{or:} \quad \mathbf{p} = \mathbf{B} \cdot \mathbf{S} + \mathbf{I} - \mathbf{S} \cdot \mathbf{f} \qquad (3\text{-}3)$$

3.2.3 *The classification function*

A comminution process normally consists of many breakage events which may operate simultaneously or consecutively or both, and selection and breakage occur within each event. However, the process may also operate so that the product from each event is subject to some size-separation operation before some fraction of it is subjected to the next breakage event.

This concept is simple to grasp in the case of jaw crushers and is shown diagrammatically in Fig. 3-1. During the crushing stroke, a lump of ore is shattered and during the reverse stroke, those fragments which are larger

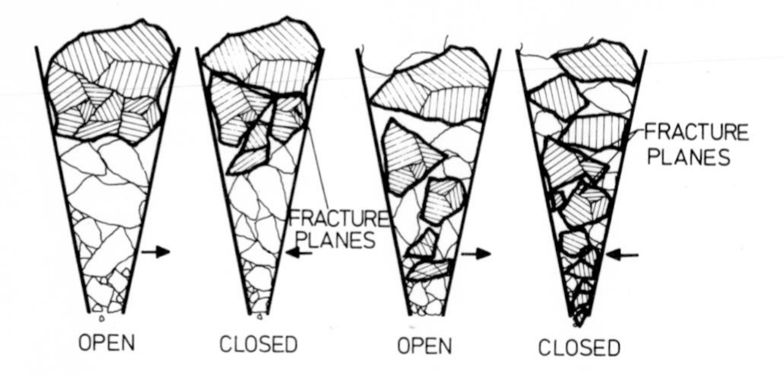

Figure 3-1. Classification occurring in a jaw crusher.

than the discharge gap are retained within the jaws for further crushing. The same type of process occurs within gyratory and roll crushers and, as will be seen, it is a particular feature of the operation of rod mills. There are indications that it occurs in almost all type of industrial grinding units though, in some cases, the occurrence may be small.

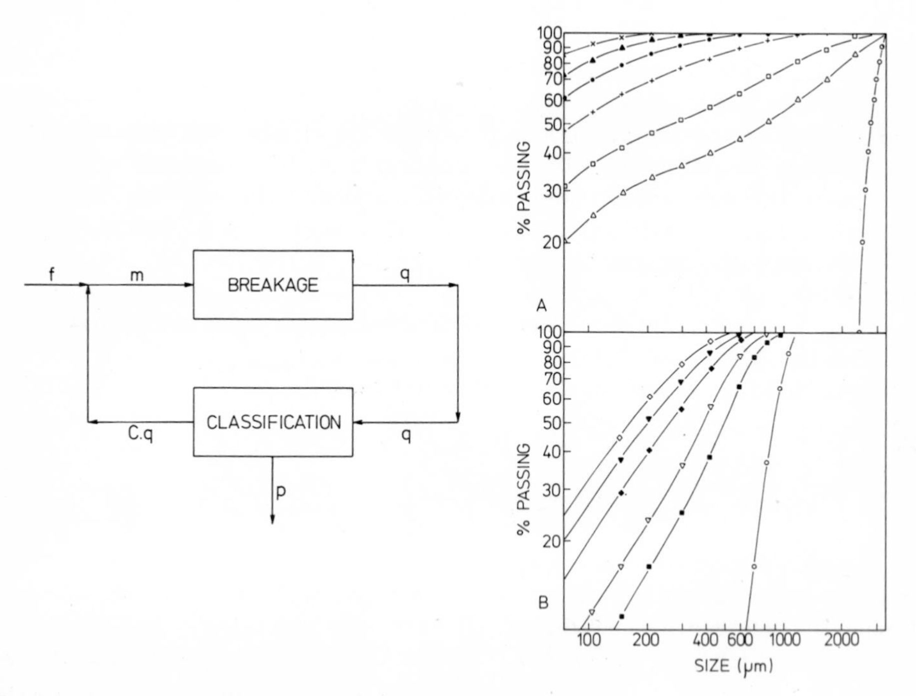

Fig. 3-2. Symbolic representation of a closed grinding circuit.

Fig. 3-3. Size distributions of products after increasing grinding times. A. Hematite in a continuous ball mill (Bush, 1967); B. anthracite in a Hardgrove mill (Klimpel, 1964). Line on right in each graph represents the feed, other lines represent the products.

A breakage-classification process operates as shown in Fig. 3-2 and the derivation of the equation which defines this process is as follows:

As a result of classification: $\mathbf{p} = (\mathbf{I}-\mathbf{C})\cdot\mathbf{q}$

or: $\mathbf{p} = (\mathbf{I}-\mathbf{C})\cdot(\mathbf{B}\cdot\mathbf{S}+\mathbf{I}-\mathbf{S})\cdot\mathbf{m}$ (3-4)

Also: $\mathbf{m} = \mathbf{f}+\mathbf{C}\cdot\mathbf{q}$

or by re-arrangement: $\mathbf{m} = \mathbf{f}+\mathbf{C}\cdot(\mathbf{B}\cdot\mathbf{S}+\mathbf{I}-\mathbf{S})\cdot\mathbf{m}$

$\mathbf{f} = [\mathbf{I}-\mathbf{C}\cdot(\mathbf{B}\cdot\mathbf{S}+\mathbf{I}-\mathbf{S})]\cdot\mathbf{m}$

or: $\mathbf{m} = [\mathbf{I}-\mathbf{C}\cdot(\mathbf{B}\cdot\mathbf{S}+\mathbf{I}-\mathbf{S})]^{-1}\cdot\mathbf{f}$ (3-5)

and from 3-4 and 3-5: $\mathbf{p} = (\mathbf{I}-\mathbf{C})\cdot(\mathbf{B}\cdot\mathbf{S}+\mathbf{I}-\mathbf{S})\cdot[\mathbf{I}-\mathbf{C}\cdot(\mathbf{B}\cdot\mathbf{S}+\mathbf{I}-\mathbf{S})]^{-1}\cdot\mathbf{f}$ (3-6)

If classification is insignificant, C approaches zero and eq. 3-6 reduces to eq. 3-3.

3.2.4 *The breakage function*

The fundamental aspect of all comminution models is the description of the product of a single breakage event. This is called the breakage function, a function that has been very difficult to determine experimentally because there is no non-destructive testing technique which will give information about the inherent breakage properties of minerals. Any technique which involves physical breakage must introduce machine characteristics as well as mineral characteristics into the feed—product transformation from which it would be hoped that the breakage function could be deduced. It has frequently been observed that there is a characteristic shape of product size distributions for particular minerals ground in given machines, and the differences which can exist are illustrated in Fig. 3-3. In this figure, the size distributions of coal after increasing periods of time in a Hardgrove mill (Klimpel, 1964), and of hematite after increasing periods of time in a continuous ball mill (Bush, 1967), are shown. In the latter case, the time scale referred to the time at which the discharge was sampled after a "tagged" sample of hematite was introduced into the mill feed. In both cases, it will be noted that the product size distribution rapidly assumed a constant shape but this was quite different for the two conditions.

Broadbent and Callcott (1956) postulated that a modification of the Rosin-Rammler equation (1933) written as:

$$B_{(x,y)} = (1-e^{-x/y})/(1-e^{-1}) \qquad (3\text{-}7)$$

gave a convenient form for the distribution of particles after breakage. The function $B_{(x,y)}$ represents the proportion of particles initially of size y which appear in size ranges smaller than x after breakage. The important concept embodied in this description of particle breakage is that the distribution obtained after breakage of a mineral particle, relative to the initial size, is independent of the initial size of the particle.

A considerable amount of laboratory work has been done to examine the validity of this form of the breakage function. The technique usually involves the use of tracer mineral to examine the breakage characteristics of individual size fractions.

Kelsall (1964) used a tracer of quartz in an environment of pure calcite to obtain the instantaneous breakage function which he defines for a single size fraction as "the size distribution of the broken material produced by statistical breakage of the particles within the ball mill when the particles formed had negligible chance to rebreak". Kelsall added an impulse of quartz tracer to the feed to a ball mill when it was operating under steady-state conditions. The first broken quartz to appear from the mill in sufficient quantities to be sized was assumed to be a measure of the instantaneous breakage function. Impulses of successive tracer size fractions showed that a common relative breakage function of the form:

$$B_{(x,x_0)} = (x/x_0)^n \tag{3-8}$$

was obtained for sizes less than 590 micrometres where n took values from 0.90 to 0.95. At sizes above 590 micrometres, the test results showed increasing deviations from the relative form. However, Kelsall attributes this to other assumptions built into the model rather than indicating that the breakage function is dependent on initial size.

The use of tracers has also involved the removal of a size fraction of natural feed to the mill and "tagging" this size fraction by irradiation, for instance, Gaudin et al. (1951), Moore (1964), Bush (1967). The radioactive fraction was then returned to the feed material and ground in a batch ball mill. The radioactive content of individual size fractions of product material was then used to follow the movement of tracer. The advantage of the use of the irradiated material is that the tracer material has the same breakage properties as the host material, and consequently allows the grinding to take place in a natural environment. The results of the work of these authors indicated that the breakage function **B** can be approximated by a step matrix, that is, the breakage of a mineral particle is independent of original size. However, it was found that because of the nature of the model used, the breakage function **B** and the function defining the probability of breakage **S** (called the selection function), were not uniquely determined. Therefore, many combinations of **B** and **S** exist, each yielding a satisfactory prediction of ball mill performance.

The two definitions of the breakage function given above differ only in relation to defining when a mineral particle is broken. Kelsall's definition considers a particle to be broken only if none of it reappears in the parent size range. Particles remaining in the parent size range are considered unbroken. The Broadbent–Callcott definition allows the distribution of products of breakage into the same size range as the parent material. For purposes of the use of the mathematical models of particle breakage, this difference in definition is not important.

The breakage function which has been discussed in this section refers to crushing breakage. In some breakage processes, particles are broken by abrasion and another type of function must be considered. This will be discussed in a later section.

3.2.5 *Repetitive breakage*

Most breakage machines operate so that sequential breakage events occur while the particles are in the machines. If v events occur and each is denoted by **X**, the products from consecutive events may be written:

$$\mathbf{p}_1 = \mathbf{X} \cdot \mathbf{f}$$
$$\mathbf{p}_2 = \mathbf{X} \cdot \mathbf{p}_1$$
$$\vdots$$
$$\mathbf{p}_v = \mathbf{X} \cdot \mathbf{p}_{v-1}$$

where the product, **p**, from the machine corresponds to $\mathbf{p}_v$ and v cycles of breakage have occurred to transform **f** to **p**.

The parameter v may be introduced to describe this concept of cycles of breakage and the process may be described by the equation:

$$\mathbf{p} = \left[\prod_{j=1}^{j=v} \mathbf{X}_j\right] \cdot \mathbf{f} \qquad (3\text{-}9)$$

where $\mathbf{X}_j$ is the breakage event occurring during the jth cycle. A cycle may be related to some aspect of operation of the machine, for instance, in the case of a tumbling mill, it may be one revolution or a unit of time. Incomplete cycles may occur and may readily be handled mathematically. If the events are *identical*, the process may be described by the equation:

$$\mathbf{p} = \mathbf{X}^v \cdot \mathbf{f} \qquad (3\text{-}10)$$

3.3 KINETIC MODEL

In this approach to the formulation of grinding models, comminution is considered as a rate process. The kinetic models of grinding have been expressed both in terms of continuous functions and as discretized distributions. In discretized form, the kinetic models bear close resemblance to the matrix models discussed in the previous sections.

The basic equation defining the rate of breakage of particles has been given by Loveday (1967) as the simple first-order kinetic equation:

$$\mathrm{d}W(D)/\mathrm{d}t = -\mathrm{k}(D) \cdot W(D) \qquad (3\text{-}11)$$

where $W(D)$ = weight of particles of size D, and $\mathrm{k}(D)$ = rate constant for size D.

By defining the breakage function as the continuous function $B(D, D_0)$

and writing the mass balance for the size fraction $W(D, t)\mathrm{d}D$ (where $W(D, t)$ represents the overall size distribution at time t), Loveday obtained the integro-differential equation:

$$\mathrm{d}[W(D, t)\mathrm{d}D]/\mathrm{d}t = \int_D^\infty [\mathrm{k}(D_0) \cdot W(D_0, t) \cdot [B(D, D_0)\mathrm{d}D]\,\mathrm{d}D_0 - \mathrm{k}(D) \cdot W(D, t)\mathrm{d}D] \quad (3\text{-}12)$$

to describe the time rate of change of sizing of the contents of a batch mill. In this description of the breakage process, the functions $\mathrm{k}(D)$, $W(D)$, and $B(D, D_0)$ are all continuous distributions with respect to size D. Loveday notes that the major problem associated with eq. 3-12 is the complexity of its solution and derives an approximate analytic solution. Other solutions to the batch grinding equation have also been published—Reid (1965), Kapur (1971).

The main difficulty in the application of the continuous distribution model to practical problems in comminution is that of obtaining a satisfactory definition of the continuous function for the distribution of particle sizes. Varying curvature in conventional plots of sizing distributions for crusher, rod mill and ball mill products, requires that a multiparameter distribution be employed. Loveday has compared a number of possible distributions and concludes that at least a three-parameter distribution is required.

Because of the problem of determining suitable equations to represent particle-size distribution, it has been convenient to use size-discretized forms of the grinding equations. When the batch grinding equation is discretized with respect to size, the net rate of change of material x in a size range i may be obtained by a simple mass balance and, for the case of the batch mill, has been given by Horst and Freeh (1970) as:

$$\mathrm{d}x_i/\mathrm{d}t = -\mathrm{k}_i \cdot x_i + \sum_{j=1}^{i=1} a_{ij} \cdot \mathrm{k}_j \cdot x_i \qquad i = 1, 2, \ldots, m$$

The k_i values in these equations are the first-order breakage rate constants and the elements a_{ij} are the inter-size flow coefficients. Horst and Freeh note that this equation may be written in matrix form as:

$$\mathrm{d}\mathbf{X}_t/\mathrm{d}t = -(\mathbf{I} - \mathbf{A}) \cdot \mathrm{K} \cdot \mathbf{X}_0 \quad (3\text{-}13)$$

where the inter-size flow coefficients **A** and the rate constants K perform similar functions to the breakage matrix **B** and selection matrix **S** in the matrix model. **X** is the size distribution.

3.4 CONTINUOUS FLOW SYSTEMS

The adaptation of the batch grinding models to describe continuous flow systems essentially requires the addition of suitable terms to the model to describe the mixing and flow characteristics of the system. Again, these characteristics of the system can be represented either by continuous distributions or by a discretized form. The continuous and discretized distribution models are also sometimes referred to as "distributed" and "lumped" parameter models, respectively.

The description of mixing and flow involves a description of the longitudinal and transverse movement of the pulp within the mill. The transverse movement of the pulp is usually considered relatively unimportant in a description of the flow characteristics since this movement has no direct effect on the residence time of the pulp in the mill. Residence time is important and should be considered in conjunction with the axial movement of the pulp in the mill.

Descriptions of the mixing characteristics can be obtained as a continuous function of mill length. However, it is generally convenient to consider a more convenient "lumped"-parameter form of the model. In these models, the mixing characteristics of the system are usually approximated by the combination of a pure delay and a certain number of perfectly mixed segments connected in series. Varying flow characteristics of real systems can then be reproduced by varying the number of perfectly mixed segments used, and by variation of the degree of mixing between adjacent segments. Models of this type have been used by Kelsall et al. (1968) and by Horst and Freeh (1970). The latter authors also considered the possibility of a size-dependent diffusion coefficient to describe the transport of material in the axial direction of the mill. In lumped-parameter form this diffusion coefficient is equivalent to a classification effect at the end of each perfectly mixed segment, and is synonymous with the classification function used by Callcott and Lynch (1964) in the rod mill model. If no classification exists, then the diffusion coefficient is independent of size and becomes a single value describing the mass rate of discharge of pulp from the segment.

3.5 PERFECT MIXING MODEL

The simplest case of multi-segment models used by Kelsall et al. (1968) and Horst and Freeh (1970) is the perfect mixing mill model, that is, a model containing only one segment. This model, when expressed in matrix form could rightly belong to either of the two categories previously discussed. However, the matrix form of this model combines some of the better aspects of both approaches and, therefore, might be considered separately.

Whiten (1974), using the concept of a mill being represented by one perfectly mixed segment, showed that the rate of change of contents of the mill due to breakage and flow could be written:

$$\frac{\partial \mathbf{s}}{\partial t} = (\mathbf{A} \cdot \mathbf{R} - \mathbf{R}) \cdot \mathbf{s} + \mathbf{f} - \mathbf{p}$$

or:

$$\frac{\partial s}{\partial t} = (A \cdot R - R - D) \cdot s + f \qquad (3\text{-}14)$$

The matrices A, R and D are the appearance, breakage-rate and discharge-rate functions. They are essentially the breakage, selection and size-dependent diffusion (or classification) functions, respectively, used by other authors. The perfect mixing model is particularly appropriate for some comminution operations and use of a different set of symbols for essentially the same functions avoids confusion between models.

Whiten showed that the model is particularly easy to manipulate using simple matrix methods for simulation and for the calculation of the model parameters. These may be obtained by examining the solution at steady state which gives:

$$(D + R - A \cdot R) \cdot s = f$$

Steady-state simulation is then carried out by solving this set of triangular simultaneous equations for s, and then the predicted product is obtained simply as:

$$p = D \cdot s \qquad (3\text{-}15)$$

The product may also be calculated directly from:

$$p = D \cdot R^{-1} \cdot (D \cdot R^{-1} + I - A)^{-1} \cdot f \qquad (3\text{-}16)$$

For purposes of the calculation of model parameters, provided a form is assumed for the breakage function A, then $D \cdot R^{-1}$ may be obtained from:

$$p = D \cdot R^{-1} \cdot (I - A)^{-1} \cdot (f - p) \qquad (3\text{-}17)$$

and, by substitution:

$$R \cdot s = (I - A)^{-1} \cdot (f - p) \qquad (3\text{-}18)$$

If the mill contents can be determined, then it is possible to obtain directly the values of R and D using eqs. 3-17 and 3-18. The values of the matrix D vary with feed rate for a constant volume mill. If the values of D are independent of size then the parameter $D \cdot R^{-1}$ is effectively $k \cdot R^{-1}$ where k is a constant. In this case, the size distribution of the mill contents is the same as that of the product.

The advantage of this perfect mixing mill model is that it combines the simplicity and ease of calculation of the matrix approach with the corresponding convenience of the directly time-dependent solution obtained by the kinetic approach. The same form of the model is used both for steady-state and dynamic simulation. The steady-state solution may be obtained directly without iterating or integrating to the steady state. On the other hand, the full dynamic response may be obtained from the model using numerical integration techniques on the finite-difference equation. This model is discussed in Appendix 1.

3.6 OTHER APPROACHES TO SIZE-REDUCTION MODELLING

The other main approaches to constructing models of comminution machines have generally been either on the basis of energy—size-reduction relationships or using purely empirical techniques. The energy—size-reduction relationships have already been discussed, but the comment should be made that recently at least one attempt has been made to determine the relationship between the energy studies and the more popular kinetic approach. Kapur (1971) has used a general energy—size-reduction relationship and concluded that there is a very simple relationship between the exponent n in the energy—size-reduction equation:

$$\mathrm{d}E = -K \cdot \mathrm{d}x/x^n$$

and the expression for the rate function or selection function which is used in grinding kinetics. Further studies in the combination of both the energy—size-reduction relationship and kinetic approach may produce a better overall model since a knowledge of the energy consumed in grinding is still very important, particularly in mill design.

Empirical studies of grinding-mill behaviour have also been carried out, for example, by Kelly (1970). These studies generally utilise a statistical approach which involves the collection of a series of data values planned according to a factorial design. These data are then analysed using standard regression and analysis of variance techniques to determine those variables which have a significant effect on performance criteria. The empirical approach is of limited value in most applications for two reasons. Firstly, models developed on an empirical basis do not usually have very satisfactory extrapolation behaviour. This means, for example, that an empirical model developed on a pilot-plant-scale mill may not generally be used to predict the performance of full-scale mills. Secondly, while an empirical model developed on the full-scale mill could be validly applied, it is not generally possible to implement a series of experiments which constitute an experimental design. This is because large-scale ball mills are invariably operated in closed circuit with some form of classifier, so that the feed conditions to the mill are determined by classifier characteristics and are not available for independent manipulation.

NUMERICAL EXAMPLES

The numerical examples given refer to the matrix model since this is the model which is most commonly used in the remainder of the text.

Example 3-1. In the equation for a breakage process $\mathbf{p} = \mathbf{X} \cdot \mathbf{f}$, the values of the breakage matrix **X** and the feed size distribution **f** are shown in Table 3-VI.

TABLE 3-VI

Numerical values in the breakage matrix and feed size distribution

X							**f**
0.15	0	0	0	0	0		25
0.20	0.15	0	0	0	0		21
0.15	0.20	0.15	0	0	0		14
0.10	0.15	0.20	0.15	0	0	and	8
0.10	0.10	0.15	0.20	0.15	0		5
0.10	0.10	0.10	0.15	0.20	0.15		3
							24

What will be the size distribution of the product from the process?

Solution:

Product in size range 1 = $0.15 \cdot 25$ = 3.75
2 = $0.20 \cdot 25 + 0.15 \cdot 21$ = 8.15
3 = $0.15 \cdot 25 + 0.20 \cdot 21 + 0.1 \cdot 14$ = 10.05
4 = ... = 9.65
5 = ... = 9.05
6 = ... = 8.65

Residue = 50.70

Example 3-2. In the equation for a breakage process $\mathbf{p} = (\mathbf{B} \cdot \mathbf{S} + \mathbf{I} - \mathbf{S}) \cdot \mathbf{f}$, the values of the breakage function **B** and the feed size distribution **f** are the same as those given for **X** and **f** in Example 3-1. The probability of breakage of particles in successive size ranges is as follows:

$\langle 1.0, 0.70, 0.50, 0.35, 0.25, 0.18 \rangle$

What will be the size distribution of the product from the process?

Solution:

The solution is given in Table 3-VII.

Example 3-3. For the process discussed in Example 3-2, what will be the size distributions of the products after 2, 3 and 4 identical stages of breakage? What will be the product after a non-integral stage, for example, 3.7 stage?

Solution:

The products from successive stages of breakage are shown in Table 3-VIII.

Product after 3.7 stages = 0.7 · product (fourth stage) + 0.3 · product (third stage)
= $\langle 0.03, 1.55, 5.34, 8.53, 10.00, 10.50, 64.05 \rangle$

TABLE 3-VII

Method of calculation for a process involving both selection and breakage

Feed (f)	Probability of breakage (S)	Particles broken (S · f)	Particles unbroken [(I − S) · f]	Product from broken particles (B · S · f)	Total product [(B · S + I − S) · f]
25	1	25	0	3.75	3.75
21	0.70	14.7	6.3	7.21	13.51
14	0.50	7.0	7.0	7.74	14.74
8	0.35	2.8	5.2	6.53	11.73
5	0.25	1.25	3.75	5.77	9.52
3	0.18	0.54	2.46	5.42	7.88
24					38.87

TABLE 3-VIII

Products from successive stages of breakage

Feed	Product (1st stage) = feed (2nd stage)	Product (2nd stage) = feed (3rd stage)	Product (3rd stage) = feed (4th stage)	Product (4th stage)
25	3.75	0.56	0.08	0.01
21	13.51	6.22	2.63	1.08
14	14.74	10.93	7.23	4.53
8	11.73	11.50	9.88	7.95
5	9.52	10.75	10.58	9.75
3	7.88	9.82	10.50	10.50
24	38.87	50.22	59.10	66.18

Example 3-4. For the process which was discussed in Example 3-2, it will be assumed that a classification mechanism is operating within the mill in such a manner that no particles in ranges 1 or 2 can leave the mill in the product. That is, the process is operating according to the equation:

$$\mathbf{p} = (\mathbf{I} - \mathbf{C}) \cdot (\mathbf{B} \cdot \mathbf{S} + \mathbf{I} - \mathbf{S}) \cdot [\mathbf{I} - \mathbf{C} \cdot (\mathbf{B} \cdot \mathbf{S} + \mathbf{I} - \mathbf{S})]^{-1} \cdot \mathbf{f}$$

What will be the size distribution of the product at steady state?

Solution:
The solution is given in Table 3-IX. Alternatively, a matrix inversion procedure may be used.

Example 3-5. For the process discussed in Example 3-4, what will be the build-up of ore in the coarse size fractions in the mill due to the internal classification?

Solution:
During every increment of time of breakage, 100 units by weight of ore enter the mill. The size distribution of the feed, and the breakage, selection and classification matrices which apply during that time, are as follows:

TABLE 3-IX

Product from a breakage process which includes classification

1 Feed	2 Product after breakage	3 Retained for further breakage	4 Final product from repeated breakage of the first two size fractions	5 Portion of (2) not retained for further breakage	6 Final product (4) + (5)
(f)	[(B · S + I − S) · f]	[C · (B · S + I − S) · f]		[(I − C) · (B · S + I − S) · f]	
25	3.75	3.75	0	0	0
21	13.51	13.51	0	0	0
14	14.74	0	4.04	14.74	18.78
8	11.73	0	2.98	11.73	14.71
5	9.52	0	2.13	9.52	11.65
3	7.88	0	2.13	7.88	10.01
24	38.87	0	5.98	38.87	44.85

f	:	25	21	14	8	5	3	24
B (first column)	:	0.15	0.20	0.15	0.10	0.10	0.10	
S (diagonal)	:	1.0	0.70	0.50	0.35	0.25	0.18	
C (diagonal)	:	1.0	1.0	0	0	0	0	

Although the process is continuous it may be regarded for purposes of illustration as a sequential process in which the feed enters the mill, then breakage occurs, then classification and discharge of broken product from the mill occur.

The calculated mill performance after sequential increments of time is as follows:

Time increment No.1:

Size range	1	2	3	4	5	6	Residue
Feed (units of mass)	25.0	21.0	14.0	8.0	5.0	3.0	24.0
Product after breakage	3.75	13.51	14.74	11.73	9.52	7.88	38.87

Mass retained in mill due to classification: 17.26 units

Mass discharged: 82.74 units

Time increment No.2:

Feed (units of mass)	28.75	34.51	14.0	8.0	5.0	3.0	24.0
Product after breakage	4.31	19.72	17.19	13.52	10.84	9.20	42.48

Mass retained in mill due to classification: 24.03 units

Mass discharged: 93.23 units

Time increment No.3:

Feed (units of mass)	29.31	40.72	14.0	8.0	5.0	3.0	24.0
Product after breakage	4.40	22.36	18.15	14.23	11.33	9.69	43.87

Mass retained in mill due to classification: 26.76 units

Mass discharged: 97.27 units

Equilibrium is reached when the mass discharged for increments of time equals the mass entering, that is, equals 100 units. The build-up of ore in the coarse size fractions due to internal classication may be calculated in this manner.

If the ore becomes harder or softer, **B** will vary and change will occur in the mass retained.

CHAPTER 4

Mathematical models of some industrial size-reduction machines

Breakage within any comminution unit may be described by the equation:

$$\mathbf{p} = \mathbf{X} \cdot \mathbf{f}$$

where **X** may result from an individual process or a series of repetitive processes. The problem is to define **X** quantitatively in such a manner that it defines the process and may be represented by a single-valued parameter to which process variables may be related. Because of the importance of **X** in models of size-reduction machines it will be discussed in more detail before machine models are discussed.

The data available about a comminution process in a size-reduction machine are:

(1) the operating characteristics of the machine, that is, diameter, length and rotational speed in a tumbling mill, or gap size in a crusher;

(2) the mass flow rate of feed to the machine;

(3) the sizing analyses of the feed to and product from the machine;

(4) the values of the elements in the **X** matrix; and

(5) the function defining the retention time of the ore within the mill.

Particular types of tests are required to obtain information about items 4 and 5 but these may be carried out. The data are limited with respect to the breakdown of the **X** matrix into its components for the following reasons.

(1) Sizing distributions cannot be determined experimentally as continuous curves but only as a series of points through which a curve may eventually be drawn.

(2) These points are derived from the mass fraction of the ore retained on a series of sieves whose apertures decrease in a selected geometric ratio. This ratio is normally $2^{-0.5}$ which means that theoretically the ratio of the volumes of the largest to the smallest particles retained on any one sieve approaches 2.82. The larger particles retained on a sieve may be broken severely as a result of single or multiple breakage events and daughter fragments may still remain on the same sieve.

(3) When particles retained on a particular sieve are subjected to a breakage process and the mass fraction retained on that same sieve after breakage is determined, it is not possible to distinguish between broken fragments and unbroken particles.

In order to allot unique values to elements in the breakage and selection functions which comprise the matrix **X** it must be possible to distinguish between broken and unbroken particles. As the geometric ratio between the sieves decreases, the approximation that breakage of a particle necessarily involves removal from a sieve interval becomes more valid, but it is probable that this ratio must become very small, perhaps less than 0.01, before the validity of the approximation can be accepted. With present experimental techniques, it is not practical to obtain the required data. It may appear that it should be possible to determine a breakage function for an ore independent of the selection function. However a breakage function can only be determined by breaking, and this involves a machine and therefore a selection function. Thus it is not possible to allot a unique value of a breakage function to an ore or of a selection function to a machine.

Consequently, it is necessary to make assumptions about the form of the model and about a breakage or selection function if a useful model is to be derived. A good understanding of the operating characteristics of a process is necessary if the assumptions about the form of the model are to be soundly based.

4.1 CONE CRUSHERS

4.1.1 *Form of the model*

Some distinctive features of cone-crusher operation are:

(1) All particles which appear in the crusher product must be smaller than the open side setting of the crusher but they need not be smaller than the closed side setting of the crusher. Internal classification is important in cone crusher operation since particles larger than the open side setting are retained for re-breakage within the breakage zone until they are smaller than the open side setting.

(2) Cone-crusher design often dictates that particles in the feed which are smaller than the closed side setting pass through the crusher with little, if any, breakage. However, there are designs of cone crushers in which significant breakage of these particles does occur.

(3) The power consumed by the crusher is a function of the mass flow and the size of particles greater than the closed side setting in the crusher feed. If classification is minimised (that is more particles smaller than the closed side setting enter the breakage zone) proportionally more power will be consumed in the generation of very fine material.

Consequently, the cone crusher may be simplified to a single breakage zone with particles, as a result of classification, having a probability of entering or re-entering this breakage zone. If a particle enters the breakage zone, then it will be selected for breakage. The components of the cone-crusher

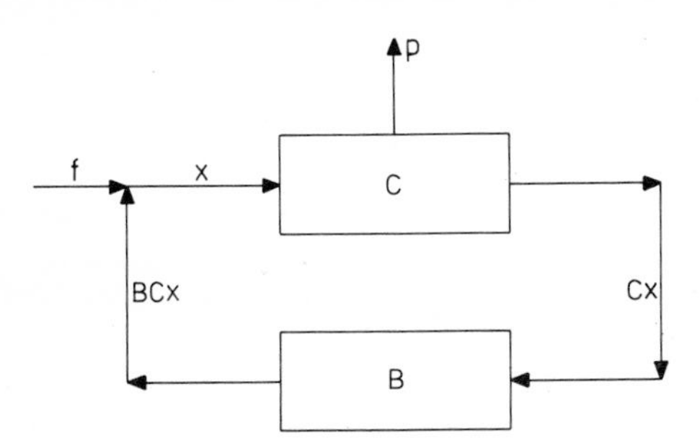

Fig. 4-1. Symbolic representation of cone crusher model.

model and the internal flows between them are shown in Fig. 4-1. The vectors **f**, **x** and **p** give the mass flow rates in each size fraction. The lower triangular matrix, **B**, gives the relative distribution of the particles after breakage and the diagonal matrix, **C**, gives the proportion of particles in any given size fraction which enter the breakage zone. Several more complicated models may be proposed for the cone crusher. However, it is considered that this model provides an adequate description of cone-crusher behaviour and that the data which are available or may be obtained from industrial operations would not be sufficient to calculate parameters for a more complex model.

Mass balances at the nodes in Fig. 4-1 give the equations:

$$\mathbf{f} + \mathbf{B} \cdot \mathbf{C} \cdot \mathbf{x} = \mathbf{x} \tag{4-1}$$

$$\mathbf{x} = \mathbf{C} \cdot \mathbf{x} + \mathbf{p} \tag{4-2}$$

Solving eq. 4-1 for **x** gives:

$$\mathbf{x} = (\mathbf{I} - \mathbf{B} \cdot \mathbf{C})^{-1} \cdot \mathbf{f} \tag{4-3}$$

The matrix $\mathbf{I} - \mathbf{B} \cdot \mathbf{C}$ is always non-singular as a unit element on the diagonal of $\mathbf{B} \cdot \mathbf{C}$ implies both no breakage and no discharge of that size fraction. Combining eqs. 4-2 and 4-3 gives:

$$\mathbf{p} = (\mathbf{I} - \mathbf{C}) \cdot (\mathbf{I} - \mathbf{B} \cdot \mathbf{C})^{-1} \cdot \mathbf{f} \tag{4-4}$$

which relates the product from the cone crusher to the cone-crusher feed. Adding eqs. 4-1 and 4-2 gives:

$$\mathbf{f} - \mathbf{p} = (\mathbf{I} - \mathbf{B}) \cdot \mathbf{C} \cdot \mathbf{x} \tag{4-5}$$

and this equation shows how the total breakage occurring in the crusher, that is, $\mathbf{f} - \mathbf{p}$, is related to the vector $\mathbf{C} \cdot \mathbf{x}$.

This model will be discussed with reference to data obtained from a series of tests which were carried out in the crushing plant of Mount Isa Mines Limited, (MIM), the flowsheet of which is given in Fig. 4-2. During these tests, crusher performance was studied at different feed size distributions,

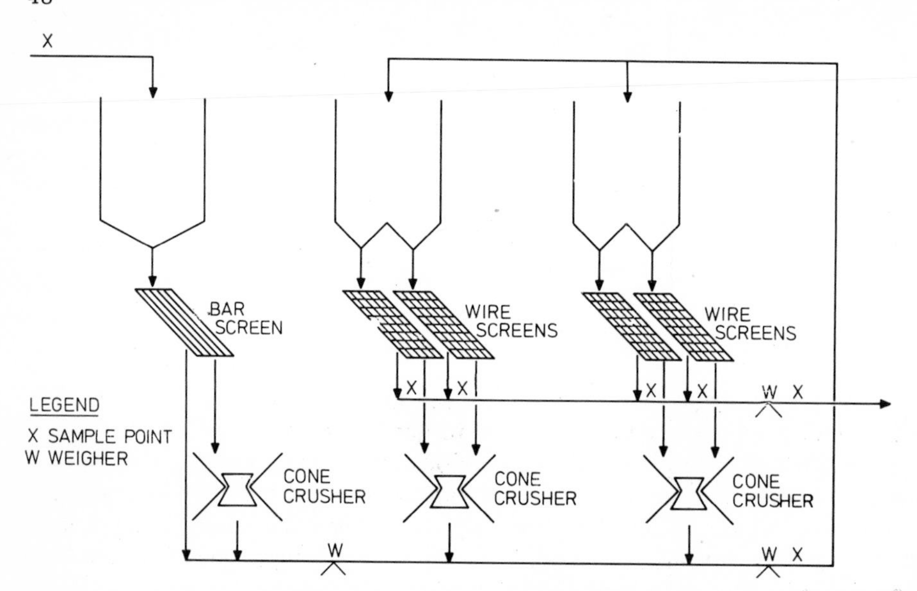

Fig. 4-2. Flowsheet of crushing plant at Mount Isa Mines Limited.

feed rates and closed side settings. These changes also produced changes in the current drawn by the crusher and a relationship was developed to predict the current for a wide range of operating conditions. The values of the parameters were determined by least squares and spline regression techniques so that the best fit between predicted and observed performance was obtained over the range of operating conditions studied.

Subsequent to the development of the model, data have been obtained for both the secondary and tertiary crushing operations at Bougainville Copper Limited. The data obtained at Mount Isa Mines Limited indicated that it was necessary for the model to incorporate separate distributions describing the coarse and fine portions of the crusher product. This has been confirmed by the data collected at Bougainville Copper Limited and the same form of the model has been successfully used to describe these crushing operations.

4.1.2 *The breakage matrix*

The observed data on the cone crusher was best explained by proposing that two modes of breakage occur in a cone crusher. In the first breakage mode particles are caught between the opposing liners of the crusher and fracture catastrophically into a small number of relatively large particles. These particles may or may not re-enter the breakage zone of the crusher model. A step matrix $\mathbf{B}_1$ which gives the product size distribution relative to the size of the original particle was used to predict the products of this breakage mode. It is calculated from the distribution:

$$B_{(x,y)} = (1 - e^{-(x/y)^u})/(1 - e^{-1}) \qquad (4\text{-}6)$$

where y is the size of the original particle and $B_{(x,y)}$ is the fraction of particles resulting from this breakage which are of size less than x. This distribution is a modification of the Rosin-Rammler distribution given by Broadbent and Callcott (1954). The value of u was calculated to be 6.0 ± 0.9 for the MIM cone crusher. The second mode of breakage described by the matrix $\mathbf{B}_2$, is the production of fine material at the points of contact between particles and the crusher liners and also between neighbouring particles. This type of breakage may account for 5—20% of the complete product size distribution. The sizing of this portion of the product is not dependent upon the size of the original particle. A Rosin-Rammler distribution is used to describe this material, that is;

$$B_{(x,x')} = 1 - e^{-(x/x')^v} \qquad (4\text{-}7)$$

where x' is a size such that a very significant portion of the product which is less than this size was produced by fine breakage. In the calculation, this distribution is altered so that the material predicted above the size of the original particle appears after breakage at the original particle size. This amount is generally small. The values calculated for the parameters in this distribution for the MIM crusher are: $x' = 3.05 \pm 0.51$ mm; $v = 1.25 \pm 0.14$.

The total breakage matrix, B, may be calculated by adding the two component matrices as follows:

$$\mathbf{B} = \alpha \cdot \mathbf{B}_1 + (1 - \alpha) \cdot \mathbf{B}_2 \qquad (4\text{-}8)$$

It was found that the value of α could be predicted from the closed side setting of the crusher:

$$\alpha = 0.8723 + 0.0045 \cdot g \pm 0.014 \qquad (4\text{-}9)$$

if g is the closed side setting in mm. The standard deviations of the coefficients in this equation are 0.15 and 0.039, respectively. This equation shows, as would be expected, that more fine material is produced as the crusher gap is decreased.

4.1.3 *The classification matrix*

The diagonal elements of the matrix, **C**, are obtained from a function of particle size $C(s)$, which gives the probability of a particle of size s entering the breakage stage of the crusher model. It is assumed that particles less than a certain size, k_1, are not broken in the crusher, that is, they are not caught between the crusher liners and are too small to be sufficiently constrained by the surrounding particles for fine breakage to occur. Thus, $C(s) = 0$ for $s < k_1$. Also, as a result of classification there exists a size, k_2, above which particles are always broken, that is, $C(s) = 1$ for $s > k_2$. Between these sizes, the function $C(s)$ is assumed to be a parabola with zero gradient at k_2 and is

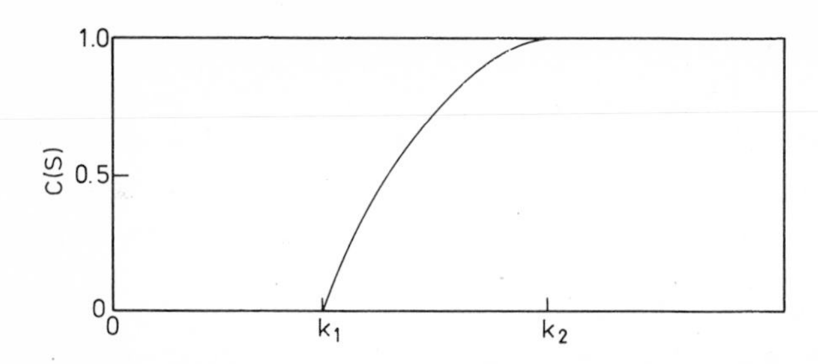

Fig. 4-3. Probability $C(s)$ of a particle entering the breakage stage (see eq. 4-10).

shown in Fig. 4-3. Consequently:

$$
\begin{aligned}
C(s) &= 0 && \text{if } s < k_1 \\
&= [(s - k_2)/(k_1 - k_2)]^2 && \text{if } k_1 < s < k_2 \\
&= 1 && \text{if } k_2 < s
\end{aligned}
\qquad (4.10)
$$

The values of the **C** matrix can be obtained as the mean values of the function $C(s)$ in the appropriate size range, that is:

$$C_i = \int_{s_i}^{s_{i+1}} C(s) \cdot ds/(S_{i+1} - S_i) \qquad (4\text{-}11)$$

where s_i and s_{i+1} are the lower and upper limits of the size interval for the ith size fraction.

The two parameters k_1 and k_2 were predicted for the MIM crusher by the equations:

$$k_1 = 0.67 \cdot g \pm 1.956 \text{ mm} \qquad (4\text{-}12)$$

$$k_2 = 1.121 \cdot g + 58.67 \cdot q + 25.4 \cdot T(t) \pm 1.8 \text{ mm} \qquad (4\text{-}13)$$

where q is the fraction plus 25.4 mm in the feed. The effect of feed tonnage on k_2 is shown in Fig. 4-4. The relation for k_1 is as would be expected. The relation between k_2 and closed side setting is reasonable and the remaining terms in the equation appear to relate to the ease of flow through the crusher, that is, large particles and low tonnages flow through the crusher more rapidly.

4.1.4 *Crusher current*

It has been seen (eq. 4-5) that the total breakage occurring in the crusher is related to the vector $\mathbf{C} \cdot \mathbf{x}$. Now this vector contains no material which does not enter the breakage stage of the crusher, and the current drawn by the crusher has been calculated using this vector. A parameter, a, which is dependent upon the size distribution of the feed to the crusher, is calculated as follows:

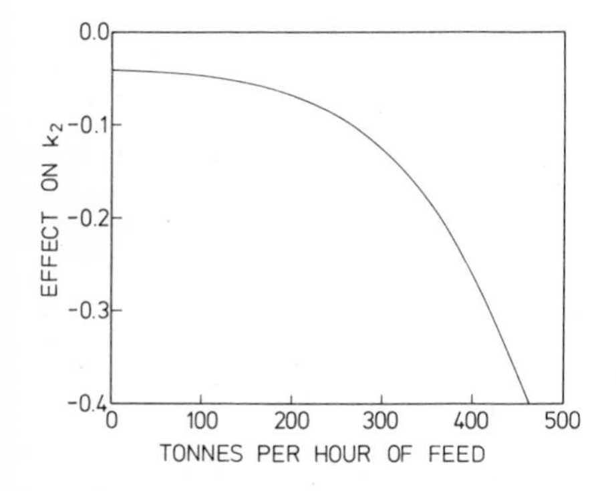

Fig. 4-4. Effect of feed rate on k_2 in the crusher model.

$$a = 25.4 \cdot \sum_{i=1}^{n} t_i/(s_i + s_{i+1}) \tag{4-14}$$

where t_i is the ith element of $\mathbf{C} \cdot \mathbf{x}$, and s_i, s_{i+1} are the lower and upper limits of the ith size fraction. For the fine-size fractions t_i equals zero. The crusher current, A, is then predicted by:

$$A = 14.2 + 0.0822 \cdot a + 0.00305 \cdot a^2 \pm 1.8 \text{ amp.} \tag{4-15}$$

4.2 ROD MILLS

4.2.1 *Form of the model*

Size distributions of rod-mill feeds and products are shown in Fig. 4-5 for hard (Mount Isa Mines Limited) and soft (New Broken Hill Consolidated Limited) ores. The important physical characteristic of rod mills is that pronounced preferential breakage of the coarser particles occurs and this is responsible for the "screening" effect which has been observed to occur within rod mills. As the ore particles pass through a rod mill they are subjected to a repetitive series of breakage processes and the "screening" effect operates so that the coarse-size ranges are progressively eliminated. Myers and Lewis (1946) demonstrated this by determining the size distribution of the ore retained in successive longitudinal sections of a continuous 2 m by 2 m rod mill after the mill was stopped. Many samples were analysed and the average size distributions in each section are shown in Fig. 4-6. The reason for this "screening" action is that the ore must pass through "screens" of decreasing aperture size as it proceeds down the mill, as is shown diagrammatically in Fig. 4-7.

Hydraulic classification also appears to influence the size distribution of particles discharged from a rod mill. An example of this is seen in the

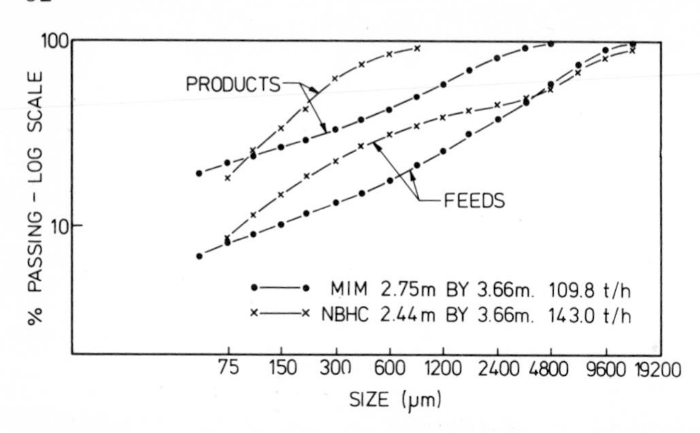

Fig. 4-5. Size distributions of rod-mill feeds and products are shown for harder (Mount Isa Mines Limited) and softer (New Broken Hill Consolidated Limited) ores. Mill sizes and feed rates are shown.

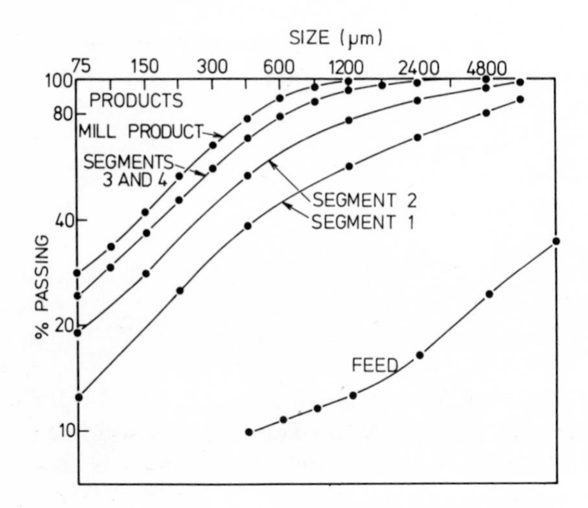

Fig. 4-6. Change in size distribution of ore in mill from feed end to discharge end.

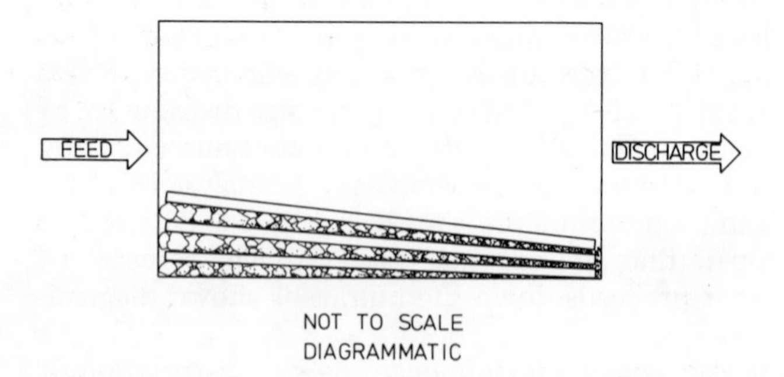

Fig. 4-7. Screening action of rods in a mill.

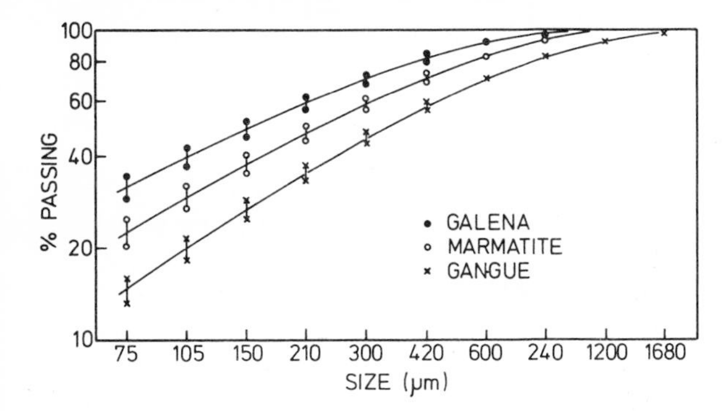

Fig. 4-8. Sizing distributions of galena, marmatite and gangue in a rod-mill product.

operation of the rod mill at New Broken Hill Consolidated Limited in which a coarsely mineralised lead—zinc ore is broken. The galena in the rod mill discharge was considerably finer than the marmatite, and both sulphide minerals were finer than the gangue. This is shown in Fig. 4-8. This difference may be due to the following reasons:

(1) For a given impact the galena and marmatite particles are broken more severely than the gangue particles.

(2) Hydraulic classification plays some part in determining which particles will be discharged from wet tumbling mills. Myers (1953) showed this in working with a slow-speed mill in which this effect would be greater than normal. This would cause the higher-specific-gravity particles in the mill product to be smaller in terms of screen sizes than those of lower specific gravity.

Consequently, a rod mill may be regarded as containing a series of grinding zones or stages in which both breakage and screening occur. Callcott and Lynch (1964) recognising this, defined a stage of breakage for a rod mill as the "interval required to eliminate the largest sieve fraction from the feed to the mill in the first instance and thereafter from the feed to successive stages". The screening or classification which occurs within each stage ensures that no particles in the maximum size range of the feed entering a stage are able to leave that stage as particles in the same size range: all must be broken into smaller size ranges. The equations which define this repetitive process, including classification, are:

$$\mathbf{p} = \left[\prod_{j=0}^{j=v} \right] \mathbf{X}_j \cdot \mathbf{f} \tag{4-16}$$

where $\mathbf{X}_j = (\mathbf{I} - \mathbf{C}) \cdot (\mathbf{B} \cdot \mathbf{S} + \mathbf{I} - \mathbf{S}) \cdot [\mathbf{I} - \mathbf{C} \cdot (\mathbf{B} \cdot \mathbf{S} + \mathbf{I} - \mathbf{S})]^{-1} \cdot \mathbf{f}$ (4-17)

$\mathbf{X}_j$ is the matrix describing the transition from the feed to the jth interface to the feed to the $(j + 1)$th interface and this represents a stage as described above. The maximum of stages is equal to the number of elements used in the feed vector, not including the undersize; for example, if a size distribution of the mill feed is represented by 9 elements, the mill operation cannot be defined by more than 9 stages because by then all of the feed will have been reduced to undersize. Several different models have been developed from eq. 4-16 to describe the changes in size distributions which are caused by rod mills, but for simplicity and utility, it is desirable to restrict the model so that one parameter only shall be sufficient to describe the total transition. The form of the model in which the matrix $\mathbf{X}_j$ is maintained constant relative to the largest size fraction of particles entering a stage has been found to be an accurate model of general application to rod mills as they are generally operated. It will be seen that the (1,1) element in $\mathbf{X}_j$ must be zero to ensure that the top element on the product is zero.

This model is an extension of the breakage-selection-classification model discussed earlier (subsections 3.2.3 and 3.2.5, and Example 3.4) but there are certain aspects of it which require further discussion. A simple assumption which may be made is that breakage operates in the first stage so that only particles in the top size range are broken but all these particles are completely broken out of the stage. This is too simple for most processes but it is useful to illustrate the argument. The first condition implies that the breakage matrix $\mathbf{X}_j$ contains elements which are other than zero in the first column only with the exception of the main diagonal, and the second condition implies that the 1,1 element in $\mathbf{X}$ is zero. The operation of the first breakage stage for a simple breakage matrix which conforms to these conditions is shown in Table 4-I.

TABLE 4-I

The first breakage stage in a process where $C_{1,1}$ is zero

$$\mathbf{X}_1 \cdot \mathbf{f}_1 = \mathbf{p}_1$$

$$\begin{bmatrix} 0 & 0 & 0 & 0 & 0 \\ 0.4 & 1 & 0 & 0 & 0 \\ 0.2 & 0 & 1 & 0 & 0 \\ 0.1 & 0 & 0 & 1 & 0 \\ 0.1 & 0 & 0 & 0 & 1 \end{bmatrix} \cdot \begin{bmatrix} 0.10 \\ 0.20 \\ 0.15 \\ 0.10 \\ 0.10 \end{bmatrix} = \begin{bmatrix} 0 \\ 0.24 \\ 0.17 \\ 0.11 \\ 0.11 \end{bmatrix}$$

$$\begin{matrix} & 0.35 & & 0.37 \end{matrix}$$

In order to keep $\mathbf{X}_2$ constant relative to $\mathbf{p}_1$, which is the feed to the second stage, as was $\mathbf{X}_1$ to $\mathbf{f}$, it is necessary to form $\mathbf{X}_2$ by deleting the final row and column of $\mathbf{X}_1$ and inserting a leading row and leading column of zeros. Thus the operation of a second breakage stage will be as shown in Table 4-II.

Products from the remaining three stages will be as shown in Table 4-III.

The characteristics of this type of model are:

TABLE 4-II

The second breakage stage in a process where $C_{1,1}$ and $C_{2,2}$ are zero

$$\mathbf{X}_2 \cdot \mathbf{f}_2 = \mathbf{p}_2 \qquad (\mathbf{f}_2 = \mathbf{p}_1)$$

$$\begin{bmatrix} 0 & 0 & 0 & 0 & 0 \\ 0 & 0 & 0 & 0 & 0 \\ 0 & 0.4 & 1 & 0 & 0 \\ 0 & 0.2 & 0 & 1 & 0 \\ 0 & 0.1 & 0 & 0 & 1 \end{bmatrix} \cdot \begin{bmatrix} 0 \\ 0.24 \\ 0.17 \\ 0.11 \\ 0.11 \end{bmatrix} = \begin{bmatrix} 0 \\ 0 \\ 0.266 \\ 0.158 \\ 0.134 \end{bmatrix}$$

Sums: $\mathbf{f}_2 = \mathbf{p}_1$: 0.37; $\mathbf{p}_2$: 0.442

TABLE 4-III

Product from the nth breakage stage where $C_{1,1}$ to $C_{n,n}$ are zero

Stage 3	Stage 4	Stage 5
0	0	0
0	0	0
0	0	0
0.264	0	0
0.187	0.293	0
0.549	0.707	1.0

(1) it provides for the progressive elimination of the top size ranges in the feed to the process as the retention time in the process is increased;
(2) for a particular process which can be described by a simple breakage matrix $\mathbf{X}_j$, one parameter is suitable to describe the results of changes in operating conditions; this parameter is v, the number of stages of breakage;
(3) change in ore type is described by change in the breakage matrix.

A critical variable in this model is the breakage matrix $\mathbf{X}_j$ which includes the parameters **B**, **S** and **C**. Change in any of these parameters will give a change in $\mathbf{X}_j$ but there is not necessarily only one combination of values for **B**, **S** and **C** which will give the same value for $\mathbf{X}_j$. These parameters cannot be uniquely determined for ores or machines but the model can still be used for simulation purposes provided that suitable assumptions are made. The approach which has been found to be suitable is to assume reasonable values for **B** and **C**, and to vary **S** to account for change in feed type or machine characteristics. These parameters are discussed further below.

4.2.2 *The breakage matrix*

The main breakage mechanism which occurs in rod mills is shatter breakage. Some breakage by attrition may occur, for instance on particles which are caught between sliding rods and the mill liner in the "lifting" zone of the rod mass, but this is a minor component of the total breakage process. It is certain that the functions which describe shatter breakage are not the same for different minerals but it is not possible to derive a unique value of a breakage function for a mineral. The alternative to using unique values of breakage functions for each mineral is to assume that the breakage function is constant and that the selection function is varied to account for differences in processes. Simulations of many rod mill grinding processes have been carried out using the model given in eq. 4-16 and assuming a constant breakage function and these simulations have been found to be accurate.

The breakage function which has been used was the function proposed by Broadbent and Callcott (1956/57), namely:

$$B(x/y) = (1 - e^{(-x/y)})/(1 - e^{-1}) \tag{4-18}$$

If the fractions, x/y, are expressed as a geometric progression then the values in the step matrix **B**, which conform to that geometric progression, may be calculated. If the ratio between successive sieves is 2^{-1}, the numerical elements in the first column of **B** are:

⟨0.1980 0.3308 0.2148 0.1225 0.0654 0.0338 0.0172 . . . . .⟩

If the ratio is $2^{-0.5}$, the numerical values are:

⟨0.1004 0.1906 0.1161 0.1361 0.1069 0.0814 0.0607 . . . . .⟩

Some other breakage function could have been chosen and provided that it had the correct form, it would have been just as suitable for use in a simulation model.

4.2.3 *The classification matrix*

A stage was defined as the interval required to eliminate the largest sieve fraction from the mill feed in the first instance and thereafter from the feed to successive stages. This requires that $C_{1,1}$ should be unity in the first stage, that is, that all particles in this size range should be retained within the breakage zone until they are reduced to less than the lower limiting size. $C_{2,2}$ must be unity in stage 2, $C_{3,3}$ in stage 3 and so on.

Values other than zero may be chosen for the other elements in stage 1 and these will then change systematically in successive stages. However, the only necessary condition is that $C_{1,1} = 1$. In practice, it has generally been found that a suitable classification matrix contains $C_{1,1} = 1$ and all other elements equal to zero in stage 1.

4.2.4 *The selection matrix*

If the values of **B** and **C** are fixed, the variations in the ore and machine characteristics are described by the selection matrix, and this is determined for a given process by a curve fitting procedure. It is generally found that the diagonal elements in S decrease with decrease in particle size, which would be expected.

Selection is one of the two variable parameters in the mathematical model of a rod mill process, the other being stages of breakage. The selection matrix incorporates a description of the ore and mill characteristics, and the number of stages of breakage is related to the feed rate, the feed size, the mill speed and other process variables.

4.2.5 *The stage breakage matrix*

The stage breakage matrix, $\mathbf{X}_j$, includes the individual matrices **B**, **S** and **C**. Change in any of these alters $\mathbf{X}_j$ and also alters the product size predicted by the model for a given feed size. The number of stages of breakage in a grinding process is a measure of the extent to which breakage proceeds and is closely linked to the retention time in the mill. It is invariably proportional to the feed rate and the relationship takes the form shown in Fig. 4-9 for constant feed size. This relationship shifts slightly, as shown in Fig. 4-9, where the feed size changes.

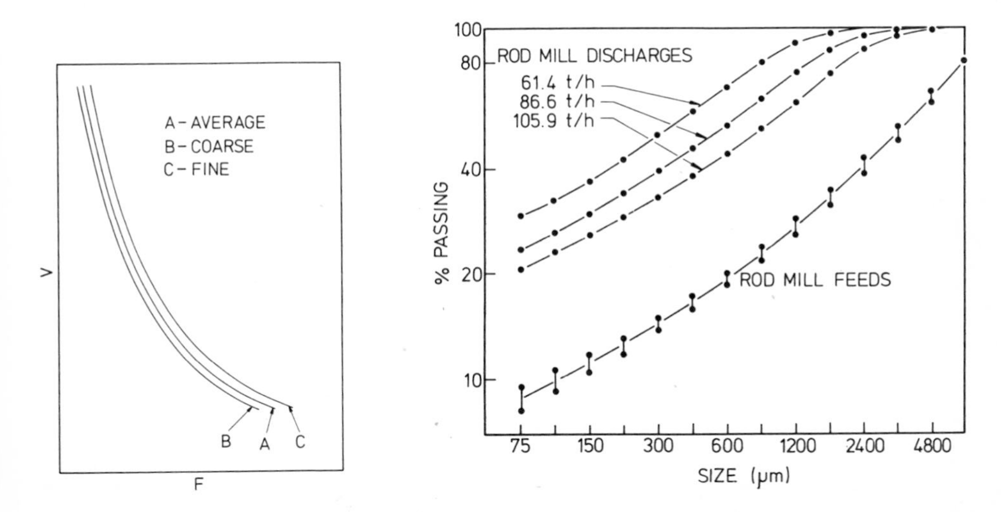

Fig. 4-9. General form of the feed rate, stages of breakage relationship.

Fig. 4-10. Size distributions of rod-mill discharges at different feed rates.

4.2.6 *Application of the model to operating mills*

In order to investigate the validity of this model over a wide range of operating conditions, a considerable amount of test work has been carried out on the 2.74 m by 3.66 m open-circuit rod mills treating Mount Isa Mines Limited copper and lead—zinc ores. The size distributions shown in Fig. 4-10

TABLE 4-IV

Numerical values in the X_j matrix

0.0000	0.0000	0.0000	0.0000	0.0000	0.0000	0.0000	0.0000
0.4126	0.1980	0.0000	0.0000	0.0000	0.0000	0.0000	0.0000
0.2677	0.3309	0.5990	0.0000	0.0000	0.0000	0.0000	0.0000
0.1527	0.2147	0.1655	0.7995	0.0000	0.0000	0.0000	0.0000
0.0815	0.1225	0.1074	0.0827	0.7995	0.0000	0.0000	0.0000
0.0421	0.0654	0.0613	0.0537	0.0827	0.7995	0.0000	0.0000
0.0214	0.0338	0.0327	0.0306	0.0537	0.0827	0.5990	0.0000
0.0054	0.0087	0.0086	0.0085	0.0164	0.0306	0.1074	0.5990

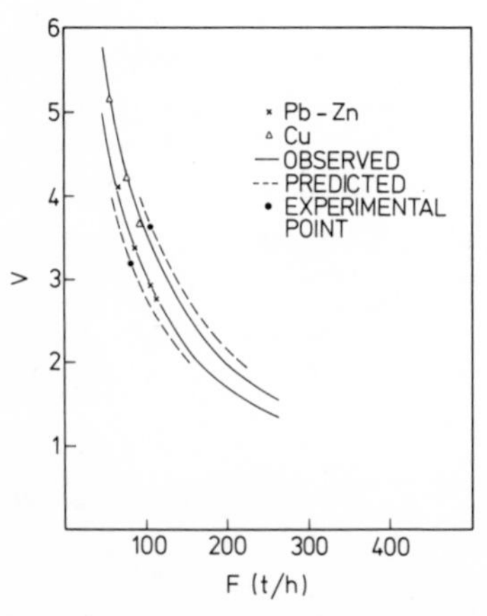

Fig. 4-11 Experimental verification of the feed rate, stages of breakage relationship. The full lines show the predicted effect of feed rate on stages of breakage at the same feed size and the dotted lines show the effect at different feed sizes (coarser for the lead—zinc ore, finer for the copper ore). The experimental points which verify the relationships are also shown.

were from some of these tests. The numerical values used in the X_j matrix in applying the model to this mill were as shown in Table 4-IV (sieve ratio 2:1).

The graphs showing the observed and calculated relationships between the feed rates and the stages of breakage are given in Fig. 4-11. The continuous lines are the F—v relationships developed from a single observation of the manner in which the ore is being broken in the mills and the individual points show the results of experimental tests. Calculated product-sizing analyses using this model were very close to the observed sizing analyses.

The technical literature contains few other results sufficiently detailed for an analysis of this type. However, the results which are available have been analysed and F—v graphs of the type shown in Fig. 4-11 apply in every case. Satisfactory agreement between the observed and calculated product-sizing analyses have also been obtained in every case (Lynch, 1959). Mitchell et al. (1954a, b) carried out extensive work on 0.76 m by 1.22 m overflow and end-peripheral discharge rod mills. T.G. Callcott (private communication) analysed these results in terms of stages of breakage and it was found that, with a maximum size of 0.038 m and the selection matrix **S** equated to **I**, the classification matrix **C** had the following elements:

1, 0.5, 0.2, 0.2, 0.25, 0.35, 0.5, 0.6.

This work of Mitchell et al. has also provided data to enable an assessment to be made concerning the effect of mill speed upon breakage. A graph showing the relationship between stages of breakage and the percent critical speed for the Allis-Chalmers mill is shown in Fig. 4-12. The relative effect of mill speed on breakage is shown in Fig. 4-13.

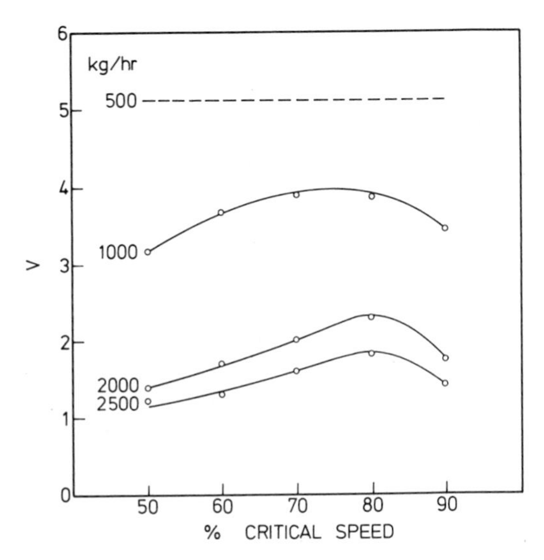

Fig. 4-12. Mill speed, stages of breakage relationship.

The following comments may be made:

(1) Where v is high, change in mill speed has little effect upon v. This agrees well with the observations on the New Broken Hill Consolidated 2.44 m by 3.66 m (Lynch, 1959), and the Tennessee Copper Company 1.83 m by 2.74 m (Myers and Lewis, 1949) rod mills.

(2) As v decreases, in this case due to a large increase in feed rate (but it could also be due to variation in the grindability of the ore at constant feed rate), change in mill speed has a progressively greater effect upon v.

(3) Change in mill speed affects the capacity of the mill to accept ore at a constant value of **f**. This is not to be confused with the effect of mill speeds upon the F—v graph.

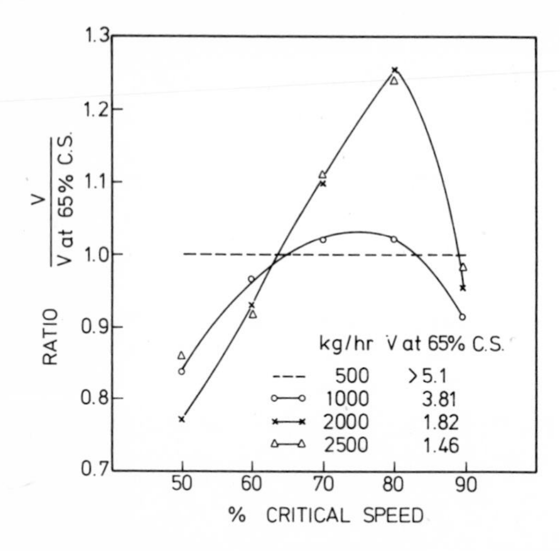

Fig. 4-13. Relative effect of change in mill speed on stages of breakage.

It has been found that, over the normal range of rod-mill operation, the following relationship is approximately correct:

$$(\text{mill feed rate}) \cdot (\text{number of stages of breakage})^{1.5} = \text{constant} \qquad (4\text{-}19)$$

The constant has been termed the mill constant, MC.

4.2.7 *Discussion*

The mathematical model of rod mills which has been discussed is not the only one which has been considered. Many others have been rejected because they failed to account in some way for observed operating behaviour. This model is not the only one suitable to describe the behaviour of rod mills; however, it is an accurate model and may be used as a mathematical description of a rod mill operating under normal conditions.

4.3 BALL MILLS

4.3.1 *Form of the model*

A graph showing feed and product size distribution for typical closed circuit overflow ball mills is given in Fig. 4-14. For grate discharge mills, the 100% passing size of the products may be affected by other variables. The following discussion will refer to overflow discharge mills unless grate mills are mentioned specifically.

The screening effect which occurs within rod mills and which led to the

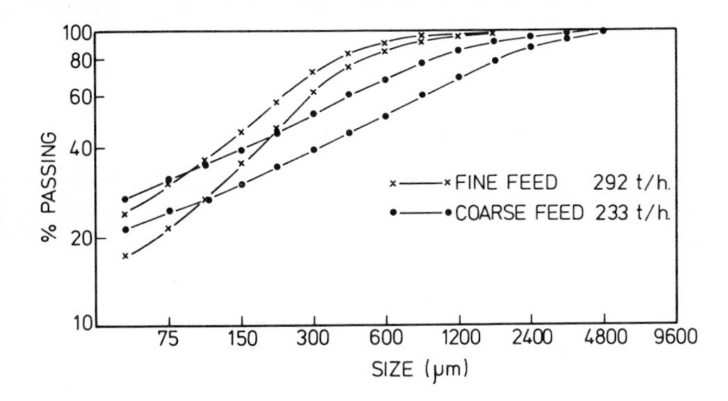

Fig. 4-14. Feed and product-size distributions for typical overflow-type closed circuit ball mills.

development of the rod mill equations is not a major feature of ball mill behaviour. Particles in the coarser size ranges in the feed to a ball mill may appear in the same size ranges in the mill product, although breakage of these particles may have occurred during passage through the mill. This contrasts with the behaviour of rod mills. The assumption can be made that the breakage process occurring within any one vertical increment in a continuous cylindrical ball mill is the same as that occurring within any other vertical increment, although the products of breakage from successive increments change because the feeds to these increments change. Successive repetitive processes in ball mills may be regarded as identical, as distinct from those in rod mills, and the process can be described by the following equations:

$$\mathbf{p} = (\mathbf{X}_j)^v \cdot \mathbf{f} \tag{4-20}$$

$$\mathbf{X}_j = \mathbf{B} \cdot \mathbf{S} + \mathbf{I} - \mathbf{S} \tag{4-21}$$

The breakage matrix for the jth stage, $\mathbf{X}_j$, does not change for any other stage. In this case also for any given ore ground in the particular circuit, the ball mill operation is defined entirely by the parameter v. The definition of v for a ball mill is not identical with v for a rod mill.

A stage of breakage within ball mills may be defined as the interval in which all particles in the maximum size range in the feed to that stage have a unit probability of being broken, and this is equivalent to writing unity in the (1,1) element in S. However, v and the remaining elements in S are still dependent upon **B**, which is not known absolutely. Numerical values in $\mathbf{X}_j$, which make this matrix suitable for process analysis and simulation on the one hand and comparable for different grinding operations on the other, may be determined by a process of iterative calculations.

4.3.2 *Application of the model to operating mills*

The model given in eqs. 4-20 and 4-21 has been used in the simulation of many ball-mill circuits using the following approach:

(1) assume that a Broadbent-Callcott breakage function as described in eq. 4-18 is applicable; (2) assume that the 1,1 element in S has a value of 1.0; (3) assume that the product (mill feed rate)·(number of cycles of breakage) is a constant, MC.

If the size distributions of the mill feed and product are known for one feed rate, all elements in S and the value of MC may be determined by a least-squares fitting. All information is then available to calculate the product for any change in feed rate or feed size distribution using eq. 4-20.

4.3.3 *Mixing in ball mills*

Comments in this section may be applied to rod mills as well as ball mills.

A ball mill may be considered as a mixer in which breakage is taking place. Danckwerts (1953) outlined various types of mixing that could occur. He considered that two basic mixing patterns exist, piston-flow and perfect mixing, but that these may be modified by limited longitudinal mixing in the first case and dead space in the second. Material that passes through a mixer with piston flow has a single value of residence time within the unit, whereas material that travels through a mixer with any other type of mixing pattern has a range of residence times. The size analysis of material in a batch mill is continuously changing. Continuous grinding systems operating in equilibrium have constant additions of material to and removal of material from the mill while the size analysis of particles in the mill is constant. If a mill is considered as a perfect mixer, the size analysis of the product will be that of the material throughout the mill. Under any other conditions of mixing this will not be so.

A grinding mill may be considered as consisting of a number of transverse sections, each of which contains the same amount of material. The material in the final segment of such a representation is equivalent to the product from the mill. During one cycle of mixing, some of the material in each segment will travel towards the product end of the mill, some will remain in the same segment, while the rest will be back-mixed towards the feed end of the mill. Material from each segment will be mixed in a similar manner. There will be, however, an overall forward movement of material to the discharge end of the mill. Once material appears in the product segment of the mill, it is lost from the system and cannot be subjected to any back-mixing.

The longer a group of particles is retained in the mill the greater is the breakage which occurs. If the retention time distribution is known, the

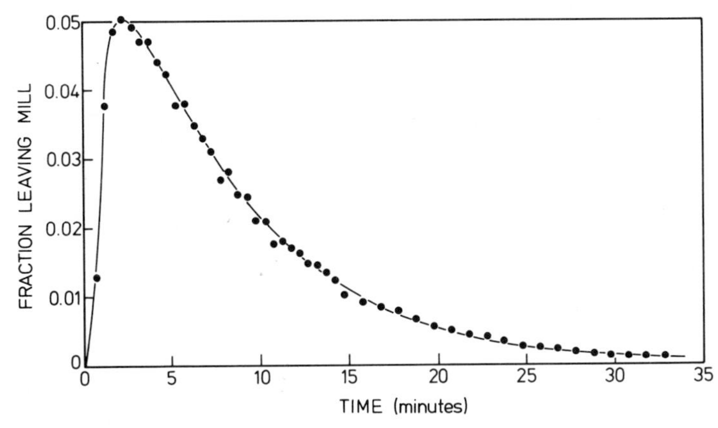

Fig. 4-15. Typical distribution of residence times of pulp in a continuous ball mill. An impulse of tracer was introduced into the mill feed at zero time and the fractions leaving the mill after successive time periods are shown on the y-axis.

fractions of the pulp leaving the mill which have been in the mill for different and known periods of time may be calculated. The total product size distribution is the weighted average of the size distributions of the broken fractions. In the perfect mixing model of Whiten (1972), it is assumed that the mill is a single perfectly mixed segment and the calculations are simplified. A typical distribution of residence time for a continuous ball mill is shown in Fig. 4-15. If steady-state behaviour of the mill is to be studied then the distribution of retention times need not be considered, but if the dynamic behaviour is to be studied, this must be considered.

4.3.4 *The behaviour of ball mills of different diameters*

It has been noted that the energy consumed per tonne of ore for a given size reduction is higher for very large ball mills than for smaller ball mills. Thus metallurgists at Bougainville Copper Limited, Panguna, (private communication) reported a higher apparent work index for ore broken in a 5.34 m diameter by 6.41 m long ball mill than for the same ore broken in a 1.71 m by 1.43 m mill. There are few comparative data available concerning breakage of the same ore in mills of different diameters, however, a qualitative appreciation of the reason for this could be obtained by deriving the parameters in the mill models for the small and large mills at Panguna. The selection functions for these mills are shown in Fig. 4-16, (Lees, 1975).

The important indication from Fig. 4-16 is that there is a noticeable decrease in the values in the selection function as particle size increases with the very large mill and that this does not seem to occur with the smaller mill. The larger mill, as it is presently operated, appears to be less able to cope with the coarser particles than the smaller mill. Much work remains to be

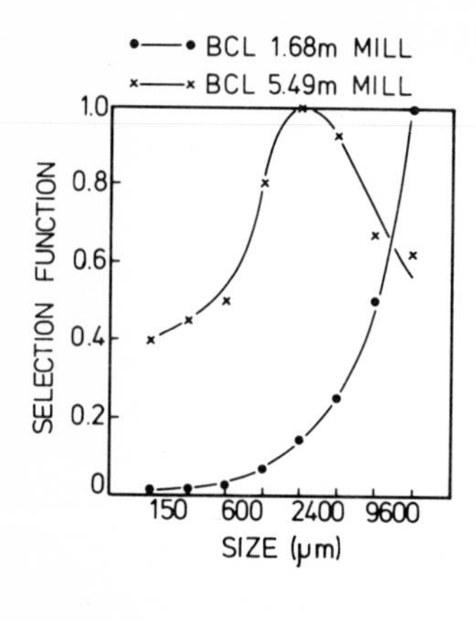

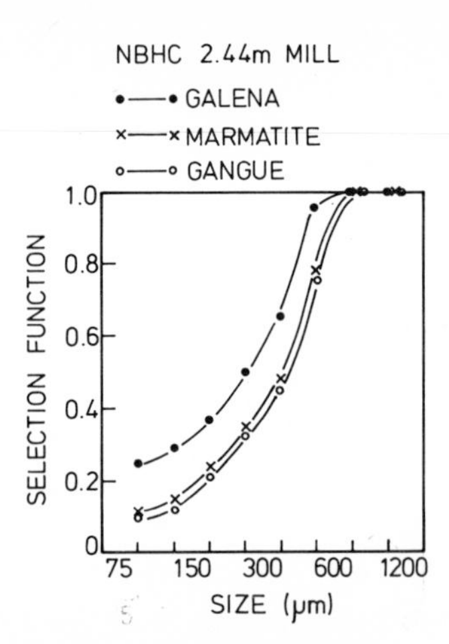

Fig. 4-16. Selection functions for ball mills of different diameters showing reduction in probability of breakage of larger particles as mill size increases.

Fig. 4-17. Selection functions for different minerals in the same mill. Elements in the selection function are shown on the *y*-axis.

done to determine the size of particle and the diameter of mill at which this inefficiency starts to occur. The solution to the efficient use of large-diameter mills may be found in reducing the size of the new feed to the circuit as the mill diameter increases.

The problem is particularly evident in the size range 4700—9400 μm but its occurrence above this size may be obscured by the fact that classification within the mill becomes very significant as particle size increases. The correct model to use may be the complete model:

$$\mathbf{p} = (\mathbf{I} - \mathbf{C})\cdot(\mathbf{B}\cdot\mathbf{S} + \mathbf{I} - \mathbf{S})\cdot[(\mathbf{I} - \mathbf{C})\cdot(\mathbf{B}\cdot\mathbf{S} + \mathbf{I} - \mathbf{S})]^{-1}\cdot\mathbf{f}$$

where **C** is related to the ore characteristics such as particle shape or specific gravity.

4.3.5 *The behaviour of a mixture of minerals in a ball mill*

Frequently the purpose of comminution is to liberate minerals before concentration. The efficiency of the concentration process is influenced by the degree of liberation and the sizing analysis of the valuable mineral. If this efficiency is to be kept at a maximum while variations occur in the valuable mineral content of the ore, it is useful to have some knowledge of the breakage behaviour of the different minerals. This will be discussed with reference to the Broken Hill, Australia, orebody in which the chief economic minerals are galena, marmatite, and tetrahedrite, and the chief gangue minerals are quartz, rhodonite, bustamite, garnet, fluorite, and calcite. The economic minerals are coarse grained and liberation from constituent gangue minerals

TABLE 4-V

Sizing analyses of flotation products at Broken Hill

Company	% wt. minus 240-mesh B.S.S.		
	Lead concentrate	Zinc concentrate	Residue
North Broken Hill Limited	87.4	56.6	35.7
Broken Hill South Limited	80.0	53.1	37.6
The Zinc Corporation Limited	81.3	45.7	24.7
New Broken Hill Consolidated Limited	69.7	40.0	24.2

is possible at a relatively coarse size. There is considerable variation in the sizing analyses of the flotation products as is shown in Table 4-V. This Table shows that preferential grinding of the valuable minerals, in particular of the galena particles, occurs.

The behaviour of the minerals in the New Broken Hill Consolidated Limited (NBHC) ball mill—rake classifier grinding circuit was studied over a wide range of operating conditions. All size fractions of all samples taken during the investigation were assayed for lead and zinc, and these were converted to galena and marmatite by assuming that galena contains 86.6% lead and marmatite 56.6% zinc. The assumption was made that all particles of galena and marmatite are completely liberated; this is not correct but it is an approximation which enables some insight to be gained into the behaviour of the individual minerals. The best fit values of the selection functions for galena, marmatite and gangue, calculated from the mill feeds and products, are shown in Fig. 4-17. It will be noted that galena and marmatite break more rapidly than the gangue minerals and this can be represented in a model by appropriate changes in the selection functions.

The breakage properties of minerals are not the only characteristics which affect their sizing distributions in the closed grinding circuit products. Their classification behaviour also has a large effect on this and this is discussed with respect to hydrocyclones in Chapter 6.

4.4 AUTOGENOUS MILLS

The term autogenous milling means a process in which the size of the constituent pieces of a supply of rock is reduced in a tumbling mill purely by the interaction of the pieces, or by the interaction of the pieces with the mill shell, no other grinding medium being employed. The definition thus covers both "run-of-mine" and "pebble" milling, the only difference from the mathematical modelling viewpoint being that the feed to the first has a continuous, and the second a non-continuous, size distribution. Autogenous milling differs fundamentally from non-autogenous milling in two respects.

(1) Size reduction occurs by two main modes, namely the detachment of material from the surface of larger particles (referred to as "abrasion") and disintegration of smaller particles (called "crushing"). Abrasion and crushing breakage overlap on the size scale. This contrasts with non-autogenous milling, in which only crushing breakage is regarded as significant.

(2) The grinding parameters of the autogenous mill load are not independent of the mill feed; the load is continually generated from the feed, and its parameters therefore depend directly on those of the feed.

These two characteristics must be specifically included in the model of the autogenous mill.

4.4.1 *Form of the model*

Because of the importance of the load in autogenous milling, the type of model adopted must include the load. The "perfect mixing model" contains the mill load as a parameter and is suitable for describing the behaviour of autogenous mills. The important equations are repeated here for convenience. The perfect mixing model is written:

$$\mathrm{d}\mathbf{s}/\mathrm{d}t = (\mathbf{A}\cdot\mathbf{R}-\mathbf{R})\cdot\mathbf{s}+\mathbf{f}-\mathbf{p} \tag{4-22}$$

The relationship between contents and product is:

$$\mathbf{p} = \mathbf{D}\cdot\mathbf{s} \tag{4-23}$$

where $\mathbf{D}$ (the discharge matrix) is a diagonal matrix giving the discharge rates of each component of s. Consequently:

$$\mathrm{d}\mathbf{s}/\mathrm{d}t = (\mathbf{A}\cdot\mathbf{R}-\mathbf{R}-\mathbf{D})\cdot\mathbf{s}+\mathbf{f} \tag{4-24}$$

and, for steady-state condition:

$$(\mathbf{A}\cdot\mathbf{R}-\mathbf{R}-\mathbf{D})\cdot\mathbf{s}+\mathbf{f} = 0 \tag{4-25}$$

If $\mathbf{A}$ is known or assumed, and $\mathbf{f}$, s and $\mathbf{p}$ are known, both $\mathbf{R}$ and $\mathbf{D}$ can be computed. If s, the mill contents, is not known, $\mathbf{R}$ and $\mathbf{D}$ cannot be separated but a combined parameter, $\mathbf{D}\cdot\mathbf{R}^{-1}$, can be calculated if $\mathbf{f}$ and $\mathbf{p}$ are known (see eq. 4-28). There are two types of models which may be used to simulate autogenous mills: a model which includes s, the mill contents, and which should be used where complete simulation is required, and a model which does not include s and which may be used to predict the product size for change in feed rate or feed size. The first model is a much better model to use because the mill load is a critical variable in autogenous milling but the experimental work required to obtain the data to use this model is considerable.

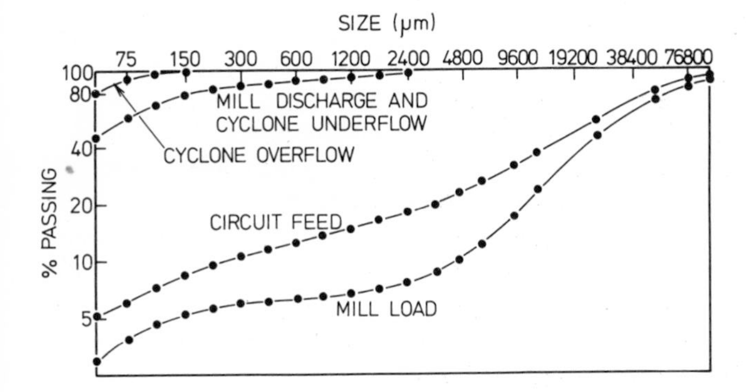

Fig. 4-18. Typical size distributions of particles in pulp streams in the autogenous mill—hydrocyclone circuit at the Warrego concentrator of Peko Mines Limited.

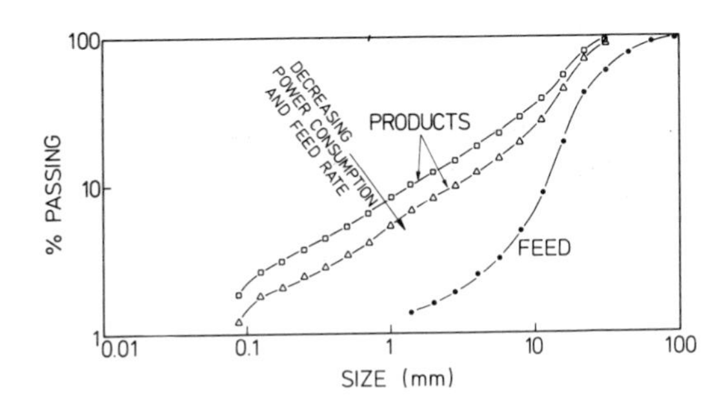

Fig. 4-19 Effects of changes in operating variables on autogenous grinding.

4.4.2 *Mill load*

In autogenous milling, the mill load is the critical operating variable because any change in feed characteristics is reflected in a change in the mill load. An increase in mill load is accompanied by an increase in power draft to a maximum value but this decreases as the load is increased further. Maximum grinding throughput is obtained at maximum power draft. Any change in feed addition rate causes change in the mill load and in the product size. Typical size distributions of the solid particles in pulp streams in an autogenous mill–hydrocyclone circuit are shown in Fig. 4-18, and the nature of the relationship between circuit feed rate, circuit feed size, mill power and mill load for an autogenous mill–hydrocyclone circuit is shown in Fig. 4-19.

4.4.3 *The breakage matrix*

Abrasion and crushing breakage occur simultaneously in the autogenous mill and both must be included in the breakage matrix. Abrasion applies mainly in the coarser sizes, crushing mainly in the finer sizes and there is a transition zone between the two. When pieces evenly distributed over a single size fraction are subjected to an abrasion breakage "event", all the pieces will be slightly reduced in size, some near the lower limit of the fraction will pass into the next smaller fraction, and the material abraded from the original pieces will report to a number of size ranges further down the size scale. From a consideration of the relative mean weights of particles in successive $\sqrt{2}$ size fractions, Wickham (1972) postulated that, if the particles in a single size interval lose a proportion x of their weight as the result of an abrasion breakage event, then $0.354 \cdot x$ (that is $(1/\sqrt{2})^3 \cdot x$) will appear in the next smaller $\sqrt{2}$ interval, and the remaining $0.646 \cdot x$ will form the detritus, which is spread over a number of considerably smaller size intervals. This concept may be used to describe abrasion breakage.

To describe crushing breakage, the following modification of the Rosin-Rammler equation proposed by Broadbent and Callcott (1956) will be used:

$$\mathbf{B}(x, y) = (1 - e^{(-x/y)^u})/(1 - e^{-1}) \tag{4-26}$$

In describing rod and ball mill operations, the exponent u was allotted the value of unity and the resulting model was found to be accurate, but in the case of autogenous mills, the best value of u was found to be 2. Two other parameters which must be known before the complete breakage function matrix can be set up are the upper and lower limits of the transition zone between abrasion and crushing breakage. The upper limit, that is, the size above which no crushing breakage occurs, has been named the crushing limit; similarly, the lower limit, the size below which no abrasion breakage occurs, has been designated the abrasion limit.

The crushing and abrasion limits may be determined by visual examination of the size fractions in the load. By such examination, it is possible to fix the size interval in which the sharp edges and conchoidal faces characteristic of crushing breakage first appear to any significant extent among the smooth fragments generated by abrasion. The upper size of this interval is taken to be the crushing limit. Through successively finer fractions, it has been noted that the proportion of fragments resulting from crushing steadily increase at the expense of those resulting from abrasion, until the latter disappear completely. The upper size of the first interval to contain no abraded fragments is taken as the abrasion limit. Generally, the transition from all-abrasion to all-crushing occupies six $\sqrt{2}$ size intervals. Within the abrasion-to-crushing transition zone, a simple linear transition of the type:

$$\mathbf{B} = \alpha \cdot \mathbf{B}_1 + (1 - \alpha) \cdot \mathbf{B}_2 \tag{4-27}$$

is used, where $\mathbf{B}_1$ is the abrasion breakage function, $\mathbf{B}_2$ is the crushing function, and α is the proportion of the distance (in size-interval terms) across the transition zone.

The regions over which crushing and abrasion breakage apply have been

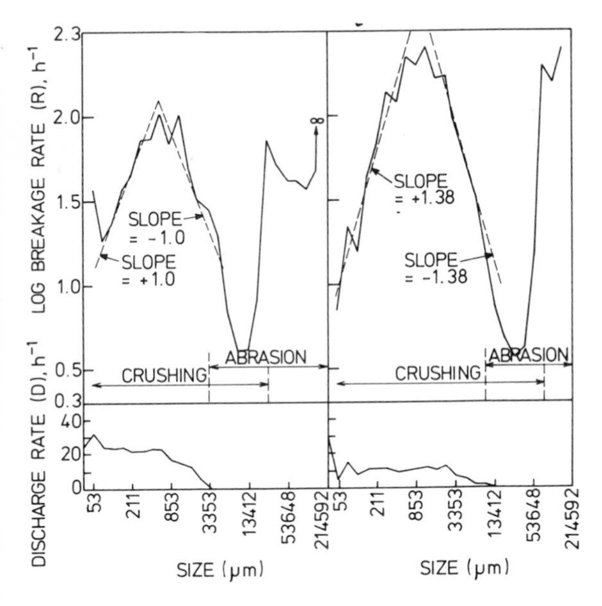

Fig. 4-20. Typical breakage and discharge rate functions for the Cobar (left hand side) and Warrego (right hand side) autogenous mills. Regions over which crushing and abrasion breakage apply are shown.

found experimentally for siliceous ores to be as shown in Fig. 4-20. While a breakage matrix of the type given in eq. 4-27 must be used where precise simulation is required, it is difficult to obtain data to determine the limits of the transition zone and the exact nature of α. The simpler approach, which is less accurate but which may not involve major errors, is to assume an abrupt changeover from one form of breakage to another. The assumption that abrasion breakage applies above 12700 μm and crushing breakage below this size has been found to be reasonable.

4.4.4 *The breakage rate*

If the mill load, **s**, and the product size, **p**, are known, the discharge rate matrix, **D**, can be calculated. If the feed size, **f**, and the breakage matrix, **A**, are also known, the breakage rate, **R**, can be calculated. The characteristics of the breakage rate and discharge rate matrices have been studied experimentally using this approach. The units of each matrix are h^{-1}. Typical breakage rate and discharge rate functions are shown in Fig. 4.20. It has been found that rock wear, or abrasion breakage, is dependent on particle weight down to about 5 cm, then dependent on particle surface area. This is shown in Fig. 4-21. Consequently, the breakage rate function will be constant down to the changeover point and will then increase exponentially with decreasing size until the crushing limit is reached. If the pebble wear in

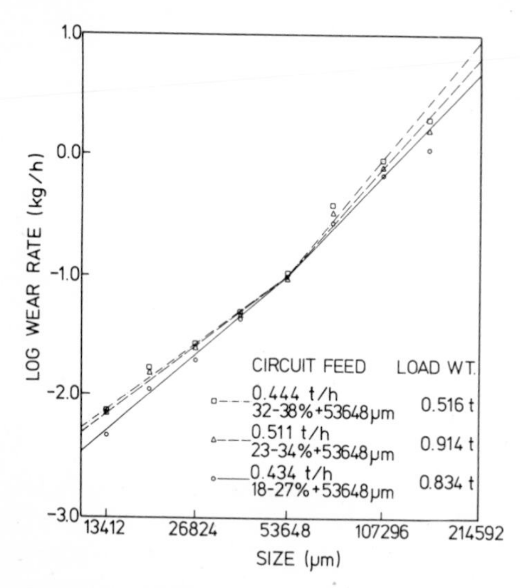

Fig. 4-21. Rate of pebble wear related to particle size.

any one size interval can be predicted and the changeover point is known, the breakage rates for all sizes above the crushing limit follow automatically.

The crushing breakage portion of the breakage rate curve can be simulated with fair accuracy by two intersecting straight lines of slope $+A$ and $-A$, respectively, where A is approximately 1.0. The implication is that the probability of crushing breakage increases with decreasing size to a maximum and thereafter decreases with decreasing size. The problem of simulating the crushing breakage rate function then becomes the problem of determining the particle size at which the peak rate occurs and the magnitude of the peak. The size at which the peak breakage rate occurs requires further study as it appears to be related to both the ore type and the mill. The evidence so far is that it is in the range 700—900 μm. The magnitude of the peak was found to be controlled by three factors: (1) the net mill energy input per unit weight of ore fed to the mill (new feed plus circulating load), (2) the weight of dry load, and (3) the concentration in the pebble portion of the circuit feed of the "most effective" pebble size.

4.4.5 *The discharge-rate matrix*

It will be seen from Fig. 4-20 that the discharge-rate function includes a portion which is independent of particle size and a portion which is size-dependent. The problem of simulating the discharge function of the mill can be reduced to two elements: (1) determining the magnitude of the plateau value of the function $D_{\max}$, and (2) determining the maximum size to which the plateau extends (size $D_{\max}$). The end of the plateau is then connected to the series of zero values commencing at the discharge aperture size by means of an S-curve of the form:

$$D(I) = D_{max} \cdot (x - b)^2 \cdot (2 \cdot x - 3 \cdot a + b)/(b - a)^3$$

where x = log size at $D(I)$, a = log size D_{max}, and b = log aperture size of mill discharge.

Analysis of test results has shown that D_{max} is controlled by two factors: (1) the pulp density of the mill discharge (decreasing linearly with increasing density), and (2) the size distribution of the load. Size D_{max} depended only on the discharge pulp density, increasing with this quantity.

For the prediction of steady-state results, use can be made of the combined parameter $\mathbf{D} \cdot \mathbf{R}^{-1}$ which can be calculated from $\mathbf{f}$ and $\mathbf{p}$ by the equation:

$$\mathbf{D} \cdot \mathbf{R}^{-1} = (\mathbf{I} - \mathbf{A}) \cdot \mathbf{p} \cdot (\mathbf{f} - \mathbf{p})^{-1} \tag{4-28}$$

This function was computed for tests carried out on several autogenous mills and the general form of the results is given in Fig. 4-22. When the diagonal elements of each $\mathbf{D} \cdot \mathbf{R}^{-1}$ matrix were plotted against size on a log-log basis for one mill, it became apparent that these functions followed the relationship:

$$\mathbf{D} \cdot \mathbf{R}^{-1} = P \cdot \mathbf{X} \tag{4-29}$$

where $\mathbf{X}$ was a typical $\mathbf{D} \cdot \mathbf{R}^{-1}$ matrix chosen as a standard and where P was effectively a multiplier. P was determined so as to minimise the sum of squares of errors between the observed and predicted mill products. Fig. 4-23 shows a linear relationship between the parameter P and tonnage throughput/amp drawn by the mill. With this method for predicting P, and hence $\mathbf{D} \cdot \mathbf{R}^{-1}$, and with the breakage function given in Table 4-VI, simulations of the mill behaviour can be carried out. For the prediction of total operating conditions

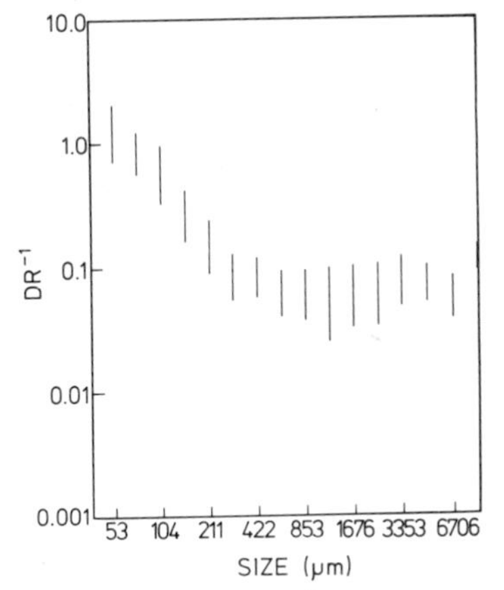

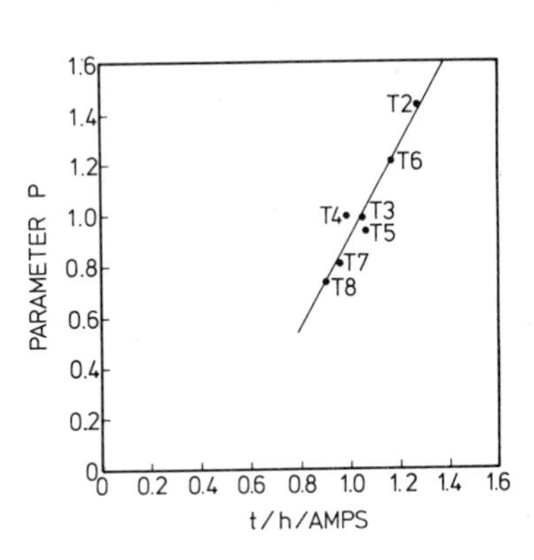

Fig. 4-22. The $\mathbf{D} \cdot \mathbf{R}^{-1}$ function for the rests on the Warrego mill.

Fig. 4-23 Linear relationship between the multiplier of $\mathbf{D} \cdot \mathbf{R}^{-1}$ and the tonnage throughput per amps drawn by the autogenous mill.

TABLE 4-VI

Abrasion and crushing breakage functions used in the autogenous mill model

Size (microns)	Abrasion breakage function (1st column)	Crushing breakage function (1st column)
152400	0.9000	
101600	0.0352	
76200	0.0002	
50800	0.0007	
38100	0.0011	
25400	0.0018	
19400	0.0022	
12700	0.0026	
9520	0.0026	0.0836
6350	0.0026	0.2070
4780	0.0026	0.1640
3327	0.0026	0.1300
2326	0.0026	0.0114
1651	0.0026	0.0811
1168	0.0026	0.0611
833	0.0026	0.0440
589	0.0026	0.0329
417	0.0026	0.0237
300	0.0026	0.0163
212	0.0026	0.0124
150	0.0026	0.0088
106	0.0026	0.0063
75	0.0033	0.0044
53	0.0042	0.0032
37.5	0.0050	0.0022

including the mill load, use must be made of eqs. 4-23 and 4-25. If **f**, **p** and **s** are known for one condition of operation, the discharge matrix **D** can be calculated from eq. 4-23 and the breakage-rate matrix **R** from eq. 4-25. The effect of change in operating conditions on **D** and **R** cannot yet be defined accurately but there is an indication (Gault, 1975) that eq. 4-30 is valid for many conditions:

$$\mathbf{D}/F = \text{constant} \tag{4-30}$$

where F is the feed rate.

By combining eqs. 4-29 and 4-30, it will be seen that the breakage rate **R** may vary systematically with the feed rate but experimental verification of this is not yet available. It must be recognised that the relationship between **D** and **R** and the operating conditions must be determined before the model can be used for circuit design and control purposes.

4.4.6 *Partial (or semi) autogenous milling*

It is common practice to add a small mass of steel grinding media to an autogenous mill to increase the rate of breakage of the finer particles. Mathematically, the effect of this is to increase the values of the elements in the lower part of the **R** matrix and this may have some effect on the elements in the upper parts also. Experimental data to determine these effects are lacking. The simulation of partial autogenous mills may be carried out in the same way as for pure autogenous mills.

NUMERICAL EXAMPLES

Example 4-1: cone-crusher model. The breakage matrix for a particular crushing operation is given below. The matrix is made up from both coarse and fine components and is not a step matrix because the production of fines is not dependent upon original particle size. The size distribution of the feed is given by f for the sizings shown in mm.

$$\mathbf{B} = \begin{bmatrix} 0.64 & & & \\ 0.15 & 0.64 & & \\ 0.10 & 0.17 & 0.66 & \\ 0.05 & 0.13 & 0.20 & 0.71 \end{bmatrix}$$

	weight(%)	size(mm)
f =	20.2	102/51
	33.3	51/25.5
	18.7	25.5/12.8
	15.8	12.8/6.4
	12.0	

The classification matrix which has been found to be applicable is:

$$\mathbf{C} = \begin{bmatrix} 1 & & & \\ 0 & 0.8 & & \\ 0 & 0 & 0.4 & \\ 0 & 0 & 0 & 0 \end{bmatrix}$$

Determine the size distribution of the crusher product and the amperage drawn by the crusher given that eqs. 4-14 and 4-15 are applicable. From eq. 4-4:

$$\mathbf{p} = (\mathbf{I}-\mathbf{C})\cdot(\mathbf{I}-\mathbf{B}\cdot\mathbf{C})^{-1}\cdot\mathbf{f}$$

Solution:

The problem can be solved without inversion of the matrix $(\mathbf{I}-\mathbf{B}\cdot\mathbf{C})$:

put $\mathbf{x} = (\mathbf{I}-\mathbf{B}\cdot\mathbf{C})^{-1}\cdot\mathbf{f}$, therefore: $(\mathbf{I}-\mathbf{B}\cdot\mathbf{C})\cdot\mathbf{x} = \mathbf{f}$

First calculate $\mathbf{B}\cdot\mathbf{C}$ by matrix multiplication:

$$\mathbf{B}\cdot\mathbf{C} = \begin{bmatrix} 0.64 & & & \\ 0.15 & 0.64 & & \\ 0.10 & 0.17 & 0.66 & \\ 0.05 & 0.13 & 0.20 & 0.71 \end{bmatrix} \cdot \begin{bmatrix} 1 & & & \\ 0 & 0.8 & & \\ 0 & 0 & 0.4 & \\ 0 & 0 & 0 & 0 \end{bmatrix} = \begin{bmatrix} 0.64 & & & \\ 0.15 & 0.51 & & \\ 0.10 & 0.14 & 0.26 & \\ 0.05 & 0.10 & 0.08 & 0 \end{bmatrix}$$

Now calculate $(\mathbf{I}-\mathbf{B}\cdot\mathbf{C})$:

$$\mathbf{I} - \mathbf{B} \cdot \mathbf{C} = \begin{bmatrix} 0.36 & & & \\ -0.15 & 0.49 & & \\ -0.10 & -0.14 & 0.74 & \\ -0.05 & -0.10 & -0.08 & 1 \end{bmatrix}$$

Therefore:

$$\begin{bmatrix} 0.36 & & & \\ -0.15 & 0.49 & & \\ -0.10 & -0.14 & 0.74 & \\ -0.05 & -0.10 & -0.08 & 1 \end{bmatrix} \cdot \begin{bmatrix} x_1 \\ x_2 \\ x_3 \\ x_4 \end{bmatrix} = \begin{bmatrix} 20.2 \\ 33.3 \\ 18.7 \\ 15.8 \end{bmatrix}$$

Now, solve for **x** element by element, for example:

$0.36 \cdot x_1 = 20.2, \quad x_1 = 56.11$
$0.15 \cdot 56.11 + 0.49 \cdot x_2 = 33.3, \quad x_2 = 85.16$

Similarly: $x_3 = 48.96, \quad x_4 = 31.04$

Now, $\mathbf{p} = (\mathbf{I} - \mathbf{C}) \cdot \mathbf{x}$

Therefore;

$$\begin{bmatrix} 0 & & & \\ 0 & 0.2 & & \\ 0 & 0 & 0.6 & \\ 0 & 0 & 0 & 1 \end{bmatrix} \cdot \begin{bmatrix} 56.11 \\ 85.16 \\ 48.96 \\ 31.04 \end{bmatrix} = \begin{bmatrix} 0 \\ 17.03 \\ 29.38 \\ \underline{31.04} \end{bmatrix}$$

$$22.55$$

The vector **x** is that which is defined in eq. 4-3 and so $\mathbf{C} \cdot \mathbf{x}$, necessary for the determination of the amperage, can be calculated:

$$\mathbf{C} \cdot \mathbf{x} = \begin{bmatrix} 56.11 \\ 68.13 \\ 19.58 \\ 0 \end{bmatrix}$$

Now, using eq. 4-14:

$$a = (56.11/153 + 68.13/76.5 + 19.58/38.3) \cdot 25.4 = 44.92$$

Using eq. 4-15:

$$\text{Amperage} = 14.2 + 0.0822 \cdot 44.92 + 0.00305 \cdot 44.92^2 = 24.04 \text{ amp.}$$

Example 4-2: rod-mill model. The feed and product size distributions for copper ore ground in a 2.74 m by 3.66 m rod mill were as shown in Table 4-VII.

The feed rate was 106 t/h. What would be the product size distribution at a feed rate of 61 t/h.

TABLE 4-VII

Feed and product size distribution for copper ore ground in a rod mill

Micrometres	Feed	Product
+ 19020	1.0	0
19020/9510	2.8	0
9510/4760	31.0	0.6
4760/2380	22.6	12.6
2380/1190	14.2	25.2
1190/595	8.6	18.1
595/298	5.4	10.3
298/149	4.1	7.5
149/75	1.7	4.8
75/0	8.6	20.9

Solution:

The breakage equations for a rod mill are:

$$\mathbf{p} = \left[\prod_{j=0}^{j=v} \mathbf{X}_j\right] \cdot \mathbf{f}$$

and:

$$\mathbf{X}_j = (\mathbf{I} - \mathbf{C}) \cdot (\mathbf{B} \cdot \mathbf{S} + \mathbf{I} - \mathbf{S}) \cdot [\mathbf{I} - \mathbf{C} \cdot (\mathbf{B} \cdot \mathbf{S} + \mathbf{I} - \mathbf{S})]^{-1}$$

Elements in first row of 9 × 9 breakage matrix are:
⟨0.1980 0.3308 0.2148 0.1225 0.0654 0.0338 0.0172 0.0083 0.0043⟩

Elements on main diagonal of 9 × 9 classification matrix are:
⟨1.0 0 0 0 0 0 0 0 0⟩

Elements on main diagonal of 9 × 9 selection matrix (calculated) are:
⟨1.0 0.8 0.25 0.2 0.25 0.5 0.5 0.5 0.5⟩

Product sizing analyses calculated after successive stages of breakage are shown in Table 4-VIII.

TABLE 4-VIII

Product size distributions after successive stages of breakage in a rod mill

Micrometres	Stages			
	1	2	3	Observed
+ 19020	0	0	0	0
19020/9510	1.0	0	0	0
9510/4760	26.3	9.9	0	0.6
4760/2380	22.2	25.0	13.0	12.6
2380/1190	14.8	18.8	24.3	25.2
1190/595	8.4	11.5	17.0	18.1
595/298	6.6	7.4	10.9	10.3
298/149	5.4	6.2	7.2	7.5
149/75	3.2	4.5	5.5	4.8
75/0	12.2	16.7	22.1	20.9

Calculated stages at breakage = 3.0 because there is good agreement between the observed and calculated products at this point.

Calculated mill constant = mill feed rate • (stages of breakage)$^{1.5}$ = 106 • $3^{1.5}$ = 550.

The parameters which describe the operation of the mill have now been calculated and may be used to predict the mill performance at the altered operating conditions.

At 61 t/h, the predicted stages of breakage = (MC/feed rate)$^{0.67}$ = (550/61)$^{0.67}$ = 4.3.

The calculated product size distribution after 4.3 stages of breakage is shown in Table 4-IX.

TABLE 4-IX

Size distribution of rod-mill product

Micrometres	Product (calculated)
+ 19020	0
19020/9510	0
9510/4760	0
4760/2380	0
2380/1190	11.3
1190/595	21.7
595/298	18.1
298/149	12.0
149/75	7.6
75/0	29.3

Example 4-3: ball-mill model. A circuit grinding copper ore consists of a 2.74 m × 3.66 m rod mill followed by two 3.20 m × 3.05 m ball mills in parallel, each operating in closed circuit with a hydrocyclone. The size distributions of each ball mill feed product, at a flow rate of ore through each ball mill at 170 t/h were as shown in Table 4-X.

TABLE 4-X

Feed and product size distributions for copper ore ground in a ball mill

Tyler mesh	Micrometres	Feed	Product
+ 4	+ 4699	0.1	0
4/8	4699/2362	2.6	0.6
8/14	2362/1168	10.4	2.9
14/28	1168/589	15.2	8.3
28/48	589/295	21.1	19.0
48/100	295/147	19.6	22.8
100/200	147/74	10.6	14.7
− 200	74/0	20.4	31.7

What is the calculated ball-mill product size distribution if the circuit operating conditions are changed so that the flow rate of ore through each ball mill is 208 t/h and the feed size distribution becomes:

⟨1.0 7.2 14.7 18.0 19.9 14.9 8.1 16.2⟩

Solution:

The breakage equation for a ball mill is:

$$p = (B \cdot S + I - S) \cdot f$$

Elements in first row of 7 × 7 breakage matrix are:
⟨0.1980 0.3308 0.2148 0.1225 0.0654 0.0338 0.0172⟩

Elements on main diagonal of 7 × 7 selection matrix (calculated) are:
⟨1.0 1.0 1.0 0.9 0.5 0.33 0.33⟩

Product sizing analysis calculated after one stage of breakage is shown in Table 4-XI.

TABLE 4-XI

Product size distribution after one stage of breakage in a ball mill

Tyler mesh	Micrometres	Product (calculated)	Product (observed)
+ 4	+ 4699	0	0
4/8	4699/2362	0.5	0.6
8/14	2362/1168	2.9	2.9
14/28	1168/589	8.2	8.3
28/48	589/295	18.9	19.0
48/100	295/147	22.6	22.8
100/200	147/74	14.9	14.7
− 200	74/0	32.0	31.7

Calculated stages of breakage = 1.0.
Calculated mill constant = mill feed rate • stages of breakage = 170 • 1.0 = 170.
Calculated stages of breakage at the increased flow rate of 208 t/h = 170/208 = 0.82.

The calculated product size distribution after 0.82 stages of breakage for the coarse feed is shown in Table 4-XII.

TABLE 4-XII

Product size distribution for the coarse feed at the higher feed rate

Tyler mesh	Micrometres	Product (calculated)
+ 4	+ 4699	0.4
4/8	4699/2362	2.8
8/14	2362/1168	7.1
14/28	1168/589	12.6
28/48	589/295	20.5
48/100	295/147	19.4
100/200	147/74	12.2
− 200	74/0	25.0

Example 4-4: ball-mill model. In this example, the mill parameters S and v for operating ball-mill circuits will be calculated.

Case No. 1:
Ore: magnetite/silica
Mill size: 4.00 m by 8.92 m (overflow)
Top size range: 3302—1651 μm
Circuit: fresh feed to hydrocyclone
Sizing distributions: Table 4-XIII.

TABLE 4-XIII

Sizing distributions for magnetite—silica ore in a ball-mill—hydrocyclone circuit

Tyler mesh	Circuit feed	Cyclone			Ball-mill discharge
		feed	OF	sands	
+ 8	0.1			nil	nil
8/10	0.4			0.3	nil
10/14	1.0	nil		0.2	nil
14/20	1.2	0.4		0.2	0.1
20/18	1.6	0.3		0.3	0.1
28/35	2.2	0.3		0.6	0.2
35/48	2.9	0.9	nil	1.2	0.7
48/65	4.7	1.7	0.1	2.1	1.5
65/100	8.1	4.7	0.3	5.7	4.9
100/150	9.3	8.9	0.8	9.9	9.3
150/200	12.8	21.6	2.6	25.4	24.6
200/270	14.1	30.9	13.8	33.5	32.0
− 270	41.6	30.3	82.4	20.6	26.6

Solution:
Cyclone sands is the feed to the ball mill; product sizings: 0.5 series screens:

calculated = 0 0.2 0.4 2.0 14.1 56.5 26.8

observed = 0 0.1 0.3 2.2 14.2 56.6 26.6

Calculated S = 1.0 1.0 1.0 0.66 0.2 0.085; calculated $v = 1.0$

Case No. 2:
Ore: galena/marmatite/silica
Mill size: 2.46 m by 2.77 m (overflow)
Top size range: 4820—2410 μm
Circuit: fresh feed to rake classifier
Sizing distributions: Table 4-XIV.

Solution:
Classifier sands is the feed to the ball mill; product sizings: 0.5 series screens:

calculated = 0.1 0.8 4.6 20.4 31.2 24.7 18.2

observed = 0 0.9 4.7 20.5 31.0 25.1 17.8

Calculated S = 1.0 1.0 0.9 0.5 0.2 0; calculated $v = 1.0$

TABLE 4-XIV

Sizing distributions for galena—marmatite—silica ore in a ball-mill—rake-classifier circuit

BSS mesh	Circuit feed	Classifier		Ball-mill discharge
		OF	sands	
+ 5	0.1		0.1	nil
5/7	0.4		0.3	nil
7/10	1.7		0.9	0.2
10/14	4.2		2.4	0.7
14/18	7.7		5.0	1.5
18/25	9.8		7.4	3.2
25/36	11.8	4.5	11.4	7.5
36/52	11.2	6.6	15.3	13.0
52/72	10.0	10.7	15.5	15.7
72/100	9.1	13.2	13.0	15.3
100/150	8.1	14.6	10.5	13.5
150/240	8.0	15.8	7.8	11.6
− 240	17.9	34.6	10.4	17.8

TABLE 4-XV

Sizing distributions for magnetite—silica ore in a ball-mill—hydrocyclone circuit

Tyler mesh	Circuit feed	Ball-mill discharge *	Cyclone			Ball-mill feed (calculated)
			feed	OF	sands	
+ 4	1.1					0.2
4/6	2.4		0.1		0.1	0.4
6/8	6.5	0.2	0.1		0.4	1.3
8/10	11.1	0.9	0.4		0.8	2.3
10/14	13.5	1.2	1.1		1.1	2.9
14/20	13.2	1.8	1.8		1.8	3.5
20/28	10.6	2.2	2.3		2.3	3.5
28/35	7.8	4.1	3.4	0.1	3.5	4.1
35/48	5.8	6.0	4.8	0.1	5.4	5.5
48/65	4.9	7.9	7.8	0.2	6.8	8.2
65/100	3.4	8.9	9.7	0.3	10.5	9.5
100/150	3.4	12.5	12.9	2.3	15.0	13.3
150/200	2.3	11.5	11.8	4.1	13.3	11.7
− 200	14.0	42.8	43.8	92.9	37.0	33.6

* The ball-mill discharge enters a magnetic separator and only the concentrate enters the cyclone. The cyclone sands return to the ball-mill.

Case No. 3:
Ore: magnetite/silica
Mill size: 3.05 m by 3.05 m
Top size range: 9400—4700 μm
Circuit: fresh feed to ball mill
Sizing distributions: Table 4-XV.

Solution:
Cyclone sands plus circuit feed is the feed to the ball mill; product sizings: 0.5 series screens:

calculated = 0 0.4 1.6 4.3 9.9 16.8 23.7 43.3

observed = 0 0.2 2.1 4.0 10.1 16.8 24.0 42.8

Calculated S = 1.0 1.0 1.0 0.85 0.4 0.3 0.25; calculated v = 1.0

Example 4-5: autogenous-mill model (the mass and size distribution of the mill load are not known). The size distributions of the feed to and discharge from a rock mill grinding a sulphide ore are given in Table 4-XVI. The abrasion and crushing breakage functions, which together comprise the total breakage function, are also given. The feed rate to the mill was 68.5 t/h.

TABLE 4-XVI

Feed and product sizing distributions, and breakage functions, for an autogenous mill

Micro-metres	Mill feed	Mill product	Abrasion breakage matrix	Crushing breakage matrix
152400	3.85	0	0.9000	
101600	2.57	0	0.0352	
76200	0	0	0.0002	
50800	0	0	0.0007	
38100	0	0	0.0011	
25400	0	0.06	0.0018	
19400	0	0.07	0.0022	
12700	0	0.08	0.0026	
9520	0	0.18	0.0026	0.0836
6350	21.54	7.87	0.0026	0.0270
4780	14.18	5.29	0.0026	0.1640
3327	12.63	5.09	0.0026	0.1300
2326	8.69	3.81	0.0026	0.1114
1651	7.30	3.02	0.0026	0.0811
1168	5.25	3.52	0.0026	0.0611
833	3.29	3.63	0.0026	0.0440
589	2.31	3.95	0.0026	0.0329
417	2.03	5.14	0.0026	0.0237
300	1.65	5.63	0.0026	0.0163
212	1.27	4.78	0.0026	0.0124
150	1.19	4.40	0.0026	0.0088
106	1.61	5.73	0.0026	0.0063
75	1.25	4.20	0.0033	0.0044
53	1.22	4.37	0.0042	0.0032
38	1.32	4.53	0.0050	0.0022
0	6.85	24.66		
PC solids	74.46			
Solids mass	68.50			
Water mass	23.50			

What will be the product size distribution if the feed rate is increased to 100 t/h at the same size distribution?

Solution:
The sequence of calculations is:
(1) Calculate $\mathbf{D}\cdot\mathbf{R}^{-1}$ from the equations:

$$\mathbf{R}\cdot\mathbf{s} = (\mathbf{I}-\mathbf{A})^{-1}\cdot(\mathbf{f}-\mathbf{p}), \text{ and: } \mathbf{p} = \mathbf{D}\cdot\mathbf{R}^{-1}\cdot\mathbf{R}\cdot\mathbf{s}$$

in which $\mathbf{D}\cdot\mathbf{R}^{-1}$ is the only unknown parameter. The second equation can be derived from eq. 4-23.
(2) Scale the $\mathbf{D}\cdot\mathbf{R}^{-1}$ parameter to the new feed rate using the equation:

$$\mathbf{D}\cdot\mathbf{R}^{-1} = \mathbf{p}\cdot\mathbf{X}$$

as was discussed in eq. 4-29 and Fig. 4-23.
(3) Calculate the new product size distribution from the equations given in step 1. Equations 4-28 and 4-23 may be rearranged to give:

$$(\mathbf{I}-\mathbf{A})\cdot\mathbf{R}\cdot\mathbf{s} = \mathbf{f}-\mathbf{p} \tag{4.31}$$

This equation may be solved directly for $\mathbf{R}\cdot\mathbf{s}$ provided that the inverse of $(\mathbf{I}-\mathbf{A})$ exists. However, it is more convenient to solve this set of triangular simultaneous equations for $\mathbf{R}\cdot\mathbf{s}$ by using a back substitution technique. The following example explains how to calculate $\mathbf{R}\cdot\mathbf{s}$ by this method, and then $\mathbf{D}\cdot\mathbf{R}^{-1}$ by:

$$\mathbf{D}\cdot\mathbf{R}^{-1} = \mathbf{D}\cdot\mathbf{s}\cdot(\mathbf{R}\cdot\mathbf{s})^{-1} = \mathbf{p}\cdot(\mathbf{R}\cdot\mathbf{s})^{-1}$$

Eq. 4-31 may be expanded as shown in Table 4-XVII.

TABLE 4-XVII.

Expansion of eq. 4-31

$$\begin{bmatrix} (1-a_{11}) & 0 & 0 & \cdots & 0 \\ -a_{21} & (1-a_{22}) & 0 & & 0 \\ -a_{31} & -a_{32} & (1-a_{33}) & & 0 \\ & & & & \vdots \\ -a_{n1} & -a_{n2} & -a_{n3} & \cdots & (1-a_{nn}) \end{bmatrix} \cdot \begin{bmatrix} (r\cdot s)_1 \\ (r\cdot s)_2 \\ \\ \\ (r\cdot s)_n \end{bmatrix} = \begin{bmatrix} (f_1-p_1) \\ (f_2-p_2) \\ \\ \\ (f_n-p_n) \end{bmatrix}$$

$(1-a_{11})\cdot(r\cdot s)_1 = (f_1-p_1)$

that is: $(r\cdot s)_1 = (f_1-p_1)/(1-a_{11})$

similarly: $-a_{21}\cdot(r\cdot s)_1 + (1-a_{22})\cdot(r\cdot s)_2 = (f_2-p_2)$

and: $(r\cdot s)_2 = [f_2-p_2+a_{21}\cdot(r\cdot s)_1]/(1-a_{22})$

The general solution may be written as:

$$(r\cdot s)_i = \left[f_i-p_i+\sum_{j=1}^{i-1} a_{ij}\cdot(r\cdot s)_j\right] \Big/ (1-a_{ii})$$

The method for calculating successive elements in $\mathbf{R}\cdot\mathbf{s}$ from the original data is as follows:

$$(r\cdot s)_1 = [(3.85-0)\cdot 68.5]/[100\cdot(1-0.9)] = 26.3725$$

$(r \cdot s)_2 = [(2.57 - 0) \cdot 68.5/100 + 0.0352 \cdot 26.3725]/(1 - 0.9) = 26.8876$

Similarly, the remaining elements of the **R·s** matrix may be determined. A minor difficulty may be experienced after the 8th size fraction, when the crushing breakage elements are included in the calculations. The simple example in Table 4-XVIII illustrates how the combined breakage matrix is constructed from the abrasion and crushing breakage elements.

TABLE 4-XVIII

Combination of abrasion and crushing breakage

a_1	0
a_2	0
a_3	0
a_4	C_1
a_5	C_2
Abrasion breakage elements	Crushing breakage elements

a_1	0	0	0	0
a_2	a_1	0	0	0
a_3	a_2	a_1	0	0
a_4	a_3	a_2	C_1	0
a_5	a_4	a_3	C_2	C_1

Combined breakage matrix

It may be shown that $(r \cdot s)_9 = 0.1010$. Therefore:

$(d \cdot r^{-1})_9 = p_9 \cdot (r \cdot s)_9^{-1} = (0.18 \cdot 68.5)/(100 \cdot 0.1010) = 1.22$

When **R·s** and $\mathbf{D} \cdot \mathbf{R}^{-1}$ are calculated from the original data, the new value of $\mathbf{D} \cdot \mathbf{R}^{-1}$ may be calculated from Fig. 4-23, and the new product size distribution may then be calculated.

The values of **R·s** and $\mathbf{D} \cdot \mathbf{R}^{-1}$ at a fresh feed rate of 65 t/h, and $\mathbf{D} \cdot \mathbf{R}^{-1}$ and the predicted product size distribution at a fresh feed rate of 100 t/h are given below. The standard $\mathbf{D} \cdot \mathbf{R}^{-1}$ matrix which is used to calculate the new $\mathbf{D} \cdot \mathbf{R}^{-1}$ are given in Table 4-XIX.

Example 4-6: autogenous-mill model (the mass and size distribution of the mill load are known). The size distributions of the feed to and discharge from a 5.11 m by 5.18 m autogenous mill grinding a magnetite—silica ore in closed circuit with a hydrocyclone and of the mill load, are given below. The abrasion and crushing breakage functions which together comprise the total breakage function are given in Table 4-XX.

Mill feed rates were 216 t/h of ore and 58.6 t/h of water, and the weight of the mill load was 70 tonnes. The breakage function given in Table 4-VI is assumed to apply to this operation. What will be the size distributions of the mill load and mill discharge, and the weight of the mill load, if the mill feed rate is increased to 300 t/h at the same solids content?

TABLE 4-XIX

Procedure for calculation of a rock-mill discharge sizing distribution (excluding mill load)

Micro-metres	$R \cdot s$	$D \cdot R^{-1}$ (old)	$D \cdot R^{-1}$ (stand.)	$D \cdot R^{-1}$ (new)	Rock-mill discharge (calculated)
152400	26.3725	0	0	0	0
101600	26.8876	0	0	0	0
76200	9.5172	0	0	0	0
50800	3.5884	0	0	0	0
38100	1.7605	0	0	.0	0
25400	1.0530	0.0390	0.0100	0.0140	0.03
19400	1.0887	0.0440	0.0300	0.0421	0.07
12700	1.3377	0.0410	0.0400	0.0561	0.10
9520	0.1010	1.2206	1.2357	1.7338	0.20
6350	10.4342	0.5166	0.9257	1.2988	12.79
4780	9.2175	0.3931	0.3372	0.4731	5.62
3327	9.7997	0.3558	0.3252	0.4563	5.65
2326	9.2052	0.2835	0.2995	0.4202	4.64
1651	9.8210	0.2106	0.3434	0.4818	5.01
1168	8.8030	0.2739	0.4587	0.6436	5.46
833	7.7085	0.3225	0.5831	0.8181	5.38
589	6.7927	0.3983	0.7179	0.0073	5.28
417	5.5231	0.6374	1.0860	1.5237	5.76
300	4.4113	0.8742	1.2304	1.7263	5.26
212	4.1183	0.7950	1.0211	1.4327	4.19
150	3.7637	0.8007	0.7312	1.3065	3.60
106	2.5366	1.5472	1.8056	2.5334	4.40
75	2.7244	1.0559	1.1426	1.6032	3.20
53	2.1157	1.4147	1.4617	2.0509	3.59
38	1.6033	1.9352	2.9430	4.1293	16.58
0					
PC solids					74.46
Solids mass					100.00
Water mass					34.30

The simplifying assumptions can be made that (1) the breakage rates remain constant, (2) the discharge rates may be scaled to the new feed rates. These are simplifying assumptions which are used to illustrate the use of the model. In practice, **D** and **R** vary according to mill load and the exact relationships are still to be determined.

Solution:
The sequence of calculations is:

(1) calculate **D** from the equation $\mathbf{p} = \mathbf{D} \cdot \mathbf{s}$ in which **p** and **s** are known;

(2) calculate **R** from the equation $(\mathbf{A} \cdot \mathbf{R} - \mathbf{R} - \mathbf{D}) \cdot \mathbf{s} + \mathbf{f} = 0$ in which **R** is the only unknown parameter;

(3) calculate the new values in **D** of the increased feed rate according to the equation:

$$\mathbf{D}_{\text{new}}(I) = \frac{\text{vol. flow of new feed}}{\text{vol. flow of old feed}} \cdot \mathbf{D}_{\text{old}}(I)$$

(4) calculate the new product size distribution from the equation:

TABLE 4-XX

Feed, mill load and product sizing distributions and breakage functions for an autogenous mill

Micro-metres	Mill feed	Mill product	Mill content	Abrasion breakage matrix	Crushing breakage matrix
152400	2.47	0	3.04	0.9000	
101600	3.16	0	8.19	0.0352	
76200	4.67	0	9.66	0.0002	
50800	4.25	0	9.77	0.0007	
38100	4.10	0	13.54	0.0011	
25400	3.72	0	13.26	0.0018	
19400	2.98	0	9.47	0.0022	
12700	2.25	0	5.62	0.0026	
9520	2.77	1.08	2.60	0.0026	0.0836
6350	2.56	0.87	1.40	0.0026	0.2070
4780	3.01	1.62	0.99	0.0026	0.1640
3327	2.76	1.60	0.65	0.0026	0.1300
2326	2.30	1.42	0.34	0.0026	0.1114
1651	2.24	1.22	0.37	0.0026	0.0811
1168	1.38	0.92	0.24	0.0026	0.0611
833	1.64	1.16	0.32	0.0026	0.0440
589	1.34	0.99	0.29	0.0026	0.0329
417	2.29	1.73	0.56	0.0026	0.0237
300	1.98	1.80	0.46	0.0026	0.0163
212	3.47	3.56	0.95	0.0026	0.0124
150	4.91	5.33	1.46	0.0026	0.0088
106	9.43	10.33	3.90	0.0026	0.0063
75	10.77	12.67	2.48	0.0033	0.0044
53	6.45	10.87	6.01	0.0042	0.0032
0	13.09	42.83	4.40		
PC solids	78.66		93.33		
Solids mass	216.00		70.00		
Water mass	58.60		5.00		

$$\mathbf{p} = \mathbf{D} \cdot \mathbf{R}^{-1} \cdot (\mathbf{I} - \mathbf{A})^{-1} \cdot (\mathbf{f} - \mathbf{p})$$

in which **p** is now the only unknown parameter;

(5) calculate the new size distribution and mass of the mill load from the equation $\mathbf{p} = \mathbf{D} \cdot \mathbf{s}$.

The calculated values are shown in Table 4-XXI.

TABLE 4-XXI

Procedure for calculation of a rock-mill discharge sizing distribution (including mill load)

Micro-metres	D (old)	R	D (new)	Mill discharge (calculated)	Mill content
152400	0	25.07	0	0	3.25
101600	0	15.18	0	0	8.75
76200	0	19.46	0	0	10.32
50800	0	20.28	0	0	10.44
38100	0	14.65	0	0	14.47
25400	0	14.26	0	0	14.17
19400	0	17.56	0	0	10.12
12700	0	24.91	0	0	6.00
9520	1.28	4.96	1.78	1.38	2.56
6350	1.92	7.82	2.66	1.10	1.36
4780	5.05	12.40	7.02	1.96	0.92
3327	7.60	20.99	10.56	1.93	0.60
2326	12.89	43.53	17.91	1.71	0.32
1651	10.18	45.95	14.14	1.49	0.35
1168	11.83	70.52	16.44	1.12	0.22
833	11.19	57.21	15.54	1.40	0.30
589	10.54	66.21	14.64	1.20	0.27
417	9.54	37.74	13.25	2.06	0.51
300	11.34	43.26	15.75	2.11	0.44
212	11.57	22.49	16.07	4.02	0.82
150	11.27	14.45	15.65	5.82	1.23
106	8.17	5.16	11.35	10.82	3.14
75	15.76	6.87	21.89	12.92	1.95
53	5.58	1.36	7.75	10.45	4.45
0	30.04		41.72	38.49	3.04

CHAPTER 5

Fundamentals of size separation

5.1 INTRODUCTION

Size separation of comminuted particles is carried out: (1) by screening, and (2) by classification. In screening, particles are separated according to their size and shape while in classification particles are separated according to their differences in size, shape and density because these properties jointly affect their movements through fluid media. These separation techniques are considered in this chapter.

5.2 SIZE SEPARATION BY HYDRAULIC CLASSIFICATION

5.2.1 *Movement of particles through fluids*

Separation of a group of particles of different sizes, shapes and specific gravities into fractions of a more homogeneous nature is achieved by allowing them to settle in a fluid (liquid, air or gas) which is at rest or in motion. Here, the separation occurs due to the differences in the rates of travel of particles (with different physical properties) in the media. A particle settling in a fluid under the influence of gravity experiences a resistance and attains a constant maximum velocity (commonly called the terminal settling velocity) when the resistance becomes equal to the effective weight of the particle. In the case of spheres this terminal velocity (V_m) can be calculated from the equation:

$$V_m = \frac{4\pi}{3k} \cdot \mu^{(n-2)} \cdot \frac{(\delta - \delta')}{\delta'^{(n-1)}} \cdot g \cdot d^{(3-n)} \tag{5-1}$$

where μ = viscosity of the fluid, δ = density of the solid, δ' = density of the fluid, g = gravitational acceleration, d = radius of *sphere*, and n and k are functions of particle radius. The popular laws of Stokes and Newton can be considered as special cases of eq. 5-1.

Stokes' Law:

$$V_m = \frac{4\pi}{3k} \cdot \frac{(\delta - \delta') \cdot g \cdot d^2}{\mu} \tag{5-2}$$

This is when $n = 1$ and $k = 6$ and applies for small solids falling in water (say, quartz spheres up to 0.005 cm radius) under laminar or viscous conditions of flow (that is, the lines of flow are smooth and unbroken) where the speeds of travel are low.

Newton's Law:

$$V_m = \left[\frac{4\pi}{3k} \cdot g \frac{(\delta - \delta') \cdot d}{\delta'}\right]^{0.5} \tag{5-3}$$

This is when $n = 2$ and $k = \left(\frac{\pi}{2}\right) \cdot Q$, where Q is the drag coefficient and applies for large solids falling in water (say, quartz spheres with radius greater than 0.25 cm) where the velocities are high and the flow is turbulent. Streamline flow of liquid round a particle is limited to conditions where Reynolds Number $\left(\frac{V_m \cdot d \cdot \delta'}{\mu}\right)$, Re, is less than 1.0 and the flow is fully turbulent only if Re is greater than 2000. Most applications in mineral processing involve the range of conditions which is intermediate between laminar and turbulent flow and the theory for these conditions is not fully developed.

Under ideal conditions, eqs. 5-1 to 5-3 may be used to predict the movement of particles in fluids and, with appropriate modifications, to account for factors such as: (1) irregularities in the shapes of particles, and (2) interference to the fall of a particle by the presence of other particles and the walls of the container in which separation is being affected.

In practice, however, these factors cannot be defined very accurately and the flow regimes encountered are so complex that the equations are unsuitable for the analysis and modelling of industrial processes. Consequently, another approach to modelling must be used.

The entry of a particle into any fraction by classification is due to:

(1) the physical properties of the particle such as size, shape and specific gravity;

(2) the physical properties of the fluid, such as density, solids content, consistency or viscosity; and

(3) the design and operating variables of the equipment in which the separation is carried out.

By proper selection and careful manipulation of (2) and (3) above, the effects of one of the properties of a particle can be minimised while the effects due to the other properties dominate. In grinding circuits physical size is the important criterion since the larger particles require further breakage while the smaller particles should be discharged from the circuit. In gravity concentration techniques such as jigging, classification conditions are selected for separation of particles into different specific-gravity groups, containing the heavy valuable material and the light gangue fraction. Separation in classifiers would be highly effective if differences in specific gravities of the individual particles of the feed of a wide size range were insignificant, while the gravity concentration techniques would be efficient if the feed particles were in a relatively close size range.

The units in which the separation of solids in fluid media is carried out can be broadly divided into two major categories: (1) classifiers, and (2) gravity concentration units. The latter do not come within the scope of the present discussion and will not be considered further.

5.2.2 *Classifiers*

The types of classifiers commonly used in industry are mechanical (rake and spiral classifiers) and centrifugal (hydrocyclones). Rake and spiral classifiers always operate wet and consist of an open settling trough which is fixed at an angle to the horizontal. The slurry is introduced into the chamber at a point near the overflow discharge end and the rate of feed is such that it will allow enough time for only the large particles to sink to the bottom. The fine particles are left to overflow with the major portion of the feed water and the settled coarse material is slowly dragged up the slope of the tank and is discharged at the opposite end of the tank. With rake classifiers (Fig. 1-9), the dragging of solids is done by the movement of a rake which consists of sweeping along the base of the tank, rising, moving backwards and then dipping into the tank to continue the cycle of operation. With spiral classifiers, the solids are carried upwards by the continuous rotation of a spiral mounted in a semi-cylindrical tank. If two or three rakes are mounted in the classifiers, they are called duplex and triplex classifiers.

A cyclone is a typical centrifugal classifier. If operated dry, it is called an air cyclone and if operated wet, it is called a hydrocyclone. Only hydrocyclones will be discussed in this book*. A hydrocyclone, shown in Fig. 5-1,

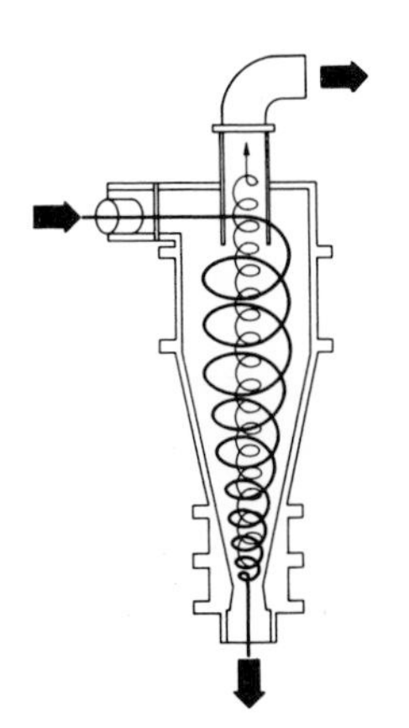

Fig. 5-1. Internal flows within a hydrocyclone.

* In the remainder of this book the term "cyclone" refers to a hydrocyclone.

consists of a cylindrical top section joined to a conical base, with overflow and underflow outlets provided at the top and bottom of the unit respectively. It contains no moving parts. The feed, in the form of a pulp, is introduced under pressure tangentially through a feed inlet positioned at the cylindrical top of the hydrocyclone and products leave the hydrocyclone along the central axis at right angles to the direction of entry. At the points of exit, the outlet streams are moving in opposite directions. The flows within a hydrocyclone are shown in Fig. 5-1.

Classification of solid particles of different weights contained in the inlet stream occurs as a result of the conversion of the direction and velocity of flow of the inlet stream into that of the outlet streams and is due to centrifugal drag and, to an extent, gravity forces acting on the particles. A particle suspended in fluid rotating in the cyclone tends to move towards the wall of the cyclone if the centrifugal force acting on it is greater than the drag force, otherwise it tends to move radially inwards. Although short circulation may occur within the cyclone causing coarse particles to appear in the discharge from the vortex finder, solid particles in general have to pass through a zone of maximum tangential velocity before emerging through the vortex finder. This ensures that all particles will be subjected to the maximum centrifugal force before they can pass from the outer to the inner spiral and this contributes to the efficiency of cyclones in rejecting coarse particles from the fine-particle product.

Because of their simplicity, flexibility of operation, and high capacity per unit area occupied, hydrocyclones are rapidly replacing mechanical classifiers as size-separating devices in mineral processing plants. As such, in the next section, no attempt will be made to discuss mechanical classifiers and only hydrocyclones will be considered.

5.3 HYDROCYCLONES

The general performance of a hydrocyclone is influenced by both the design variables such as the cyclone dimensions, and by operating variables such as the feed pressure and the physical properties of feed solids and feed pulp. In this section, the flow regime within a cyclone is discussed qualitatively and the effects of different variables upon the capacities and solids separation characteristics of cyclones are discussed quantitatively.

5.3.1 *Flow regime in hydrocyclones*

Detailed discussions of the flow pattern in hydrocyclones have been given by various authors (Kelsall, 1952; Lilge, 1962). The general flow pattern may best be considered by resolving the linear velocity with which the fluid enters the cyclone into three components: (1) the tangential component, (2) the radial velocity component, (3) the vertical component.

Kelsall used an optical method to determine these velocity components. Some of the important conclusions from his observations were:

(1(i)) below the bottom of the vortex finder, fluid "envelopes" of constant tangential velocity, which are cylinders coaxial with the cyclone, exist; (ii) the tangential velocity (v) decreases with increase in radius (r) according to the relationship $v \cdot r^n$ = constant, until a maximum velocity is reached at a radius smaller than that of the inside wall of the vortex finder; however, the relationship becomes $v \cdot r$ = constant as the radius is further decreased;

(2) at parallel levels considerably below the bottom of the vortex finder, the radial velocity decreases with decrease in radius and becomes zero at the air core;

(3) a conical envelope of zero vertical velocity separates the flow of fluid towards the apex and the flow towards the vortex finder.

This work was extended by Ruangsak (1963) under the supervision of Kelsall in order to investigate the effect of change in fluid viscosity upon the flow pattern. Liquids of viscosities up to 20 centipoise were used and it was found that change in fluid viscosity caused a proportionate change in the velocity component within the cyclone, but did not alter the flow profiles.

Several other workers, for example, Bradley and Pulling (1959), studied the flow profiles using dye injection techniques and high-speed cinematography. These studies have resulted in improvement in cyclone design but were not comprehensive enough to be of significance either in quantifying the velocity profiles or predicting the performance of a hydrocyclone.

Rietema (1961) suggested that the residence time of particles plays an important part in the mechanism of solids separation in cyclones. From a theoretical analysis of the flow pattern in cyclones, he proposed a theory to explain this phenomenon, and he has developed correlations claimed to be useful in the design of cyclones. These studies were based upon some simple assumptions, one of them being that no hindered settling occurs in cyclones. In practice the separation in hydrocyclones is a direct result of hindered settling of the particles, and the volume concentration of solid particles in the pulp is normally high with a consequent effect upon the viscous characteristics of the pulp. This theory was not adequate as presented to explain and predict the performance of a cyclone classifying pulps with significant quantities of comminuted solids.

Lilge (1962) gave an excellent account of the paths, within a cyclone, generally taken by particles because of the different forces acting on them at various positions in the cone. Fig. 5-2, reproduced from his publication, illustrates the essential features of the particle paths in cyclones.

It is clear that the classification at high pulp densities of the feed to the cyclone diameter (Dahlstrom, 1954). With respect to the design variables, particles. Fahlstrom (1963) proposed an "hypothesis of crowding" to explain the solids separation at high densities. According to this theory, "the separation is not only due to hindered settling of particles, but also due to

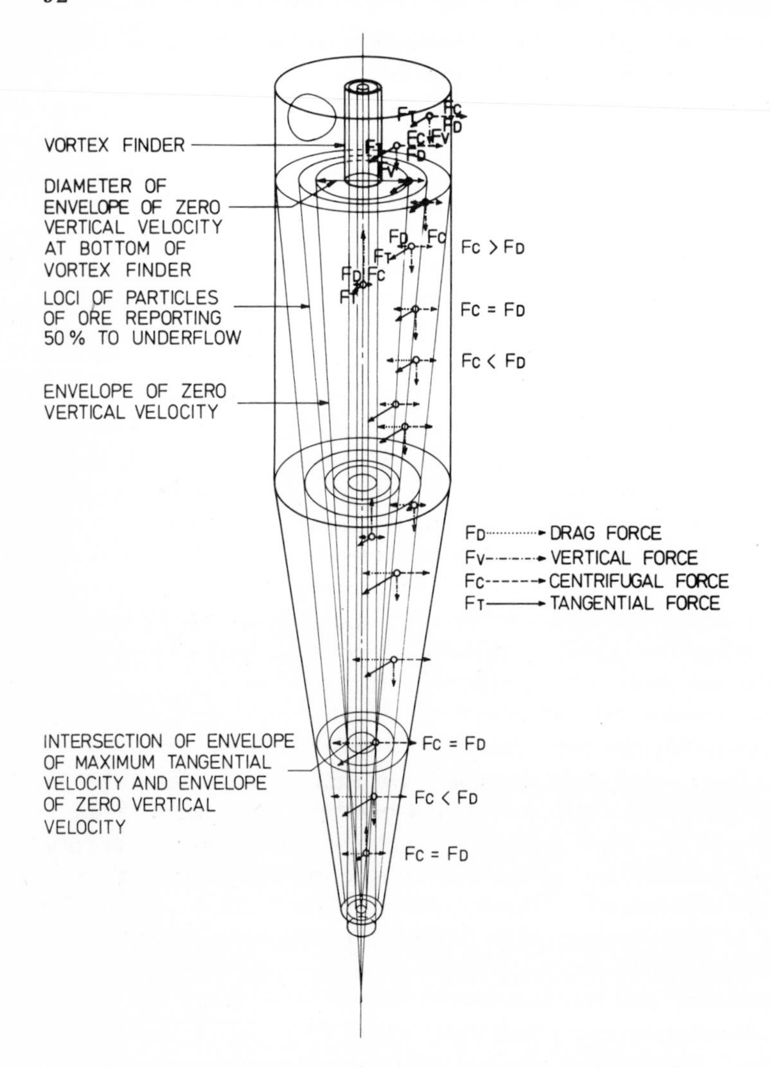

Fig. 5-2. Forces within a hydrocyclone, according to Lilge (1963).

hindered discharge through the apex". In this regard, his contribution may be regarded as an important investigation which has considered data from industrial units. It is particularly valuable in this context although in the context of a simulation model the relationships developed are of limited use.

5.3.2 *Capacities of hydrocyclones*

One of the most widely studied properties of cyclones has been their capacities. It has been generally realised that variables, such as the cyclone inlet and vortex finder (VF) dimensions, and the operating pressures, affect the capacity of a cyclone. Several correlations have been developed to relate the cyclone capacity to these variables and the more important of these are considered below.

An empirical expression, derived by Dahlstrom (1949) from his experiments, is:

$$Q/P^{0.5} = K' \cdot (\mathrm{VF} \cdot \mathrm{Inlet})^{0.9} \tag{5-4}$$

The proportionality constant K', was found to be primarily a function of included angle of the cone and minor design variables, but independent of cyclone diameter (Dahlstrom, 1954). With respect to the design variables, Moder and Dahlstrom (1952) reported different values of K' for various ratios of vortex finder and feed inlet areas. This formula has been generally found to give a reliable estimate of hydrocyclone capacity with water or pulp with a low solids content as the feed.

From studies using a 7.5-cm cyclone, Kelsall (1952) found that: (1) the capacity varies as the pressure to the power 0.416; (2) the long narrow rectangular feed openings give small volumes of flows as compared with the circular openings of the same cross-sectional area. He also observed (Kelsall, 1953) that a decrease in spigot diameter, at constant feed pressure, has negligible effect upon the total flow through the cyclone. His work was not extended to large units.

Chaston (1958), from simple hydraulic theory and a study of published results, proposed a formula for calculating the approximate capacity of a cyclone:

$$Q = K'' \cdot A \cdot P^{0.5} \tag{5-5}$$

where A = area of feed inlet, P = feed pressure and Q = flow rate of feed.

The value of the constant, K'', was found to be dependent upon the ratio of vortex finder area to feed inlet area. This expression is a simplication of Dahlstrom's (1949) equation.

5.3.3 *Efficiency curves*

The performance of a cyclone is affected by its dimensions and operating conditions. In order to develop correlations which include all of the essential variables, a method of representing the performance, either by a single number or a curve, is needed. Some of the methods that are commonly used are considered in this section.

Dahlstrom (1949) was the pioneer in conducting detailed experimental studies on hydrocyclone performance. He used the d_{50} value of an operation to represent the solid elimination efficiency of the cyclone. While it will be seen that d_{50} is a useful parameter, the representation of complete classification by a single size is not generally satisfactory, since it does not describe the full size range of particles. A study of product size analyses from a

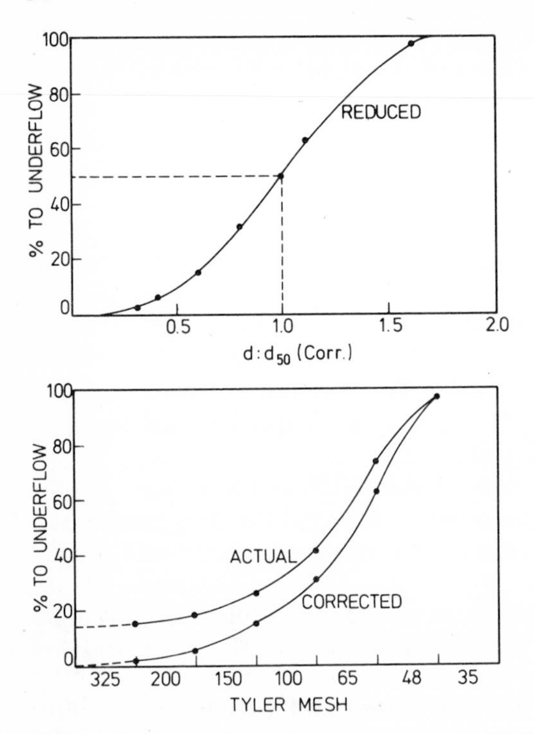

Fig. 5-3. The actual, corrected and reduced efficiency curves of a hydrocyclone.

cyclone shows that a proportion of each particle size of the feed reports to the underflow product and that one single-valued parameter is not an adequate description of the process.

A more logical approach is to plot "performance (efficiency) curves" to represent the performance of a cyclone operation, as described in the following section. The efficiency curve of a cyclone (Fig. 5-3) expresses the relationship between the weight fraction or percentage of each particle size of the feed reporting to the underflow discharge and the particle size. Actual efficiency, E_a, at a point on this curve for a given size material is given by:

$$E_a = W_u \cdot M_u / W_f \cdot M_f \qquad (5\text{-}6)$$

where W_u and W_f are the weight fractions of the given size material in the underflow and the feed ore streams, respectively, and M_u and M_f are the mass flow rates of the underflow and the feed ore streams, respectively.

The actual performance curve of a cyclone will not pass through the origin and the reason for this has been explained by Kelsall (1953). He suggested that, independent of centrifugal forces acting on the particles, if R_f is the fraction of feed fluid reporting to the underflow (flow ratio), R_f percent of all sizes of particles are discharged through the spigot. Therefore, separation due to centrifugal action, or corrected efficiency, E_c (Fig. 5-3) is given by:

$$E_c = \frac{E_a - R_f}{100 - R_f} \cdot 100 \qquad (5\text{-}7)$$

Yoshioka and Hotta (1955) observed the similarity in the shapes of performance curves for different operating conditions and cyclone sizes, and developed a method of reducing them to a single curve called the "reduced efficiency curve". This curve is obtained by plotting centrifugal efficiency or corrected weight percentage of particles reporting to the underflow against the actual size divided by corrected d_{50}. A reduced efficiency curve applicable to the actual and corrected efficiency curves is also given in Fig. 5-3.

In order to use the reduced efficiency curve to derive the actual performance curve, formulae for estimating the values of: (1) the corrected d_{50}, and (2) the flow ratio of water for any condition of cyclone operation are needed so that the cyclone performance for the new conditions of operation may be derived.

5.3.4 *Estimation of d_{50}*

The difference between the actual d_{50} and the corrected d_{50} values must be emphasised. The former is taken from the actual performance curve and the latter is derived from the corrected performance curve. The corrected d_{50} value is always greater than the actual d_{50} value and this is shown in Fig. 5-3. Because the actual d_{50} value is the result of two mechanisms, and the d_{50} (corrected), or d_{50}(c), value is due to one of these, it is appropriate to calculate d_{50}(c) before calculating d_{50} (actual). It is possible to formulate a comprehensive equation for the direct calculation of d_{50} (actual) without considering the mechanisms individually; the theoretical work of Schubert and Neese (1973) and Schubert (1975) is of interest in this respect.

Many expressions based upon theory and experiments have been given by various authors to relate the different design and operational variables of a cyclone operation to d_{50}. The more important of these are mentioned below and reference can be made to the original papers for additional details.

For the design of large cyclones, the following empirical expression was derived by Dahlstrom (1949), and has found wide application:

$$d_{50} = \mathrm{K} \cdot \frac{(\mathrm{VF} \cdot \mathrm{Inlet})^{0.68}}{Q^{0.53}} \cdot \left(\frac{1.73}{\rho_s - \rho_f}\right)^{0.5} \qquad (5\text{-}8)$$

where K, the constant, was found to be sensitive only to major changes in cyclone geometry. However, the application of this expression is restricted to feed suspensions containing solids (of density equivalent to that of quartz) up to 20% by weight and underflow volume splits of up to 15% of total flow. Matschke and Dahlstrom (1959) modified this formula for miniature size (10—40 mm in diameter) cyclones by altering slightly the constant K and the exponents.

Bradley (1958), making several assumptions concerning the nature of flow within cyclones, including particle movement according to Stokes' Law, derived a theoretical equation:

$$d_{50} = 2.7 \cdot \left[\frac{\tan \frac{\theta}{2} \cdot \eta \cdot (1 - R_{\text{f}})}{D_{\text{c}} \cdot Q \cdot (\rho_{\text{s}} - \rho_{\text{f}})} \right]^{0.5} \cdot \left[\frac{2.3 \cdot \text{VF}}{D_{\text{c}}} \right]^{n} \cdot \frac{\text{Inlet}^2}{\alpha} \qquad (5\text{-}9)$$

In their experimental investigations, Bradley and Pulling (1959) studied the "classification surface" within a cyclone and modified Bradley's original equation. They compared the value derived from this equation with the actual result obtained from experimental determinations, using perspex spheres, and the agreement between the results was found to be reasonably good. Many assumptions made in deriving the equation, and its practical limitations in application, were discussed in their original paper.

Lilge, in his (1962) paper titled "Hydrocyclone fundamentals", has suggested that a relationship exists between the "cone ratio" (underflow diameter/overflow diameter) and the volume distribution of fluid entering the cyclone. This graph is reproduced in Fig. 5-4. From this graph, it was concluded that the volume of fluid appearing in the overflow or underflow is dependent only upon the cone ratio. Lilge used this relationship and the value of the cyclone throughput to calculate the radial and vertical velocities. The fluid regime in the cyclone is also markedly affected by the fluid viscosity of a medium in the cone. In this manner, Lilge considered many variables and presented the "cone force equation":

$$(\rho_{\text{s}} - \rho_{\text{f}}) \cdot d \cdot V_{\text{T}}^2/r = C_{\text{D}} \cdot \rho_{\text{f}} \cdot V_{\text{R}}^2/2 \qquad (5\text{-}10)$$

This equation was used to estimate the d_{50} value of an operation, when V_{T} is the maximum tangential velocity, r the corresponding radius and V_{R} is the radial velocity.

This paper has been discussed extensively by various authors (vide *Trans. Inst. Min. Metall.*, 1962) and the various points considered at that time will not be repeated here.

Two factors might limit the common application of this equation:

(1) Cyclone feed pressure and the feed solids content were not shown to have an effect upon the flow distribution relationship mentioned above, but available data from industrial cyclones show that these operating variables have influence upon the discharge ratio of fluid through the cyclone outlets.

(2) As pointed out by Tarjan (1962) and emphasised by Lilge (1962), the specific gravity and apparent viscosity of the feed pulp cannot be used in the cone force equation, when there is segregation of medium in the cone. In practice, the extent to which the segregation may be neglected in cyclone-classifying pulps of high solids concentration at low operating pressures is not certain.

Lilge showed that the underflow diameter of cyclones, the properties of

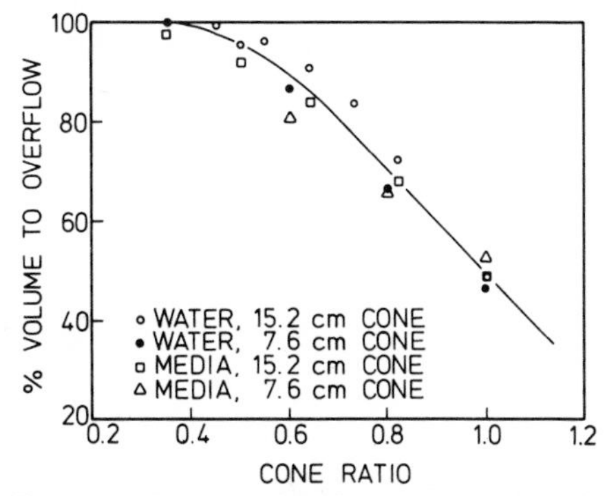

Fig. 5-4. Relationship between cone ratio and the volume distributions of fluid entering the hydrocyclone, according to Lilge (1963).

solid particles, and the apparent viscosity of the feed pulp have significant effects upon the d_{50} value of an operation.

In "Hydrocyclone fundamentals", Lilge approximated the performance curve of a cyclone to a straight line and observed that: (1) for + 150-mesh feed particles, the performance lines for various "cone ratios" were parallel on a log-normal plot with a slope of 0.02, and (2) for − 150-mesh feed particles, the performance lines for various "cone ratios" were parallel as a linear plot with an average slope of 2.0.

These comments suggest that, by knowing the d_{50} value, the actual performance curves for an operation may readily be derived using these slopes. The work of Rao (1966) showed this approximation to be invalid.

Fahlstrom (1963) derived an expression to estimate d_{50} as:

$$d_{50} = \mathrm{K} \cdot (1 - g_{\mathrm{u}})^{1/n} \qquad (5\text{-}11)$$

Fahlstrom's experimental data demonstrated the dependence of g_{u}, the quantity of solids discharged through the apex per unit time, upon the percent by weight and the size distribution of solid particles in the feed to the cyclone. By incorporating g_{u} in the equation, he has endeavoured to show the important effect of physical properties of feed pulp upon the cyclone classification operation.

Dahlstrom (1949) and Moder and Dahlstrom (1952) agree that d_{50} for a feed pulp does not vary with change in spigot diameter. Fahlstrom's (1963) publication and Lilge's (1962) findings both showed the important effect of spigot diameter upon the cyclone performance.

5.3.5 *Comments*

An excellent review of the "state of the knowledge" on hydrocyclones was given by Bradley (1965) and some important aspects of his review were discussed above. The following comments are made about the state of knowledge at the time at which this was published.

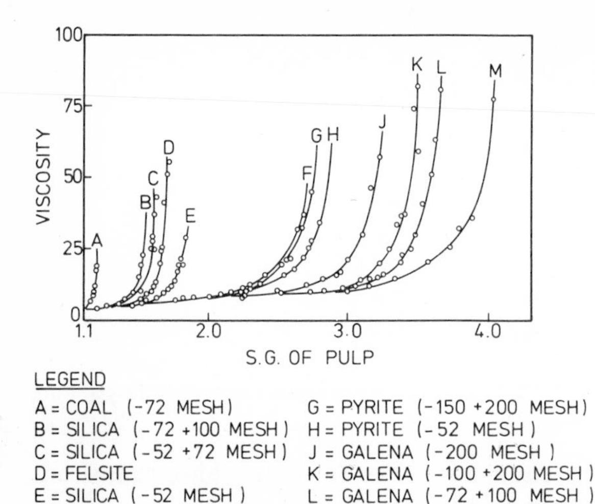

Fig. 5-5. Relationship between solids content and pulp viscosity.

(1) Most of the detailed literature on cyclones discussed the result of tests conducted on small cyclones (76 mm in diameter) operating at high shear rates and treating pulps of low solids content, with limited size distribution of homogeneous solid particles.

(2) All of the theoretical expressions were based upon many simplified assumptions. The parameters used in these equations could not be determined accurately and limited experimental results were used to support the simplified assumptions.

(3) Equations that require the measurement of apparent viscosities of feed pulps are of limited use in milling practice. Several measurements of apparent viscosities of different mill pulps have been carried out by Rao (1966) adopting the method described by Devaney and Shelton (1940). From the general shape of these curves, it is evident that the measurements of viscosities of mill pulps is not very accurate because of the sharp increase in the viscosity at high solids content, as shown in Fig. 5-5. Laboratory measurement of viscosities of small samples of mill pulps are limited in reliability because of fluctuations in the physical conditions, such as the temperature of the pulps, and the large amounts of coarse particles present in the feed to a cyclone.

(4) The pressure—throughput relationships were suitable only to estimate the capacities of cyclones treating pulps of low solids content (below 10—20% by weight). However, very little information was available to make any deductions concerning the importance of the effects of specific gravity of the medium on cyclone capacity.

(5) It had not been shown by any investigator whether the concept of the reduced efficiency curve was applicable in industrial systems in which pulps

of low moisture content and solids of wide size range and of heterogeneous nature are classified.

(6) The effects of feed size distribution and feed pulp densities upon the d_{50} value of a cyclone could not be estimated with the information available in literature. In particular, expressions to estimate the corrected d_{50} value of a cyclone operation, which is required in order to derive corrected performance curve from a reduced efficiency curve, were non-existent in literature.

(7) De Kok (1956) and Peachey (1960) observed a linear relationship between the water rate in the feed and the rate at which water reports to the overflow. However, expressions relating the water rates in these two streams to design and operating variables of a hydrocyclone which are needed in order to estimate the actual performance curve of a hydrocyclone from corrected performance curve, were not available in literature.

Detailed test programmes were carried out by Lynch and Rao (1965, 1975) in order to develop a model which could be used to estimate the performance of industrial-size hydrocyclones operating under widely varying conditions. The model which was developed will be discussed in Chapter 6.

5.4 SIZE SEPARATION BY SCREENING

Two types of screens are in common use in industrial plants, vibrating screens and wedge-wire screens. In both cases, separation is obtained by repeated presentation of particles to apertures. This is obtained with vibrating screens by "bouncing" the particles on the screens and with wedge-wire screens by causing the slurry to flow across slotted plates and taking "slices" of the flowing stream at each slot or aperature. The operating characteristics of each screen are discussed below.

5.4.1 *Vibrating screens*

The action of a vibrating screen is to present particles repeatedly to the screen surface which consists of a number of equal sized aperatures. At each presentation or trial, every particle which is capable of passing through the aperature has a probability of doing so and the higher the number of trials the higher is the probability that the particle will appear in the screen undersize.

Vibrating screens may be operated as batch or continuous units. In batch screening, the particles are placed on the screen and vibrated for a period of time, and the number of trials is related directly to the screening time. In continuous screening, the particles are fed continuously onto one end of an inclined vibrating screen and flow across the screen under the influence of gravity. The number of trials is proportional to the screen length and angle of inclination. Although continuous screening is steady-state, that is for

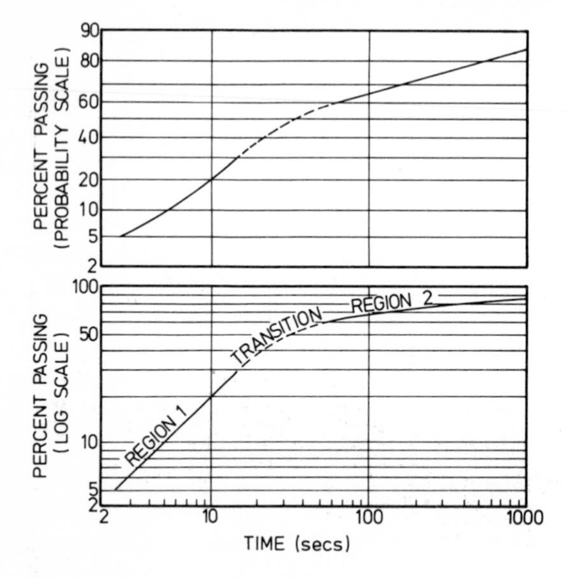

Fig. 5-6. Batch sieving. Percent passing—time relationship, according to Whitby (1958).

constant feed conditions, the oversize and undersize are constant for successive periods of time, and batch screening is non-steady-state, that is the undersize is continuously changing, there is a close analogy between them.

Whitby (1958) working with a batch sieve, showed that the time–percent passing curve could be divided into two regions with a transition region in between as shown in Fig. 5-6. Region *1* describes the screening behaviour early in the process when many of the particles retained on the screen are much smaller than the aperture size, and region *2* describes the screening behaviour later in the process when the particles entering the undersize are "near-mesh", that is approaching the size of the aperture. Whitby found that the rate at which material passed the sieve in region *1* was almost constant and could be described by the equation:

$$\text{percent passing} = a \cdot t^b \tag{5-12}$$

where $t =$ sieving time, $b = 1$ (approximately), $a =$ sieving rate constant. Region *2* was found to be represented accurately as a straight line on log probability paper. The duration of region *1* was found to depend on the material being sieved, that is on the size distribution and shape of the feed particles and Whitby showed the large variation which occurred for the region-*1* period between material readily screened, glass beads, and material difficult to screen, cake flour, though load, sieve and particle size distribution were identical. He also showed that increasing the load on the batch sieve delays the transition from region *1* to region *2* and that the transition time is directly proportional to the load.

In the case of continuous screening, region *1* exists close to the point at which the particles are fed onto the screen, and the transition zone and region *2* exist further down the screen.

5.4.2 *Wedge-wire screen*

Wedge-wire screens, the most common forms of which are known as sieve bends, were first developed in The Netherlands at the Mining Research Establishment of the Dutch State Mines in the Province of Limburg in 1953, for use as a size-separation device in coal washing plants. Since then, sieve bends have gained widespread use in such fields as mineral treatment and chemical industries, cement, potato starch and corn starch factories and in sewerage treatment plants.

The sieve bend, as used in the mineral treatment industry, consists of a 60°-segment of a circle made up of stainless steel bars which are of uniform size and wedge-shaped. The feed pulp, a mixture of solid particles and water, is introduced tangentially onto the upper surface of the screen and travels across this surface in a direction normal to the bars. On passing a slot between bars, a fraction of the feed pulp strikes the leading edge of a bar and, as a result, a small layer of pulp is removed from the main pulp stream and this passes through the slot and thus to the undersize fraction. The remaining pulp passes over this bar whereupon the process is repeated at each subsequent bar, producing a cumulative undersize product and a final oversize product. The sizing action of a sieve bend is shown in Fig. 5-7. Fontein (1954) found that the thickness of the undersize stream, which was removed at each bar, was about one half to one quarter the width of the slot. He also found that the maximum-size particle which appeared in the undersize product, was twice the thickness of the undersize removed at each bar. Thus, for a sieve bend with a 1.0-mm slot width, the undersize stream will be

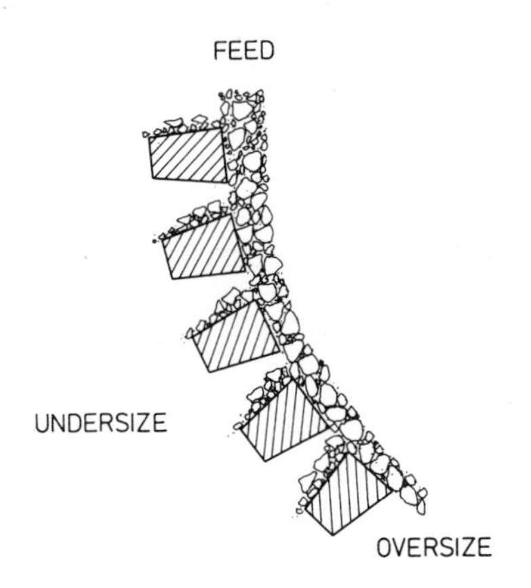

Fig. 5-7. Normal operation of a wedge-wire screen.

about 0.25 mm thick and the largest particle appearing in the undersize product will have a diameter of 0.5 mm.

Fontein (1965) showed that the thickness of pulp striking a bar was not a result of the centrifugal force acting on the pulp due to the curved nature of the sieve band. He proposed that the thickness of this layer was a direct consequence of the frictional drag acting on the pulp in contact with a bar. This caused a retardation in velocity of the pulp in contact with a bar and the resultant differing flow velocities caused a deflection of part of the pulp onto the next bar and thus to the undersize product. Fontein showed that this was correct in a series of tests in which the frictional drag at the pulp/metal interface was lowered by coating the bars with a grease. It was found that as the frictional drag decreased, the maximum-size particle appearing in the undersize product decreased and also the volume split (Qd/Qt) across the screen decreased. This confirmed his proposal concerning the mechanism of separation of wedge-wire screens.

Since the solids separation on a sieve bend is dependent on the flow velocities within the feed pulp, Fontein considered the effects of a Reynold's Number where:

$$\mathrm{Re} = V \cdot SW/Y \tag{5-13}$$

He found that the solids separation and volume split were effectively constant if the Reynold's Number was greater than a critical value which was found to be about 300. At a Reynold's Number less than the initial value, the volume split decreased markedly while the maximum-size particle appearing in the undersize increased. When this occurs, it was found that the sieve bend may be subject to blinding.

The important operating features of wedge-wire screens are:

(1) The separation of the particles occurs on the basis of the physical size of the particles. Consequently, these screens are of particular value where friable minerals of high specific gravity are being liberated in a closed grinding circuit and where over-grinding is to be avoided.

(2) The screens are subject to "blinding". This is particularly apparent when the flow rate over the screen decreases and Fontein concluded that the Reynold's Number should exceed 500 to ensure that blinding does not occur.

(3) Because of the nature of sieve bends, there is no restriction on the volume flow rate of feed pulp to them. However, at extremely high flow rates the fraction of the feed solids which is actually subjected to the sizing action of the sieve band will be only small.

(4) The screens are subject to wear.

During operation of a sieve bend, the high-velocity stream of pulp striking each bar causes wear to occur at the point of impact. This causes the bars to become rounded and, as a result, the thickness of the layer of pulp removed of each bar becomes smaller as wear progresses. Thus, the size of the largest particle appearing in the undersize product will change as time progresses,

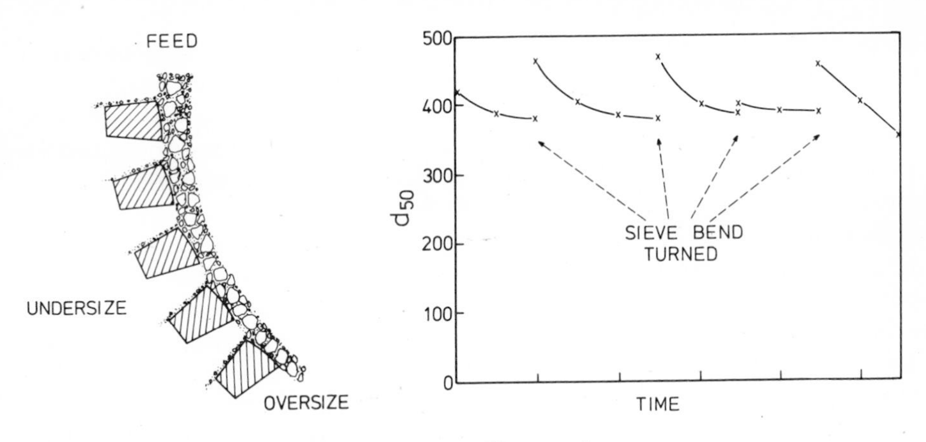

Fig. 5-8. Operation of a wedge-wire screen with worn bars.

Fig. 5-9. Influence of wear on d_{50} for a wedge-wire screen.

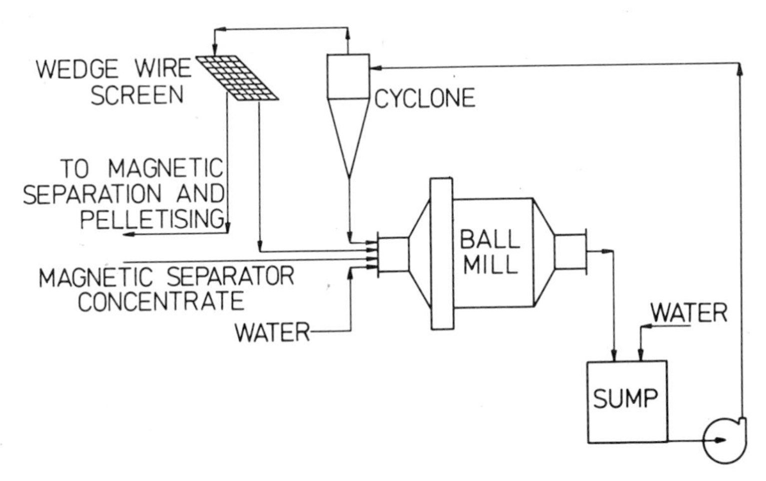

Fig. 5-10. Use of a wedge-wire screen in the grinding circuits at Erie Mining Company Limited.

although the operating conditions remain constant. The operation of a screen with worn bars is shown in Fig. 5-8. One method which is widely used to prevent excessive wear is the reversal of the direction of flow of the pulp across the screen surface. This is done by turning the screen so that the feed end becomes the oversize discharge end. This results in the minimisation of the change of the size distribution of the undersize product (Fig. 5-9).

As few quantitative data are available on the effects of wear, only experience can dictate the maximum time a sieve bend may be used under any particular circumstances before reversal of direction of flow is necessary. This time interval may vary from twenty minutes to several weeks depending on the type of ore used in the feed pulp.

Wear on a sieve bend can be useful in removing manufacturing faults on a screen surface. If at a spot on the screen surface the screening is coarser than normal, then the wear at this spot is greater than normal and thus the screening becomes finer. It should be noted that with vibrating screens, the opposite tendency occurs and irregularities on the screen cloth are amplified as wear progresses.

Following the successful use of wedge-wire screens in the form of sieve bends, flat wedge-wire screens were developed and the first commercial installation was at the Erie Mining Company. They were used on cyclone overflow streams to remove coarse light particles containing silica from a pulp containing predominantly fine magnetite as shown in Fig. 5-10.

The main difference in operation between sieve bends and flat wedge-wire screens is that centrifugal action on the pulp assists in forcing the pulp through the sieve bend, thereby keeping the apertures clear, whereas intermittent vibration is used for this purpose with flat screens. Flat wedge-wire screens are now used widely in closed grinding circuits particularly when over-grinding of heavy friable minerals such as cassiterite is to be avoided.

CHAPTER 6

Mathematical models of hydrocyclones and screens

6.1 HYDROCYCLONES

A mathematical model of a hydrocyclone which is suitable for circuit design and optimisation by simulation consists of a series of equations which relate the design and operating variables to the separation achieved. The design variables in a hydrocyclone are the vortex finder, spigot and inlet diameters, and the operating variables are the flow rate, percent solids and size distribution of the solid particles in the pulp.

Some of the theoretical attempts which have been made to develop a suitable model were discussed in Chapter 5. These attempts were of limited success due to the complexity of the flow regime and the difficulty of defining the sizes of particles as they are "seen" by the rotating flow.

However, a model has been formulated and used successfully for the analysis of hydrocyclone operation and in the simulation and control of closed grinding circuits (Lynch and Rao, 1965). This is a simple mechanistic model which is based on concepts of d_{50}(c) and the reduced-efficiency curve.

Development of this model to the point at which it could be used quantitatively required the collection of a considerable amount of data from hydrocyclones operating over a wide range of conditions. These data were used for model development, testing and modification where necessary, and for the evaluation of constants in the models. In order to obtain these data, comprehensive series of tests were conducted at the Julius Kruttschnitt Mineral Research Centre and at Mount Isa Mines Limited in test rigs designed for the purpose.

The tests at the Mineral Research Centre were carried out in 10.2, 15.2, 25.4 and 38.1-cm diameter hydrocyclones supplied by Krebs Engineers. The hydrocyclones and the fittings were used as they were received and no alterations were made to ensure geometric similarity with changes in hydrocyclone size.

Limestone of purity 99.0% was obtained locally and was ground in a ball mill to the fineness required. The three size distributions used are given in Table 6-I. Variables studied for each size range included the flow rate and solids content of the cyclone feed, the diameters of the cyclone, vortex finder and spigot, and the inlet area. Tests at Mount Isa Mines Limited were carried out on a 50.8-cm hydrocyclone and silica was used as the test material.

TABLE 6-I

Size distributions of limestone used in tests at the Julius Kruttschnitt Mineral Research Centre

Micrometres	Weight percent passing		
	coarse	medium	fine
1180	99.4	99.8	99.9
850	96.1	98.2	99.1
600	85.0	89.3	95.5
425	74.3	80.8	91.0
300	67.7	74.3	87.8
212	60.4	68.7	84.2
150	55.2	63.9	81.5
106	49.2	58.5	77.5
75	43.1	53.2	72.4
53	36.9	49.7	65.8

6.1.1 *Form of the model*

The separation process in hydrocyclones is a probability process in which particles of different sizes have different probabilities of appearing in the coarse product. The shape of a typical probability or efficiency curve was shown in Fig. 5-3. The reason that the curve does not pass through the origin was discussed in sub-section 5.3.3. The corrected and reduced efficiency curves which correspond to an actual efficiency curve were also shown in Fig. 5-3.

The two major mechanisms by which particles enter the coarse product, entrainment and classification by centrifugal action, are shown diagrammatically in Fig. 6-1 and the simulation model described below refers to these mechanisms. The coarse product discharged through the spigot consists of: (1) particles which short circuit directly into this product, and (2) particles which appear in this product as a result of the size-separation characteristics of the hydrocyclone.

The former is directly related to the mass fraction of feed water reporting to underflow and the latter may be expressed in terms of the reduced-efficiency curve and corrected d_{50} value for the process.

The hydrocyclone model consists of a series of equations which describe: (1) pressure—throughput relationship: (2) reduced-efficiency curve; (3) water flow ratio; and (4) classification size, that is, corrected d_{50}.

6.1.2 *Pressure—throughput relationship*

For a feed pulp of constant solids content and size distribution, many workers have found that the hydrocyclone throughput varies as the square root of the operating pressure according to the equation:

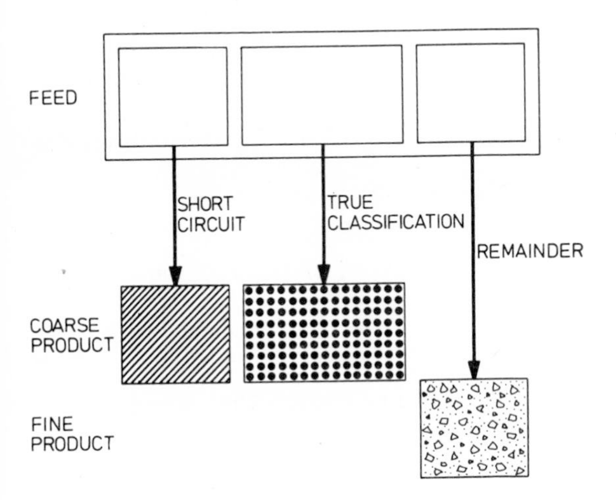

Fig. 6-1. Entrainment and classification mechanisms in a hydrocyclone.

$$Q = K \cdot (P)^{0.5} \quad (6\text{-}1)$$

It has also been found that the relationship between the vortex-finder diameter (VF) and the throughput at constant pressure is given by the equation:

$$Q = K \cdot (VF)^{1.0} \quad (6\text{-}2)$$

When the diameter of the vortex finder is appreciably larger than that of the spigot, and this is the normal case industrially, change in spigot diameter has negligible effect on the throughput.

The relationship between pressure and throughput in hydrocyclones is important in the design of the pump in hydrocyclone installations. It is necessary to know within reasonable accuracy the cyclone pressure for a given throughput so that the pump size and speed can be specified. A simple pressure—throughput relationship for a hydrocyclone with constant inlet dimensions is (Lynch and Rao, 1965):

$$Q = K \cdot VF \cdot P^{0.5} \cdot (FPW)^{0.125} \quad (6\text{-}3)$$

For a given installation, K has been found to be constant over a wide range of variation in Q, VF, P and FPW, and it may be evaluated for that installation from one set of observations. Thus, for that installation, the operating pressure may be determined for change in operating conditions, and the suitability of the pump to handle the new conditions may then be assessed.

Eq. 6-3 is valuable when an existing hydrocyclone installation is being investigated and no change in inlet area is to be made. When a new hydrocyclone installation is being designed based on data from pilot-scale units, this equation is no longer adequate. Lynch and Rao (1975) and Lynch et al. (1974, 1975) have found that for scale-up of hydrocyclones, the critical design variables are the inlet and outlet dimensions, the hydrocyclone diameter being the size of housing necessary to accommodate these fittings while maintaining a correct flow pattern.

The regression equation developed to relate pressure to throughput for cyclones with constant feed size is:

$$Q = K \cdot VF^{0.73} \cdot Inlet^{0.86} \cdot P^{0.42} \quad (6\text{-}4)$$

Where there is a large change in size distribution of the cyclone feed, the regression equation is:

$$Q = K \cdot VF^{0.68} \cdot Inlet^{0.85} \cdot Spig^{0.16} \cdot P^{0.49} \cdot (-\,53\mu m)^{-0.35} \quad (6\text{-}5)$$

A statistical analysis of the data available has shown that when all the other variables are constant the capacity of a hydrocyclone:

(1) increases linearly with increase in $\log_{10}P$; this observation is in conformity with the studies of earlier workers;

(2) increases non-linearly with increase in vortex-finder diameter. All the earlier workers have reported a linear relationship between the capacity and the vortex-finder diameter but the present analysis has shown minor errors in assuming such a relationship;

(3) increases linearly with increase in spigot diameter slightly; and

(4) increases to a certain value and then decreases linearly with increase in percent solids in the feed.

With regard to (4) there has been some disagreement about the effect of an increase in feed solids content, whether it causes an increase (Tarr, 1972) or a decrease (Lynch and Rao, 1965) in throughput at constant pressure. In order to resolve this question, Lynch, Rao and colleagues carried out a series of pressure—throughput tests on different sized cyclones at different operating conditions and a summary of some of the results is given in Fig. 6-2.

It has been found that the capacity of a hydrocyclone operating on slurries is higher than that for water alone at normal operating conditions. This partly explains the remark made by Tarr that increase in percent solids causes an increase in the capacity of a cyclone. However, as shown in Fig. 6-6-2, the capacity decreases with further increase in percent solids.

6.1.3 *Water split*

A linear relationship exists between the water in the fine product and the water in the feed over a very wide range of operating conditions. For a given feed slurry the operating variable which has the greatest influence on this relationship is the spigot diameter as shown in Fig. 6-3.

A simple equation which has been found to describe with reasonable accuracy the water split over a wide range of operating conditions for constant size distribution of particles in the cyclone feed is:

$$WOF = 1.07 \cdot WF - 3.94 \cdot Spig + K \quad (6\text{-}6)$$

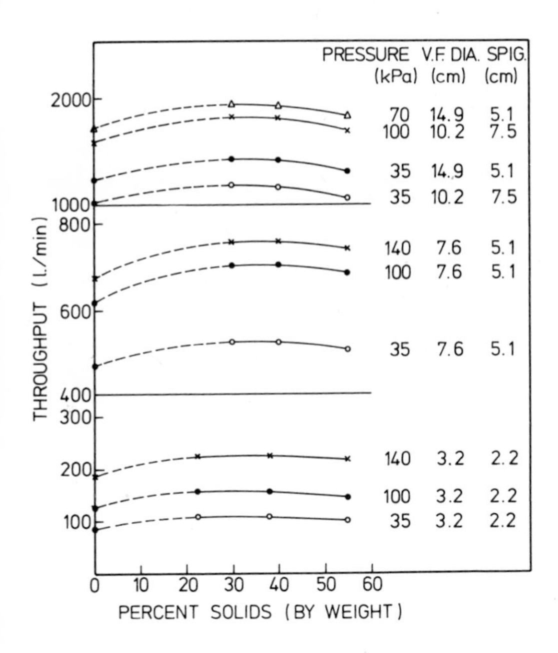

Fig. 6-2. Relationship between throughput, operating pressure and percent solids in cyclone feed.

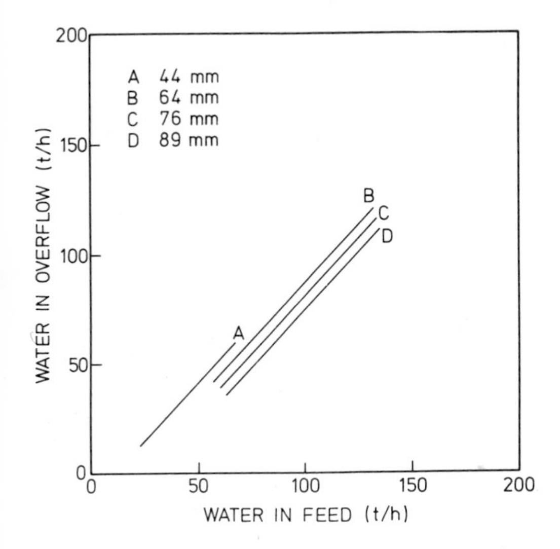

Fig. 6-3. The effect of spigot diameter on the water flow distribution in a hydrocyclone.

As in the case of eq. 6-3 for the pressure—throughput relationship, K has been found to be constant for a given installation over a wide range of variation in *WOF*, *WF* and *Spig*, and it may be evaluated for that installation from one set of observations. Thus for that installation, the water split may readily be calculated for change in spigot diameter or flow rate to the cyclone.

However, the value of this equation is limited due to three reasons:

(1) change in feed size distribution has an effect on the water split with all other conditions constant;

(2) the parameter which is required by the model is the fraction of water entering the underflow $(WF - WOF)/WF$, or R_f, rather than the mass, $WF - WOF$;

(3) it is not adequate for scale-up purposes.

Water split equations having wider application have now been determined for three size distributions of feed. The percent of feed water entering the underflow, R_f, has been related to the feed water flow rate and spigot diameter. The equations developed were of the form:

$$R_f = K_1 \cdot \frac{Spig}{WF} - \frac{K_2}{WF} + K_3 \qquad (6\text{-}7)$$

and regression techniques were used to determine the numerical values in the equations.

The equations for the individual feed sizes were:

$$\text{coarse:} \quad R_f = \frac{152.7 \cdot Spig}{WF} - \frac{213.9}{WF} + 6.67 \qquad (6\text{-}8)$$

$$\text{medium:} \quad R_f = \frac{102.2 \cdot Spig}{WF} - \frac{124.5}{WF} + 7.49 \qquad (6\text{-}9)$$

$$\text{fine:} \quad R_f = \frac{225.5 \cdot Spig}{WF} - \frac{303.3}{WF} - 7.40 \qquad (6\text{-}10)$$

and for all the data was:

$$\text{all sizes:} \quad R_f = \frac{193.0 \cdot Spig}{WF} - \frac{271.6}{WF} - 1.61 \qquad (6\text{-}11)$$

A comparison of the observed values, and the values calculated from the regression equations, is given in Fig. 6-4. The influence of the feed size on the water distribution is clearly demonstrated in Fig. 6-4.D in which the values for the coarse and fine feeds are predominantly on opposite sides of the regression line while those for the medium feed are in the centre. It will be noted that small errors, of the order of 0.1 t/h, make significant differences in calculated values of R_f for *small* flow rates and these are reflected in the apparent scatter in Fig. 6-4.

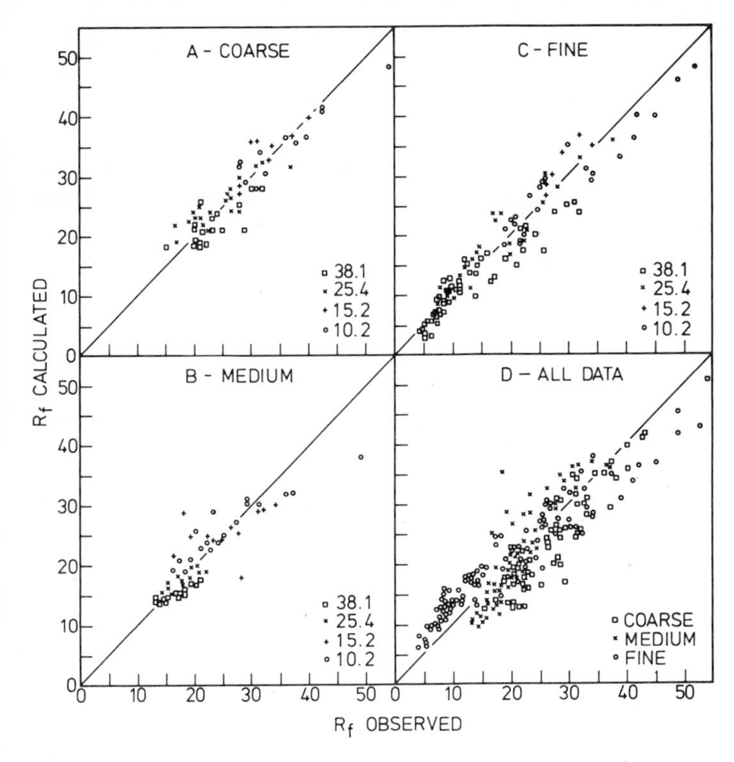

Fig. 6-4. Comparison of observed and calculated values of R_f using regression equations. Scales are in percentages. Legends refer to diameters of hydrocyclones and size of feed.

Inclusion of feed size in regression equations is difficult unless the size distributions have the same shape and can be described by a single parameter. In the present work, each feed size was described by two parameters, the percent + 35-mesh (420μm) and the percent − 300-mesh (53μm). The best regression equation for all data when these parameters were included was:

$$R_f = \frac{201.2 \cdot Spig}{WF} - \frac{268.6}{WF} - \frac{0.87 \cdot WF}{(+\,420\,\mu\text{m})} + \frac{7.85 \cdot WF}{(-\,53\,\mu\text{m})} - 6.21 \qquad (6\text{-}12)$$

This was a more accurate equation for the present data than eq. 6-11 but it cannot be regarded as a general equation for use with widely varying feed conditions. Its main importance is to demonstrate the effect of feed size on the water split.

The procedure for scale-up of the water distribution is as follows:

(1) Conduct a series of tests on a small hydrocyclone covering a wide range of R_f values. This may be done by varying the water flow rates and the spigot diameters.

(2) Derive a regression equation, relating R_f, *Spig* and *WF* using the form shown in eq. 6-7.

(3) Use the equation to predict the water distribution for any required flow rate of water in the feed. The hydrocyclone diameter will be the diameter necessary to house the spigot size required.

(4) Note that R_f is affected by the feed size distribution and the equation will lose accuracy if the feed size changes. If significant changes in feed size are expected and it is possible to carry out series of tests over the expected ranges of feed sizes, this should be done and an appropriate equation derived.

6.1.4 d_{50} *(corrected)*

For silica of constant size distribution classified in a 50.8-cm Krebs hydrocyclone, the d_{50} (corrected) or d_{50}(c) value is: (1) directly proportional to the vortex finder diameter and operating pressure; (2) inversely proportional to the spigot diameter and flow rate of water in the cyclone overflow. The relationships are shown in Fig. 6-5.

The regression relationship between log d_{50}(c) and the operating variables is:

$$\log_{10} d_{50}(\mathrm{c}) = 0.0173 \cdot FPS - 0.0695 \cdot Spig + 0.0130 \cdot VF + 0.000048 \cdot Q + \mathrm{K} \quad (6\text{-}13)$$

This equation may be used for the prediction of the effect of variables on d_{50}(c) for a particular installation operating on a siliceous ore and it is more accurate than later equations for this limited purpose. It will be noted that the sign of the throughput term in this equation is $+ve$ whereas in later equations dealing with scale-up it is $-ve$. This probably reflects the differences between the effects of changes in throughput in a single cyclone (relatively small variation) and in a range of cyclones (large variation).

However, eq. 6-13 is not adequate for scale-up work and it will not give accurate predictions where the sizing distributions of the cyclone feed change over a wide range. Consequently, from the data obtained from the tests with limestone relationships for each feed size were developed individually and then a general expression covering all data was derived. The regression equations developed were of the form:

$$\log d_{50}(\mathrm{c}) = \mathrm{K}_1 \cdot VF - \mathrm{K}_2 \cdot Spig + \mathrm{K}_3 \cdot Inlet + \mathrm{K}_4 \cdot FPS - \mathrm{K}_5 \cdot Q + \mathrm{K}_6 \quad (6\text{-}14)$$

Whenever logarithmic transformations of the dependent variables were carried out, appropriate weighting factors (Deming, 1964) have been used in developing the regression equations. The equations for the individual feed size distributions and for all data combined were as follows:

coarse: $$\log_{10} d_{50}(\mathrm{c}) = 0.0419 \cdot VF - 0.0710 \cdot Spig + 0.0467 \cdot Inlet + 0.0406 \cdot FPS - 0.00006 \cdot Q - 0.7491 \quad (6\text{-}15)$$

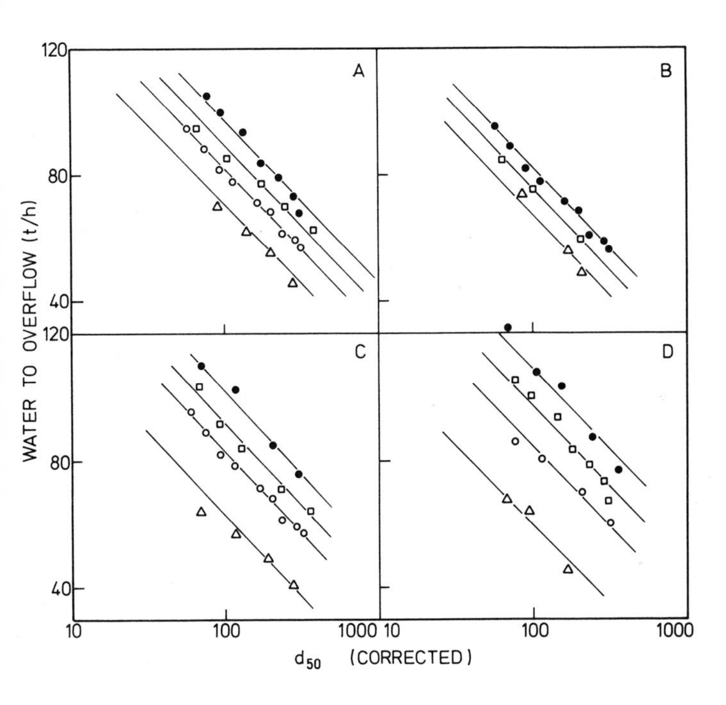

Fig. 6-5. Effect of operation variables on d_{50} (corrected).

Legend		*VF* (cm)	*SPIG* (cm)	Pressure (kpa)
	A ●	15.2	7.6	70
	□*	15.2	7.6	70
	○	14.0	6.4	56
	Δ*	14.0	6.4	56
	B ●	14.0	6.4	56
	□	14.0	7.6	56
	Δ	14.0	8.9	56
	C ●	14.0	6.4	84
	□	14.0	6.4	70
	○	14.0	6.4	56
	Δ	14.0	6.4	35
	D ●	17.1	7.6	70
	□	15.2	7.6	70
	○	14.0	7.6	70
	Δ	11.4	7.6	70

* Copper ore used in these tests. Silica used in all other tests.

medium: $$\log_{10} d_{50}(c) = 0.0637 \cdot VF - 0.0712 \cdot Spig + 0.0220 \cdot Inlet + 0.0390 \cdot FPS - 0.00008 \cdot Q - 0.4811 \quad (6\text{-}16)$$

fine: $$\log_{10} d_{50}(c) = 0.0344 \cdot VF - 0.019 \cdot Spig + 0.0513 \cdot Inlet + 0.0255 \cdot FPS - 0.00008 \cdot Q - 0.6623 \quad (6\text{-}17)$$

all data: $$\log_{10} d_{50}(c) = 0.0400 \cdot VF - 0.0576 \cdot Spig + 0.0366 \cdot Inlet + 0.0299 \cdot FPS - 0.00005 \cdot Q + 0.0806 \quad (6\text{-}18)$$

For a constant feed size distribution, equations of this form may be used to predict $d_{50}(c)$ for large throughputs from tests carried out at small throughputs. However, a change in feed size affects the accuracy of prediction and a regression equation which included the $+ 420\,\mu m$ and $- 53\,\mu m$ values for all data was as follows:

$$\log_{10} d_{50}(c) = 0.0418 \cdot VF - 0.0543 \cdot Spig + 0.0304 \cdot Inlet + 0.0319 \cdot FPS - 0.00006 \cdot Q - 0.0042 \cdot (+ 420\,\mu m) + 0.0004 \cdot (- 53\,\mu m) \quad (6\text{-}19)$$

This equation shows the nature of the effect of change in feed size on the value of $d_{50}(c)$ but it is specific to the present data.

The procedure for scale-up of the $d_{50}(c)$ value is as follows:

(1) Carry out a series of tests using a small hydrocyclone on the ore at a feed size which is fairly close to the expected size in the plant and obtain a range of $d_{50}(c)$ values.

(2) Ensure during these tests that the coarse product is flowing freely through the spigot (spray type of discharge), and that the flow rate of one stream and the solids contents and size distributions of all streams are measured accurately, and derive an equation of the form shown in eq. 6-14.

This equation will then be suitable for scale-up work, bearing in mind the approximate relationships which must exist between vortex finder, inlet, spigot, and hydrocyclone diameter. For more precise prediction when it is expected that the feed size will alter significantly the tests should be repeated over a range of feed sizes and an appropriate equation developed.

6.1.5 *Reduced-efficiency curves*

Whilst three types of efficiency curves, the actual, corrected and reduced, may be drawn for any one classification test, only the reduced-efficiency curve is suitable for use in the assessment of the performance of a cyclone as the operating conditions altered. The reduced-efficiency curve is a measure of the probability of appearance of particles in the coarse products due to centrifugal action alone and it is affected by properties of the material and by some of the characteristics of the hydrocyclone. It was found (Lynch and

Rao, 1965) that for a given operation, the reduced-efficiency curve is constant for wide changes in flow rate and solids content of the pulp and vortex-finder and spigot diameters of the hydrocyclone. However, at that time the effect on the reduced-efficiency curve of maintaining geometrical similarity of cyclones and changing cyclone diameter was not known. This is important in scale-up for design purposes but it is also important when the problem of minimising the appearance of coarse particles in the fine product is being considered. If small cyclones give a sharper reduced-efficiency curve than large cyclones, this may be a vital factor in designing a cyclone installation when the rate of production of coarse, composite particles is to be minimised.

Present investigations have shown that the reduced-efficiency curve for an ore classified in hydrocyclones remains constant for all conditions including change in hydrocyclone diameter provided that geometric similarity is maintained. Experimental results illustrating this are given in Fig. 6-6. The reduced efficiency curve for limestone is close to that for silica but, as shown in Fig. 6-7, is different from that for coal classified in a 38.1-cm Krebs hydrocyclone (D.T. Tarr, private communication). Although confirmatory tests on the behaviour of high- and low-specific-gravity minerals in hydrocyclones of different sizes are still to be carried out, the following tentative conclusions may be drawn:

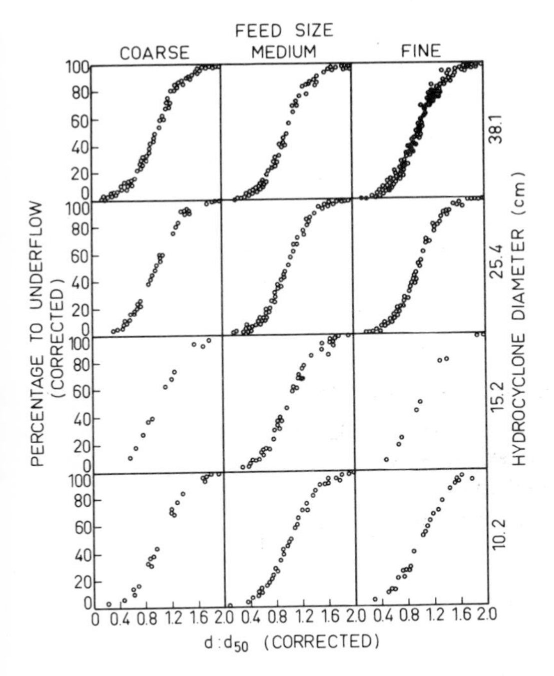

Fig. 6-6. The constancy of the reduced-efficiency curve for a single mineral, in this case silica.

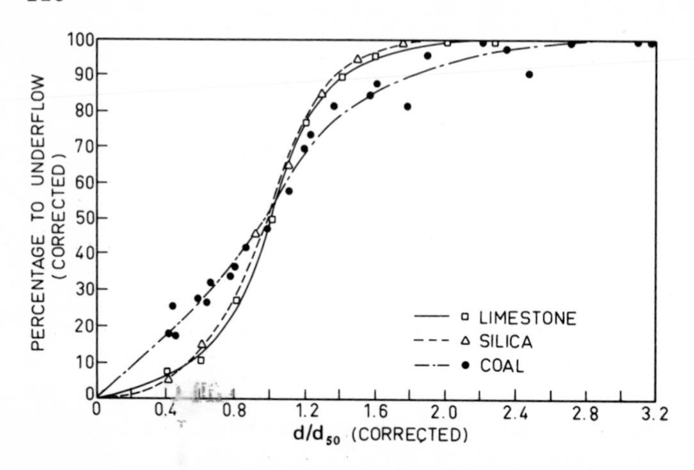

Fig. 6-7. Reduced-efficiency curves for different minerals.

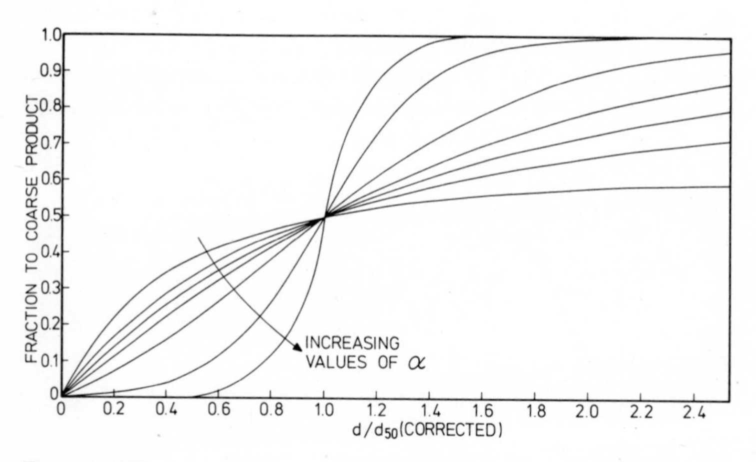

Fig. 6-8. Effect of change in α on the shape of the reduced-efficiency curve. The equation for the curve is $y = (e^{\alpha x} - 1)/(e^{\alpha x} + e^{\alpha} - 2)$.

(1) The reduced-efficiency curve for a mineral is not dependent on hydrocyclone diameter, or outlet dimensions, or operating conditions.

(2) The reduced-efficiency curve determined for a mineral on a small hydrocyclone may be used for scale-up work.

It has been found that all reduced-efficiency curves may be defined by the equation:

$$y = \frac{e^{\alpha \cdot x} - 1}{e^{\alpha \cdot x} + e^{\alpha} - 2} \tag{6-20}$$

in which α is the variable parameter which describes completely the shape of the curve. The effect of change in α on the shape of the reduced-efficiency curve is shown in Fig. 6-8.

6.1.6 *Behaviour of mixtures of minerals*

In the case of circuits which are treating ores containing mixtures of minerals with different physical properties, it is important to be able to predict the behaviour of each component in the classifier as well as in the grinding mill so that the size distribution of each component in the circuit product may be predicted. No experimental work has yet been done with mixtures of minerals in the test rigs at the Julius Kruttschnitt Mineral Research Centre (JKMRC), but data which are suitable for investigating aspects of the classification behaviour of mixtures of minerals have been collected at two plants. The plants at which the data were obtained are operated by Peko Mines Ltd. at Warrego, N.T., Australia and New Broken Hill Consolidated Limited (NBHC) at Broken Hill, N.S.W., Australia. A brief description of the ores and circuits is as follows:

Warrego concentrator. The ore is a coarse-grained magnetite—silica mixture. The grinding circuit is an autogenous mill operated in closed circuit with several Krebs 25-cm hydrocyclones in parallel. The hydrocyclone feed contained approximately 65% magnetite. The size fractions of each stream were not assayed for magnetite and consequently the behaviour of total streams only could be considered.

NBHC concentrator. The ore is a coarse-grained galena—marmatite—silica mixture. One ball mill—hydrocyclone circuit was set up as a test circuit and operated at various conditions, the performance being observed at each condition. In this case, each size fraction in each stream for each test was assayed for lead and zinc and the approximation was made that the lead was present in free galena and the zinc in free marmatite.

The reduced-efficiency curve for the Warrego data is shown in Fig. 6-9 and it will be noted that this curve has a long "tail" and does not conform to the regular shapes of the curves reported earlier. This is a feature of the reduced-efficiency curves of ores containing a mixture of minerals of differing specific gravities. However, calculations show that this type of curve for a total ore arises directly from summing the results of classification of the individual minerals which are behaving as predicted by standard reduced-efficiency curves.

A typical calculated reduced-efficiency curve for a hypothetical mixture of magnetite and silica is also shown in Fig. 6-9. It will be seen that the general shape of this curve agrees with the observed shape of the curve from the Warrego data. The "tail" is due to the fact that the d_{50}(c) for the ore is not the weighted average of the d_{50}(c) values for the components but is biassed towards the lower value. Further calculations show that the shape of the curve is affected by the size distributions of the components and their relative proportions. Similar comments may be made about the NBHC data.

The conclusion is that a composite reduced-efficiency curve may change, but the curves for the components remain constant. It is important to be

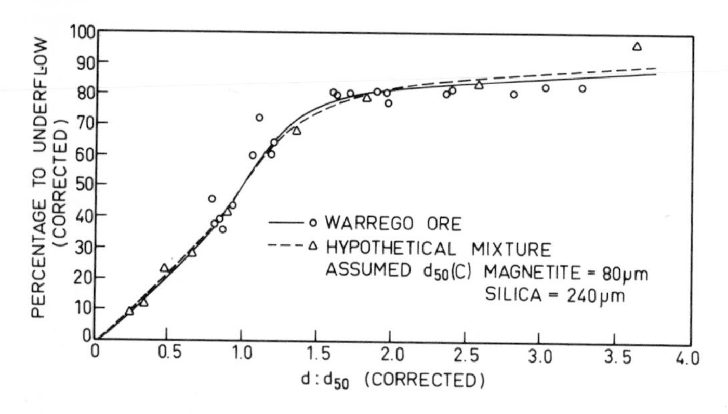

Fig. 6-9. Reduced-efficiency curves for an actual (Warrego) ore, which contains magnetite and silica, and a hypothetical magnetite—silica mixture.

able to predict d_{50}(c) for the components and the results of the NBHC tests in which the d_{50}(c) values for the ore varied between 137 and 242 μm will be discussed in this context. The Warrego data will not be discussed because of the narrow range of d_{50}(c) values.

A regression equation relating d_{50}(c) to the variables of operation for the NBHC data is as follows:

$$\log_{10} d_{50}(\text{c})\ \text{ore} = 2.2556 + 0.021 \cdot VF - 0.066 \cdot Spig + 0.0002 \cdot FPS - 0.0001 \cdot Q \quad (6\text{-}21)$$

and a comparison of the observed and calculated values is shown in Fig. 6-10. The low value of the coefficient of the *FPS* term is due to the limited range over which *FPS* is varied. The true value is probably considerably higher. The good agreement shows that this type of equation, which is similar to those derived for single minerals, is also suitable for complex ores. The final requirement is to predict d_{50}(c) for the components from known values of d_{50}(c) for the ore.

The d_{50}(c) values for the individual minerals in the NBHC tests have also been determined and the relationships between these and the d_{50}(c) values for the ore are given in Fig. 6-11.

Marlow (1973) derived the following equations for these relationships:

$$d_{50}(\text{c})\ \text{gangue} = 1.29 \cdot d_{50}(\text{c})\ \text{ore} \quad (6\text{-}22)$$

$$d_{50}(\text{c})\ \text{marmatite} = 0.75 \cdot d_{50}(\text{c})\ \text{ore} \quad (6\text{-}23)$$

$$d_{50}(\text{c})\ \text{galena} = 0.31 \cdot d_{50}(\text{c})\ \text{ore} \quad (6\text{-}24)$$

The specific gravities for the minerals and the ore are: gangue — 2.7, marmatite — 4.08, galena — 7.57, ore — 3.25. The relationship between d_{50} and the specific gravity of a mineral has been expressed (Bradley, 1965) as:

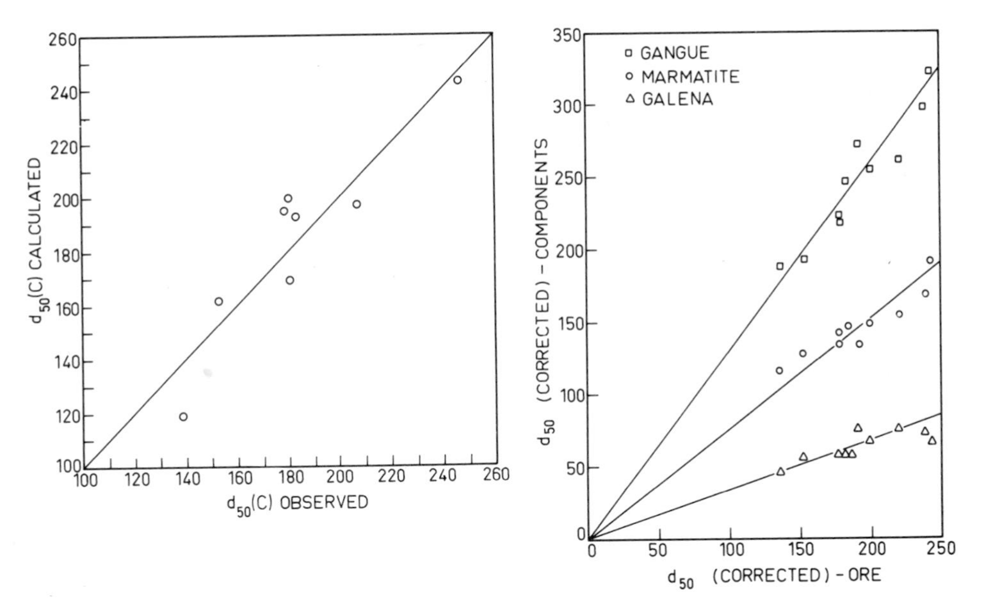

Fig. 6-10. Comparison of the observed and calculated d_{50}(corrected) values for the NBHC ore.

Fig. 6-11. Relationship between d_{50}(corrected) for NBHC ore and d_{50}(corrected) for the individual minerals.

$$d_{50} = \alpha\,(sg_{mineral} - sg_{liquid})^{-K} \qquad (6\text{-}25)$$

where K may have a value between 0.5 (Stokes' Law) and 1.0 (turbulent condition). Marlow showed that if eq. 6-25 is written:

$$d_{50} = K_1 \cdot (sg_{mineral} - 1)^{-K_2} \qquad (6\text{-}26)$$

K_1 for the individual components may be written as follows:

$$K_1 = d_{50} \cdot (sg_{mineral} - 1)^{K_2} \qquad (6\text{-}27)$$

If K_1 is constant the following equations may be written:

$$d_{50}(c)\text{ gangue} = 1.32^{K_2} \cdot d_{50}(c)\text{ ore} \qquad (6\text{-}28)$$

$$d_{50}(c)\text{ marmatite} = 0.73^{K_2} \cdot d_{50}(c)\text{ ore} \qquad (6\text{-}29)$$

$$d_{50}(c)\text{ galena} = 0.34^{K_2} \cdot d_{50}(c)\text{ ore} \qquad (6\text{-}30)$$

and by comparing these equations with eqs. 6-22 to 6-24, it will be noted that K_2 has a value very close to unity. This gives good support to the use of eq. 6-25, where the regime is fully turbulent, to predict the d_{50}(c) values for various components of a mixture of minerals if the d_{50}(c) value for the ore is known or can be predicted. This prediction can be made using an equation of the form of eq. 6-21. The requirement is that the assumption that the components are fully liberated is approximately correct.

6.1.7 *Effect of cone angle and cyclone length on cyclone performance*

A summary of the effects is given below.

(1) Increase in cyclone length increases the throughput. Increase in cone angle at shorter length of a cyclone also causes an increase in throughput, but this effect is not noticed at the full length of a cyclone.

(2) The reduced-efficiency curve is independent of the cyclone geometry.

(3) An increase in cone angle at constant length or an increase in length at constant cone angle increases the $d_{50}(c)$.

(4) Water split in a cyclone is independent of cone angle but a decrease in cyclone length decreases the fraction of water reporting to underflow.

It appears that the length of a hydrocyclone is an important variable which could be effectively used to adjust the performance characteristics of a hydrocyclone operation.

Consequently, a decrease in the length of a hydrocyclone at constant cone angle: (1) decreases the fraction of water reporting to underflow and consequently decreases the amount of fines short-circuiting to the underflow product; (2) decreases the $d_{50}(c)$, that is, decreases the amount of fines reporting to underflow.

To illustrate the overall effect of (1) and (2) above on the mass flow rates and sizing analyses of products, calculations were carried out for the 12°-cone operating at 350 l/min. The results are given in Table 6-II. It will be noted that, all other conditions being constant, a decrease in cyclone length: (1) decreases the flow rate of solids and water in underflow product, (2) decreases the fineness of underflow product, and (3) decreases the $d_{50}(c)$.

6.1.8 *Summary*

A series of equations, which may be used to define and predict the performance of high-capacity cyclones of the type which are installed in many industrial circuits, has been developed. These equations are based upon the premise that the solid particles which appear in the underflow consist of two components, as follows: (1) the representative proportion of the feed particles which are carried into the underflow by the water, and (2) the particles which appear in the underflow as a result of classification.

The particles which appear in the overflow may then be calculated from the known feed and underflow compositions. In order to derive the constants in the equations which apply to a particular operation, it is necessary to carry out a cyclone classification test on the ore, or preferably, a group of tests from which the mean values of the constants may be determined.

The sequence in which calculations are carried out to determine the performance of the cyclone for any required conditions of operation is as follows:

(1) the capacity of the cyclone;

TABLE 6-II

Effect of change in length on the performance characteristics of a 12°-hydrocyclone–simulated results

Original conditions:
Cyclone length = 100.0 cm

	Feed (t/h)	Overflow (t/h)	Underflow (t/h)	% to underflow
Water	13.10	8.40	4.70	35.9
Solids	19.60	7.25	12.35	63.0

Sizing distributions (wt. % passing):

μm	1200	850	600	425	300	212	150	106	75	53
Feed	99.8	97.9	90.4	81.5	75.0	68.0	63.7	59.2	55.2	51.8
Overflow	100.0	100.0	100.0	100.0	100.0	99.5	99.3	98.0	94.3	89.7
Underflow	99.7	96.7	84.8	70.6	60.3	49.5	42.8	36.4	32.2	29.5

d_{50}(c) (in μm) = 91
% solids by wt. in feed = 60.0; throughput = 350 l/m

New conditions:
Cyclone length = 80 cm
Feed conditions = as above
Sizing distributions (wt. % passing) calculated:

μm	1200	850	600	425	300	212	150	106	75	53
Overflow	100.0	100.0	100.0	100.0	100.0	99.5	99.3	98.5	95.6	91.1
Underflow	97.5	96.5	83.9	69.0	58.1	46.7	39.6	32.6	27.9	25.2

d_{50}(c) (in μm) = 83 (calculated)

(2) the volume flow rate of water in the feed, which is calculated from the throughput and the solids content of water in the feed;

(3) the flow rate of water in the overflow;

(4) the flow rate of water in the underflow, and the mass fraction of the water in the feed which appears in the underflow;

(5) the mass of particles of each size range which appear in the underflow due to the water;

(6) the value of d_{50}(c) for the operation;

(7) the mass flow rate of particles from each size range appearing in the underflow due to classification, calculated from the d_{50}(c) value and the reduced-efficiency curve;

(8) the total mass, sizing analysis and solids content of ore in the underflow; and

(9) the total mass, sizing analysis and solids content of ore in the overflow.

6.2 SCREENS

Although screens are used in almost all processes which involve mineral particles, there have been few attempts reported to develop comprehensive mathematical models of screen operations or to use such models for simulation studies.

Gurun (1973), in a study of the design of crushing plant flowsheets by simulation, represented the behaviour of the screen by a column vector in which successive elements described the probability of each fraction appearing in the oversize. For a particular case, the numerical values in this column vector were chosen from a large bank of equipment and performance data stored in a computer. This model is useful when predictions about screen behaviour are available and correct, but is of limited value in other circumstances and particularly when the screens are operating under a fully loaded or over-loaded condition.

The work of Whitby (1958) provides a basis for a suitable model, but this was carried out on batch screens only. While there is a close analogy between batch and continuous screening, the adaptation of Whitby's model to a continuous process for simulation purposes would require the collection of a considerable amount of special data from continuous screens and this has not yet been done.

The only model of a vibrating screen which has been shown to be useful for simulation purposes is the simple model developed by Whiten (1972b) and even this model has not yet been fully tested at all conditions of screen loading. This model is described in sub-section 6.2.1.

In the case of wedge-wire screens, the separation efficiency has been found to be influenced by the aperture size, the mass flow rate of water in the fine product, the water split and the solids content of the screen feed. The type of model which was developed to describe hydrocyclone performance has been found suitable for use in the development of a model of wedge-wire screens.

6.2.1 *Vibrating screens*

The model of the vibrating screen was developed from data collected at the Qld. Mount Isa Mines Limited in July, 1968. These data contained tests giving feed and product data for the Allis-Chalmers 1.83 m by 4.80 m single-deck vibrating screens. A theoretical model of the vibrating screen was developed using probability considerations and the parameters were found by least squares fitting to the model.

The probability of a particle of size s dropping through the screen in one trial may be calculated following the nomenclature given in Fig. 6-12.

The area of the basic separating unit of the screen (that is, hole plus wire) is $(h + d)^2$ where h is the hole size and d the wire diameter. The particle of size s must fall into an area of $(h - s)^2$ to drop through the screen without bouncing off the wire. Thus the probability of the particle dropping through the screen is:

$$[(h - s)/(h + d)]^2$$

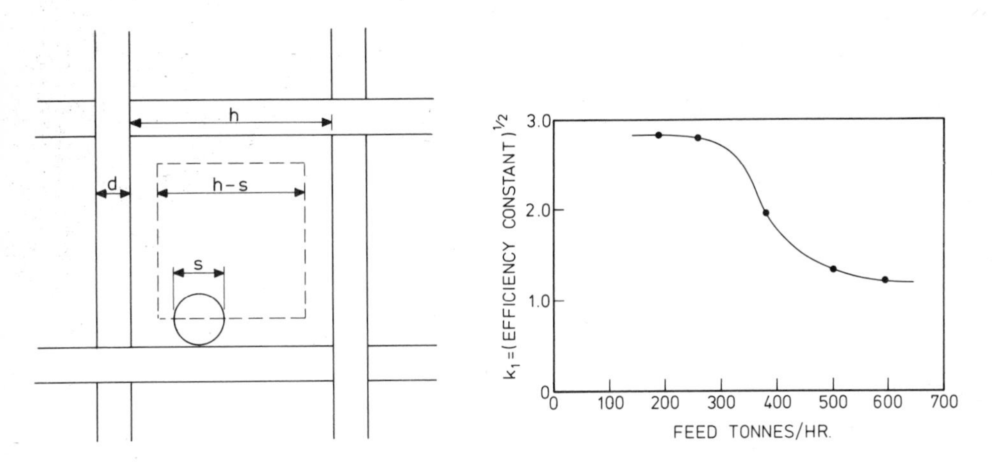

Fig. 6-12. Nomenclature for model of vibrating screen. (For legend, see text.)

Fig. 6-13. Relationship between efficiency constant and feed tonnage for a 12.7-mm square hole-wire screen.

The probability that a particle of size s does not pass through the screen in m trials is:

$$[1-\{(h-s)/(h+d)\}^2]^m$$

and this expression provides the efficiency curve for the screen.

The number of trials, m, that a particle performs in crossing the screen is considered to be proportional to: (1) an efficiency constant $(k_1)^2$, (2) the length of the screen, l, and (3) a load factor f.

Consequently:

$$m = k_1^2 \cdot l \cdot f \tag{6-31}$$

The load factor f will be unity for low feed rates and will decrease to almost zero for very large feed rates. Experimentally the load factor and the efficiency constant can be incorporated into a single load-dependent parameter. Additional extensive data were collected from Bougainville Copper Limited (Papua New Guinea) and it was found that a very good fit between predicted and experimental behaviour was obtained if the efficiency constant was allowed to vary with feed tonnage. The relationship between efficiency constant and feed tonnage is shown in Fig. 6-13 for 12.7-mm square hole wire screens. The predicted performance of a particular screen is determined by the value of the exponent m. In order to obtain the required range of values for m and at the same time have reasonable values of k_1^2 it has been found convenient to express the screen length in metres.

To use this model for the prediction of the amount of ore in size intervals an average value of the probability is required for each size interval. A method for the calculation of a reasonably accurate value is given below.

The unweighted average value of the probability is:

$$\frac{\int_{s_1}^{s_2} [1-\{(h-s)/(h+d)\}^2]^m \cdot ds}{(s_2-s_1)} \tag{6-32}$$

and the approximation:

$$[1-\{(h-s)/(h+d)\}^2]^m = e^{-m} \cdot [(h-s)/(h+d)]^2$$

may be used. Putting this approximation into eq. 6-32 and making the substitution:

$$y = \sqrt{m} \cdot (s-h)/(h+d) \tag{6-33}$$

gives:

$$\frac{\frac{h+d}{\sqrt{m}} \cdot \int_{y_1}^{y_2} e^{-y^2} dy}{s_2 - s_1} \tag{6-34}$$

and this integral may be evaluated using the approximations (Hart, 1968):

$$\int_{y}^{\infty} e^{-y^2} dy = 0.124734/(y^3 - 0.4378805 \cdot y^2 + 0.266892 \cdot y + 0.138375)$$

and:

$$\int_{0}^{\infty} e^{-y^2} dy = 0.89$$

which are accurate to 0.01. The reason for using k_1^2 instead of k_1 is associated with the evaluation of eqs. 6-32 to 6-34.

This model provides an adequate description of the screen behaviour except for the sub-mesh material of which the fraction $k_3 = 0.1$ was found to report with the coarse fraction.

6.2.2 *Wedge-wire screens*

An important feature of wedge-wire screens, particularly those which are not equipped with tapping or vibrating devices, is the rapidity with which "blinding" occurs and its influence on the separation efficiency. It may be noted that, despite the occurrence of blinding, the shape of the corrected efficiency curve remained constant.

Although wedge-wire screens can be operated at any required capacity, the lower limit of the efficient operating range may be regarded as the point at which oversize particles are not freely discharged and an arbitrary upper limit of four times the lower limit has been set. At capacities greater than this, the efficiency is low.

Increase in slot width increases the mass fraction of water appearing in the fine product. For individual slot widths the relationships for a lead—zinc ore were found to be:

(a) slot width $(SW) = 0.25$ mm; $WUS = 0.5569 \cdot WF - 0.0549 \cdot FPS + 1.159$

(b) slot width $(SW) = 0.5$ mm; $WUS = 0.8284 \cdot WF - 0.0607 \cdot FPS + 2.5906$

(c) slot width $(SW) = 1.0$ mm; $WUS = 0.9229 \cdot WF - 0.0594 \cdot FPS + 3.328$

The composite relationship of this form which took all the operating variables into account and which gave the best fit to the data over the range of conditions is:

$$WUS = 0.98 \cdot (SW)^{0.33} \cdot WF - 0.06 \cdot FPS + 2 \cdot SW \qquad (6\text{-}35)$$

This equation was developed by using a multiple linear-regression technique for the analysis of the results. It can be improved in accuracy as further results are obtained.

For a lead—zinc ore classified on wedge-wire screens of various slot widths over a range of conditions, it was found that a single reduced-efficiency curve fitted all data. This was defined by the equation:

$$y = \frac{e^{4.0 \cdot x} - 1}{e^{4.0 \cdot x} + e^{4.0} - 2}$$

It will be noted that the value of α is 4.0 which is high and shows good separation efficiency.

The variables which had the most significant effect on $d_{50}(c)$ were the slot width and the flow rate of water in the undersize product. The relationship for a lead—zinc ore was found to be:

$$\log_{10} d_{50}(c) = 1.1718 \cdot \log_{10} SW + 0.001372 \cdot WUS + 0.0029 \cdot FPS + 2.45 \qquad (6\text{-}36)$$

It is expected that this equation also will be modified as further results are obtained.

The calculation of wedge-wire screen performance may be carried out in the same manner as hydrocyclone performance. However, it is emphasised that the equation is approximate only because the data required to develop more precise general-purpose equations have not yet been obtained. It is possible that when the data have been obtained, it will be found that different sets of equations are necessary to describe the performances of coarse and fine screens.

A serious problem with wedge-wire screens is that unless they are equipped with a mechanism for minimising the occurrence of "blinding", they operate under unsteady-state conditions. This makes the collection of data for the development of accurate mathematical models difficult and limits accuracy in the application of these models.

NUMERICAL EXAMPLES

Example 6-1: Operating pressure for new conditions with the same cyclone. One set of data for a 50.8-cm diameter hydrocyclone classifying a silica water pulp is as follows:
vortex finder diameter = 14.0 cm
spigot diameter = 7.6 cm
pressure at inlet = 69 kpa
throughput = 2000 l/min
percent water in feed = 52

What will be the pressure at the inlet if the throughput is increased to 2500 l/min, the vortex finder diameter is reduced to 12 cm, and the percent water in the feed is increased from 52 to 58?

Solution:
Eq. 6-3 may be used to solve this problem, namely;

$$Q = K \cdot VF \cdot P^{0.5} \cdot (FPW)^{0.125}$$

Step 1—Calculation of K from original data:

$$K = \frac{2000}{14 \cdot 69^{0.5} \cdot (52)^{0.125}} = 10.49$$

Step 2—Calculation of new pressure:

$$P = \left[\frac{Q}{K \cdot VF \cdot (FPW)^{0.125}}\right]^2 = \left[\frac{2500}{(10.49) \cdot 12 \cdot (58)^{0.125}}\right]^2 = 143 \text{ kpa}$$

The pump for the new duty can now be specified since the new operating pressure is known.

Example 6-2: Operating pressure in scale-up calculation. One set of data for a 10.2-cm diameter hydrocyclone classifying a silica—water pulp are as follows:
vortex finder diameter = 3.3 cm
spigot diameter = 1.8 cm
inlet diameter = 2.1 cm
pressure at inlet = 120 kpa
throughput = 200 l/min
percent water in feed = 45

What will be the pressure at the inlet of a 38.1-cm hydrocyclone working on the same pulp at a throughput of 2000 l/min? The dimensions are as follows:
vortex finder diameter = 14.9 cm
spigot diameter = 7.5 cm
inlet diameter = 9.5 cm

Solution:
Eq. 6-4 may be used to solve this problem, namely:

$$Q = K \cdot VF^{0.73} \cdot Inlet^{0.86} \cdot P^{0.42}$$

Step 1—Calculation of K from original data:

$$K = \frac{200}{(3.3)^{0.73} \cdot (2.1)^{0.86} \cdot (120)^{0.42}} = 5.92$$

Step 2—Calculation of new pressure:

$$P = \left[\frac{Q}{K \cdot VF^{0.73} \cdot Inlet^{0.86}}\right]^{1/0.42} = \left[\frac{2000}{5.92 \cdot (14.9)^{0.73} \cdot (9.5)^{0.86}}\right]^{2.38} = 95.2 \text{ kpa}$$

Thus, the pump for the full-scale duty can be chosen from pressure-capacity tests on a small hydrocyclone.

The problem set in Example 6-1 can also be solved by eq. 6-4. The calculated pressure using this equation is 154 kpa which is within 8% of the pressure calculated by eq. 6-3. Eq. 6-3 is more accurate where the diameter of the cyclone remains unchanged but its range of use is limited.

Example 6-3: Water split for change in throughput. One set of data for a 38.1-cm hydrocyclone classifying a silica—water pulp is as follows:

flow rate of solids in feed = 40 t/h
percent by weight of solids in feed = 55.0
flow rate of solids in overflow = 15 t/h
percent by weight of solids in overflow = 35.3
spigot diameter = 5.1 cm

Calculate the water split if the flow rate of the feed is increased by 50% and the spigot diameter remains constant.

Solution:
Eq. 6-6 may be used to solve this problem, namely:

$$WOF = 1.07 \cdot WF - 3.94 \cdot Spig + K$$

Step 1—Calculation of flow rates:

flow rate of water in feed = $(40 \cdot 45)/55 = 32.7$ t/h

flow rate of water in overflow = $(15 \cdot 64.7)/35.3 = 27.5$ t/h

Step 2—Calculation of the constant in eq. 6-6:

$$27.5 = 1.07 \cdot 32.7 - 3.94 \cdot Spig + K \ldots K = 12.6$$

Step 3—Calculation of the new flow rate of water in the overflow:

$$WOF = 1.07 \cdot (32.7 \cdot 1.5) - 3.94 \cdot 5.1 + 12.6 = 45.0 \text{ t/h}$$

Example 6-4: New operating conditions for change in feed solids content. One set of data for a 50.8-cm diameter hydrocyclone classifying a silica—water pulp is as follows:

vortex finder diameter = 14.0 cm
pressure at inlet = 55 kpa
spigot diameter = 6.4 cm
throughput = 1888 l/min
s.g. of silica = 2.7

Sizing distributions and pulp solids contents:

Micrometres	417	295	208	147	105	75	53	38	−38	% solids (by wt.)
Feed	3.2	2.6	4.6	8.3	15.0	18.2	10.4	8.0	29.6	25.0
Overflow	0	0	0	0	0.8	8.2	15.0	14.0	62.0	13.8
Underflow	5.7	4.6	8.0	14.6	25.8	25.8	7.0	3.5	5.0	65.0

What would be the sizing distributions and mass flow rates of ore and water in each stream if the solids content of the feed is increased to 52% by weight and the throughput remained unchanged?

Solution:

Step 1—Flow rates of ore and water in the cyclone feed stream for both sets of conditions are (in t/h):

	Ore	Water
original	33.6	100.8
new	87.6	80.9

Step 2—Flow rates of ore and water in the cyclone product streams for the first set of conditions are (in t/h):

	Ore	Water
overflow	14.5	90.6
underflow	19.1	10.2

Step 3—Calculations based on water flow rates:

original: $WOF = 1.07 \cdot WF - 3.94 \cdot Spig + K$

$90.6 = 1.07 \cdot 100.8 - 3.94 \cdot 6.4 + K$

$K = 8.0$

new : $WOF = 1.07 \cdot 80.9 - 3.94 \cdot 6.4 + 8.0$

$= 69.3$ t/h

Mass fraction of water short-circuiting to underflow:

original = 0.101

new = 0.143

Step 4—Calculation of $d_{50}(c)$ and reduced-efficiency curve for original conditions. The procedure is given in Table 6-III.

$d_{50}(c)$ for original conditions = 74 μm; α in reduced-efficiency curve = 2.0

Step 5—Calculations based on $d_{50}(c)$ using eq. 6-15:

For original conditions (calculation of constant):

$$\log_{10} d_{50}(c) = 0.0173 \cdot FPS - 0.0695 \cdot Spig + 0.0130 \cdot VF + 0.000048 \cdot Q + K$$

$$\log_{10} 74 = 0.0173 \cdot 25.0 - 0.0695 \cdot 6.4 + 0.0130 \cdot 14.0 + 0.000048 \cdot 1888 + K$$

$$1.8692 = 0.4325 - 0.4448 + 0.1820 + 0.0906 + K$$

$$K = 1.6089$$

TABLE 6-III

Procedure for calculation of $d_{50}(c)$ and reduced-efficiency curve

Micrometres*	Feed (t/h)	O.F. (t/h)	U.F. (t/h)	Short circuit to U.F. (t/h)	Feed classified (t/h)	To U.F. by classification (t/h)	Correct efficiency	Reduced efficiency
495	1.09	0	1.09	0.11	0.98	0.98	100	6.78
351	0.87	0	0.87	0.09	0.78	0.78	100	4.82
246	1.54	0	1.54	0.15	1.39	1.39	100	3.37
175	2.79	0	2.79	0.28	2.51	2.51	100	2.40
124	5.04	0.12	4.92	0.50	4.54	4.42	97.4	1.70
89	6.11	1.19	4.92	0.61	5.50	4.31	78.4	1.22
61	3.51	2.18	1.33	0.37	3.14	0.96	30.6	0.84
43	2.70	2.03	0.67	0.29	2.41	0.38	15.8	0.59
	9.93	8.97	0.96	0.99	8.94	0	0	

* These are the geometric means.

For new conditions:

$$\log_{10} d_{50}(c) = 0.0173 \cdot 52 - 0.0695 \cdot 6.4 + 0.0130 \cdot 14.0 + 0.000048 \cdot 1888 + 1.6089$$

$$= 2.3363$$

$$d_{50}(c) = 217\ \mu m$$

Step 6—Calculation of product sizing distributions and mass flow rates for new conditions. Fraction entering underflow by true classification is calculated from the equation:

$$y = \frac{(e^{\alpha \cdot d/d_{50}} - 1)}{(e^{\alpha \cdot d/d_{50}} + e^{\alpha} - 2)}$$

The procedure is given in Table 6-IV.

TABLE 6-IV

Procedure for calculation of overflow and underflow sizing distributions

Micrometres	Feed (t/h)	Short circuit to U.F. (t/h)	Classified feed (t/h)	To U.F. by true classification: fract.	To U.F. by true classification: (t/h)	Total to U.F. (t/h)	Total to O.F. (t/h)	Sizings U.F.	Sizings O.F.
495	2.8	0.40	2.40	0.937	2.25	2.65	0.15	9.6	0.2
351	2.28	0.33	1.95	0.792	1.54	1.87	0.41	6.8	0.7
246	4.03	0.58	3.45	0.575	1.98	2.56	1.47	9.2	2.5
175	7.27	1.04	6.23	0.386	2.40	3.44	3.83	12.4	6.4
124	13.14	1.88	11.26	0.250	2.81	4.69	8.45	16.9	14.1
89	15.94	2.28	13.66	0.166	2.27	4.55	11.39	16.4	19.0
61	9.11	1.30	7.81	0.106	0.83	2.13	6.98	7.7	11.7
43	7.01	1.00	6.01	0.071	0.43	1.43	5.58	5.2	9.3
19	25.93	3.71	22.22	0.029	0.64	4.35	21.58	15.7	36.1

Total mass flow to underflow = 27.7 t/h
Total mass flow to overflow = 59.8 t/h

$$\text{Percent solids in underflow by weight} = \frac{27.7 \cdot 100}{27.7 + 11.6} = 70.5$$

$$\text{Percent solids in overflow by weight} = \frac{59.8 \cdot 100}{59.8 + 69.2} = 46.3$$

Example 6-5: Scale-up equations for prediction of hydrocyclone performance. Tests were carried out on the classification of silica in 10- and 15-cm hydrocyclones to obtain data for the design of a large installation. The results are shown in Table 6-V.

TABLE 6-V

Data obtained from tests on small hydrocyclones

	Measured diameters (cm)					Calculated		
Cyclone	*VF*	*Spig*	*Inlet*	*FPS*	*Q*	$d_{50}(c)$	*WF*	R_f
15	5.1	3.8	3.1	66.5	349	133.0	11.70	33.6
15	5.1	3.8	3.1	63.0	351	89.0	12.53	32.2
15	6.4	3.8	3.1	64.5	438	145.5	15.23	19.0
15	6.4	3.8	3.1	65.1	355	162.5	12.19	31.2
15	5.1	3.8	3.1	64.9	348	141.2	12.01	34.0
15	7.9	3.8	3.1	65.1	345	214.7	12.16	31.7
15	7.9	3.8	3.1	60.2	363	129.6	13.55	26.1
15	6.4	3.8	3.1	60.1	409	105.1	15.29	25.1
15	5.1	3.8	3.1	60.4	398	72.4	14.82	27.4
15	7.9	3.8	3.1	55.4	439	81.6	17.51	16.4
15	7.9	3.8	3.1	52.9	372	77.4	15.34	23.3
10	3.2	2.2	2.1	65.9	129	157.5	4.38	28.8
10	3.2	2.2	2.1	65.7	182	120.8	6.17	21.6
10	3.2	2.2	2.1	63.3	183	92.5	6.53	21.2
10	3.2	2.2	2.1	63.6	127	114.0	4.33	28.9
10	3.2	2.2	2.1	58.0	127	69.0	4.72	23.3
10	3.2	2.2	2.1	63.6	160	87.0	5.65	24.8
10	3.2	2.2	2.1	63.4	192	77.0	6.80	20.4
10	3.2	2.2	2.1	67.7	188	100.0	6.15	41.3
10	3.2	2.2	2.1	67.7	187	113.0	6.12	24.5
10	3.2	2.2	2.1	58.3	143	68.0	5.49	25.0

What will be the water split and the $d_{50}(c)$ value for a 38-cm hydrocyclone at the following operating conditions?

Operating condition	*VF*	*Spig*	*Inlet*	*FPS*	*Q* (l/min)
1	14.9	6.2	9.5	64.5	2100
2	10.2	6.2	9.5	64.5	1400

Solution:
The regression equations for the calculation of R_f and $d_{50}(c)$ are as follows:

$$R_f = \frac{K_a \cdot Spig}{WF} - \frac{K_b}{WF} + K_3$$

$$\log_{10} d_{50}(c) = K_c \cdot VF - K_d \cdot Spig + K_e \cdot Inlet + K_f \cdot FPS - K_g \cdot Q + K_h$$

The first step is to calculate the coefficients in these regression equations. A problem is that there is limited variation only in *Spig*, *Inlet* and *FPS* which would cause serious inaccuracies with predictions of large-scale performance. Until further information is obtained about the coefficients of the variables the procedure which should be followed is to set the values of the coefficients of the variables to the values given in eqs. 6-11 and 6-18 and use regression techniques to calculate the independent coefficients.

Consequently, the equations in which K' and K'' are to be found from the data, are:

$$R_f = \frac{193 \cdot Spig}{WF} - \frac{272}{WF} + K' \quad \text{and:}$$

$$\log_{10} d_{50}(c) = 0.0400 \cdot VF - 0.0576 \cdot Spig + 0.0366 \cdot Inlet + 0.0299 \cdot FPS - 0.00005 \cdot Q + K''$$

K' and K'', determined from the data by regression, were 2.2 and 0.0517. For the proposed operating conditions the calculated values of *WF* were 75.0 t/h and 50.0 t/h, respectively, for conditions 1 and 2:

$$R_f \text{ for } 1 = \frac{193 \cdot 6.2}{75} - \frac{272}{75} + 2.2 = 14.5; \quad R_f \text{ for } 2 = \frac{193 \cdot 6.2}{50} - \frac{272}{50} + 2.2 = 20.7$$

$$\log_{10} d_{50}(c) \text{ for condition } 1 = 0.0400 \cdot 14.9 - 0.0576 \cdot 6.2 + 0.0366 \cdot 9.5 + 0.0299 \cdot 64.5 - 0.00005 \cdot 2100 + 0.0517 = 2.4618$$

$$d_{50}(c) = 290\ \mu\text{m}$$

$$\log_{10} d_{50}(c) \text{ for condition } 2 = 0.0400 \cdot 10.2 - 0.0576 \cdot 6.2 + 0.0366 \cdot 9.5 + 0.0299 \cdot 64.5 - 0.00005 \cdot 1400 + 0.0517 = 2.3088$$

$$d_{50}(c) = 204\ \mu\text{m}$$

Example 6-6: Behaviour of an ore containing a mixture of minerals. A galena—sphalerite ore is divided into coarse and fine fractions in a hydrocyclone operating under the conditions given below. What will be the operating conditions of the hydrocyclone if an extra 20 t/h of water is added to the feed? It may be assumed that all particles are fully liberated. Original operating conditions for a 38.1-cm diameter hydrocyclone operating on a galena—sphalerite ore:

Diameters:		Specific gravity:	
Vortex finder	= 14.9 cm	Ore	= 3.3
Inlet	= 9.5 cm	Galena	= 7.6
Spigot	= 6.2 cm	Sphalerite	= 4.1
		Gangue	= 2.7

Mass flow of ore in cyclone feed = 146.4 t/h
Mass flow of water in cyclone feed = 57.8 t/h
Volume flow rate of cyclone feed = 1603 l/min

Calculated flow rates (t/h) of each component in each stream are given in Table 6-VI.

TABLE 6-VI

Flow rates of individual mineral in a lead—zinc ore in each stream of a hydrocyclone

	Feed			Overflow			Underflow		
Micrometres	ore	galena	sphalerite	ore	galena	sphalerite	ore	galena	sphalerite
+ 840	10.6	0.6	1.9	0	0	0	10.6	0.6	1.9
840/600	9.2	0.7	2.2	0.4	0	0	8.6	0.7	2.2
600/420	12.7	1.0	3.4	2.1	0	0	10.6	1.0	3.3
420/295	17.8	1.6	5.0	4.0	0	0.2	13.8	1.6	4.8
295/208	19.5	2.1	5.7	4.5	0	0.8	15.0	2.1	5.0
208/149	18.4	2.7	4.9	7.2	0.1	1.8	11.2	2.6	3.1
149/105	17.9	3.5	4.6	8.2	0.3	2.2	9.7	3.2	2.3
105/74	10.8	2.4	3.0	5.9	0.6	1.8	4.9	1.8	1.2
− 74	29.4	6.3	9.2	19.2	3.5	6.2	10.2	2.9	3.0
Total	146.4	21.0	39.8	51.2	4.5	13.0	94.9	16.5	26.8
Water	57.8			39.9			17.9		
Calculated $R_f = 31.0\%$									

Solution:
Step 1—The points on the corrected efficiency curves, the $d_{50}(c)$ values, and the values of α for each mineral are given in Table 6-VII.

TABLE 6-VII

Calculated points on the corrected efficiency curves, the $d_{50}(c)$ values and the values of α for each mineral for a lead—zinc ore classified in a hydrocyclone

Corrected efficiency curve (μm)	Galena	Sphalerite	Gangue	Ore
+ 840	1.0	1.0	1.0	1.0
840/600	1.0	1.0	0.91	0.84
600/420	1.0	0.96	0.65	0.76
420/295	1.0	0.94	0.51	0.67
295/208	1.0	0.82	0.53	0.67
208/149	0.95	0.47	0.29	0.37
149/105	0.88	0.28	0.18	0.34
105/74	0.65	0.14	0.05	0.21
− 74	0.21**	0.02	0	0.05
α	1.9	3.1	1.9	—
$d_{50}(c)$ micrometres*	67	182	337	225

* For notes, see p. 133.

* The $d_{50}(c)$ values for galena, sphalerite and gangue using the observed $d_{50}(c)$ value for the ore and eqs. 6-28 to 6-30 are 77, 164 and 297 μm, respectively. The differences between the observed and the calculated values may be attributed to experimental error.
** This value is significantly greater than zero because some of the coarse particles in this size range enter the underflow by classification because of the high s.g. of galena.

Step 2—Calculation of constants in the cyclone equations (eqs. 6-6 and 6-21).

Water split: $WOF = 1.07 \cdot WF - 3.94 \cdot Spig + K_1$

$$39.9 = 1.07 \cdot 57.8 - 3.94 \cdot 6.2 + K_1$$

$$K_1 = 2.5$$

$d_{50}(c)$: The coefficient of FPS will be assumed to be 0.020 rather than 0.0002 for the reason discussed earlier (p. 118).

$$\log_{10} d_{50}(c) \text{ ore} = K_2 + 0.021 \cdot VF - 0.066 \cdot Spig + 0.020 \cdot FPS - 0.0001 \cdot Q$$

$$\log_{10} 225 = K_2 + 0.021 \cdot 14.9 - 0.066 \cdot 6.2 + 0.020 \cdot 71.6 - 0.0001 \cdot 1603$$

$$K_2 = 1.1768$$

Step 3—Calculation of water split, $d_{50}(c)$ for new conditions:
New mass flow of water in cyclone feed = 77.8 t/h
Mass flow of ore in cyclone feed = 146.4 t/h (as before)
New volume flow rate of cyclone feed = 1943 t/h
New percent solids in cyclone feed = 65.3
New water split:

$$WOF = 1.07 \cdot 77.8 - 3.94 \cdot 6.2 + 2.5 = 61.3 \text{ t/h}$$

$$R_f = 16.3/77.8 = 21.0\%$$

New $d_{50}(c)$ for the ore:

$$\log_{10} d_{50}(c) \text{ ore} = 1.1768 + 0.021 \cdot 14.9 - 0.066 \cdot 6.2 + 0.020 \cdot 65.3 - 0.0001 \cdot 1943$$

$$d_{50}(c) \text{ ore} = 156\ \mu\text{m}$$

New $d_{50}(c)$ for each mineral (from eqs. 6-30, 6-31 and 6-32):
galena = 53 μm; marmatite = 114 μm; gangue = 206 μm.
The calculations of the flow rates and sizing distributions of the individual minerals in the products are given in Table 6-VIII (steps 4 and 5).

The effect of adding extra water to the cyclone feed is to increase the percentage of −200 mesh in the overflow from 37.3 to 45.1 due to changes in the distributions of galena, sphalerite and gangue.

Example 7: Vibrating screens. A 12.7-mm square hole wire screen was operated at a feed rate of 257 t/h. The screen is 6.4 m long and 2.4 m wide with a wire diameter of 2.5 mm. The efficiency constant has been found to be 2.79^2 at this feed rate when the screen length used in calculations is given in metres. The sizing distributions of the feed and undersize are given in Table 6-IX, and 149 t/h of undersize were produced. Use the screen models to calculate a predicted tonnage and size distribution of the product. Assume the fraction of the sub-mesh which reports in the oversize is 0.1.

TABLE 6-VIII

Calculation of the cyclone underflow and overflow flow rates and compositions

Step 4. Calculation of underflow composition:

	Feed			Short-circuiting (21.0%)			True classification		
Micrometres	gangue	galena	sphalerite	gangue	galena	sphalerite	gangue	galena	sphalerite
+ 840	8.1	0.6	1.9	1.7	0.1	0.4	6.4	0.5	1.5
840/600	6.3	0.7	2.2	1.3	0.1	0.5	5.0	0.6	1.7
600/420	8.3	1.0	3.4	1.7	0.2	0.7	6.3	0.8	2.7
420/295	11.2	1.6	5.0	2.4	0.3	1.1	7.2	1.3	3.9
295/208	11.7	2.1	5.7	2.5	0.4	1.2	5.6	1.7	4.1
208/149	10.8	2.7	4.9	2.3	0.6	1.0	3.5	2.1	3.0
149/105	9.8	3.5	4.6	2.1	0.7	1.0	2.1	2.7	2.0
105/74	5.4	2.4	3.0	1.1	0.5	0.6	0.7	1.5	0.9
− 74	13.9	6.3	9.2	2.9	1.3	1.9	0.7	1.6	0.9

Step 5. Calculation of both product streams:

	Underflow				Overflow			
Micrometres	gangue	galena	sphalerite	ore	gangue	galena	sphalerite	ore
+ 840	8.1	0.6	1.9	10.6	0	0	0	0
840/600	6.3	0.7	2.2	9.2	0	0	0	0
600/420	8.0	1.0	3.4	12.4	0.3	0	0	0.3
420/295	9.6	1.6	5.0	16.2	1.6	0	0	1.6
295/208	8.1	2.1	5.3	15.5	3.6	0	0.4	4.0
208/149	5.8	2.7	4.0	12.5	5.0	0	0.9	5.9
149/105	4.2	3.4	3.0	10.6	5.6	0.1	1.6	7.3
105/74	1.8	2.0	1.5	5.3	3.6	0.4	1.5	5.5
− 74	3.6	2.9	2.8	9.3	10.3	3.4	6.4	20.1
Total	55.5	17.0	29.1	101.6	30.0	3.9	10.8	44.7
Water		16.3				61.5		

Calculated $R_f = 21.0\%$

TABLE 6-IX

Sizing distributions of the feed and undersize for a vibrating screen

Sieve aperture	%retained	
(μm)	feed	undersize
45250/22630	9.16	0
22630/16000	10.17	0
16000/11314	19.67	1.01
11314/5660	26.89	40.92
5660/2830	11.53	19.99
2830/1000	8.60	14.87
1000/250	6.31	10.88
Sub-mesh	7.67	12.33

Solution:
Eq. 6-36 must first be evaluated:

$$h + d/\sqrt{m} = \frac{(12.7 + 2.5) \cdot 1000}{(6.4 \cdot 2.79^2)^{1/2}} = 2154.2$$

(Note: This is not dimensionally correct, as for an efficiency constant of 2.79^2, screen length must be given in m).

Values of y (see eq. 6-33) must be determined for evaluation in eq. 6-34. A value of y corresponding to a sieve aperture of 12700 μm is also calculated as $-12700/+11314$ μm material does have a probability of passing through the screen.

Values of $\int_y^\infty e^{-y} \cdot dy$ are then determined using the approximate integration formula given previously and the fact that the function is symmetrical. These values are given in Table 6-X.

TABLE 6-X

Evaluation of $\int_y^\infty e^{-y} \cdot dy$ for various sieve apertures

Sieve aperture	y	$\int_y^\infty e^{-y} \cdot dy$
12700	0.0	0.89
11314	−0.6434	0.3156
5660	−3.268	$3.99 \cdot 10^{-3}$
2830	−4.582	$1.41 \cdot 10^{-3}$
1000	−5.431	$8.38 \cdot 10^{-4}$
250	−5.780	$6.92 \cdot 10^{-4}$

This allows calculation of $\int_{y_1}^{y_2} e^{-y} \cdot dy$ and evaluation of eq. 6-34 gives the probability

$E(S)$ that particles of a given size fraction do not pass through the screen. The probability that they do pass through is then $1 - E(S)$ and the evaluation of this is given in Table 6-XI.

TABLE 6-XI

Calculation of the probability that particles pass through a screen aperture

Interval	$\int_{y_1}^{y_2} e^{-y^2}$	$E(S)$	$1 - E(S)$
12700/11314	0.5744	0.8961	0.1039
11314/5660	0.31161	0.1187	0.8813
5660/2830	0.00258	0.00196	0.99804
2830/1000	0.00057	0.00067	0.99934
1000/250	0.00015	0.0	1.0

A fines factor of 0.1 is used, therefore 0.9 of the sub-mesh material passes to the undersize. The amount of feed material in the 12700/11314 size interval is determined by linear interpolation.

The predicted tonnage and size distribution of the undersize are then calculated and are given in Table 6-XII.

TABLE 6-XII

Calculated size distribution of the screen undersize

Sieve aperture	Predicted tonnage	Perc. retained
16000/11314	1.98	1.33
11314/5660	60.90	41.01
5660/2830	29.57	19.91
2830/1000	22.09	14.88
1000/250	16.22	10.92
Sub-mesh	17.74	11.95
Total	148.5	100.0

CHAPTER 7

Mathematical simulation of operating circuits

7.1 INTRODUCTION

Information about any mineral processing circuit, such as its metallurgical efficiency or the values of parameters in the simulation models of the processing units in the circuit, requires information about the flow rates and compositions of the streams entering and leaving the circuit. In most circuits flow measurements are made on feed and product streams and occasionally one or more of the internal streams. Flow rates of the remaining streams are calculated from other measured characteristics such as assays or sizing distributions of samples collected at appropriate points. Investigation of circuit efficiency using simulation techniques involves: (1) calculation of complete circuit material balances from incomplete raw plant data; (2) calculation of model parameters from the completed set of plant data; (3) circuit simulation on a digital computer followed by optimisation studies.

The data completion calculations in particular are complex and a complete exposition of the techniques which may be used, including their merits and failures, would be the subject of a separate monograph. Treatment of the three sets of calculations will be at an elementary level in this monograph and references will be given for use if detailed treatments are required.

7.2 MATERIAL BALANCE CALCULATIONS

The problem is that samples cut from streams flowing within circuits may be in error in that they may not be truly representative of the streams from which they were cut and the analytical techniques used in processing the samples are also subject to error. Flow rates, solids contents or mineral compositions of streams measured on-line may also be in error.

The problem will be considered by examining the separation of a feed to a processing unit into two products and then by examining more complex problems. The underlying assumption is that "true" values of stream components and flow rates exist. If an experiment, or a sampling campaign, were repeated, the average of these repeated measurements would approach the true values as the number of repetitions increased. This approach is impractical in most cases because the "true" values change with time since a mineral processing plant cannot be expected to remain at steady state for very long. However, an alternative approach is usually available. Most sampling campaigns produce sufficient information for the calculation of many estimates of material flow rates in each stream. For example, if the size distributions of the feed to and products from a classifier are known the proportion of total feed entering each product may be estimated from each size fraction. If these estimates coincide, no further analysis is necessary because the data are already self consistent. This will rarely occur with experimental data. Analytical techniques are available which use the many estimates produced by an experiment to estimate the "true" values and provide an indication of the accuracy of the estimate.

The techniques discussed in section 7.2 are illustrated in Examples 7-1 to 7-3 and reference to these examples while reading the section may help to clarify the concepts involved.

7.2.1 *Flow rate calculation from single-component assays of each stream*

The simplest and most common case is a two-product separation. The analytical techniques to be described are simplified by using a reasonably compact notation.

There are three types of variable:

(1) the true values (denoted by an asterisk*) which are self-consistent but cannot be derived from the data;

(2) the adjusted values (denoted by a bar —) which are numerically self-consistent and can be derived from the data; and

(3) the experimental values (no superscript) which are known to a certain accuracy but which are not self-consistent.

Let A^*, B^* and C^* be the true flow rates of the feed to and products from a separation process. Components of the streams are designated a_i^*, b_i^* and c_i^* respectively where $i = 1$ to n. The components a_i^*, b_i^*, c_i^* are of the same type for a particular value of i but different types are possible within each stream. For instance, in a chalcopyrite flotation process a_1^*, b_1^*, c_1^* could be fractional copper assays and a_3^*, b_3^*, c_3^* could be the fraction less than 53 μm. This process may be written:

$$A^*(a_i^*) \longrightarrow \begin{cases} B^*(b_i^*) \\ C^*(c_i^*) \end{cases}$$

Since these values are exact, the mass balance is exact in flows and component flows. The following equations may be written:

$$A^* = B^* + C^* \tag{7-1}$$

$$a_1^* \cdot A^* = b_1^* \cdot B^* + c_1^* \cdot C^* \tag{7-2}$$

$$a_i^* \cdot A^* = b_i^* \cdot B^* + c_i^* \cdot C^* \tag{7-3}$$

Dividing by A^* in eq. 7-1:

$$1 = B^*/A^* + C^*/A^* \tag{7-4}$$

if:

$$\beta^* = B^*/A^* \tag{7-5}$$

by combining eqs. 7-3 and 7-5:

$$a_i^* = \beta^* \cdot b_i^* + (1 - \beta^*) \cdot c_i^* \tag{7-6}$$

or:

$$\beta^* = (a_i^* - c_i^*)/(b_i^* - c_i^*) \tag{7-7}$$

If β^* and any one of A^*, B^* and C^* are known the other two can be calculated.

Eqs. 7-6 and 7-7 are the basis of the three product formulae often used to calculate component recovery. Eq. 7-7 implies that β^* may be calculated from a single set of component measurements if the experimental accuracy is perfect. This is only true if the compositions of the products are not identical. β^* can only be calculated from eq. 7-7 if separation occurs and the importance of the degree of separation can be shown by considering two extreme cases.

(1) Perfect separation:

$$1(a_1^* a_2^*) \longrightarrow \begin{cases} \beta^*(b_1^* b_2^*) \\ (1-\beta^*)(c_1^* c_2^*) \end{cases}$$

where $b_1^* = c_2^* = 1$, $b_2^* = c_1^* = 0$.

Then $\beta^* = a_1^*$ or $\beta^* = 1 - a_2^*$ from eq. 7-7. In this limiting case only two streams need be analysed to obtain all the required information.

Consider a perfect copper-silica separation at 100 t/h:

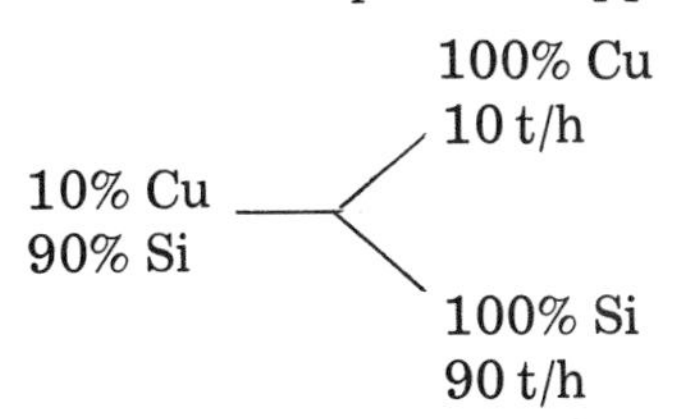

In this case β^* equals 0.1 and can be calculated from two of the three streams.

(2) Perfect splitting: In this case $a_i^* = b_i^* = c_i^*$ for all i and eq. 7-7 yields $\beta = 0/0$. In this case even a perfect sampling procedure provides no information about the flow proportions because no separation is occuring.

Real processes lie between these two limiting cases and the information which may be derived depends on which limiting case is approached by the process.

Only the very simple form of eq. 7-6 allows the simple solution which is given in eq. 7-7. More complex cases require more complex methods of solution.

Some alternative methods will be considered for the simple case and will then be extended to more complex cases.

Consider eq. 7-6 for any estimate β_0 of the true value β^*. If the estimate does not coincide with the true value, the material balance will *not* add up to zero but will differ by an amount or error Δ_i in each component:

$$\Delta_i = a_i^* - \beta_0 \cdot b_i^* - (1 - \beta_0) \cdot c_i^* \tag{7-8}$$

The range of reasonable estimates of β^* is from zero to one which will generate errors ranging from $(a_i^* - c_i^*)$ to $(a_i^* - b_i^*)$ as shown in Fig. 7-1. The value of β^* at zero error is given by eq. 7-6 as before. Alternatively, the square of the error:

$$\Delta_i^2 = [a_i^* - \beta_0 \cdot b_i^* - (1 - \beta_0) \cdot c_i^*]^2 \tag{7-9}$$

may be considered. This equation defines the curve shown in Fig. 7-2 which is generated by considering a single component (for example, a_2, b_2, c_2) and different β estimates.

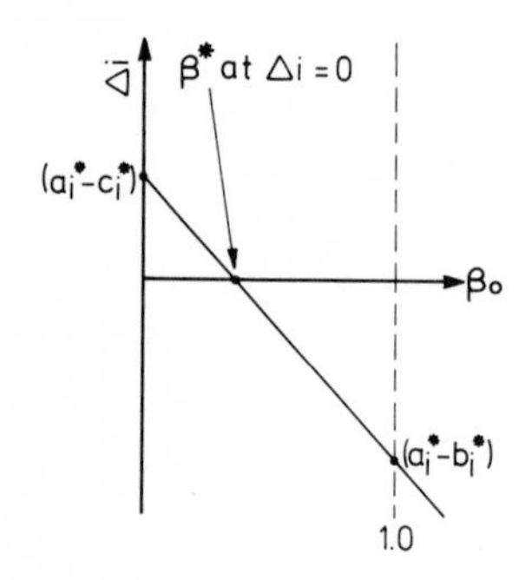

Fig. 7-1. Mass flow error in component i at a junction for any estimate β_0 of the true split β_0^*.

The value of β^* can be found from the point of contact of the curve with β_0 axis. More importantly, β^* can be found from the *minimum* or point of zero slope of the error-squared functions. There are several methods of locating this minimum point which are described in the references.

If only one set of experimental data (a, b, c) is available, no redundant information is present, eq. 7-7 provides an estimate of β^* and the data are consistent with this estimate.

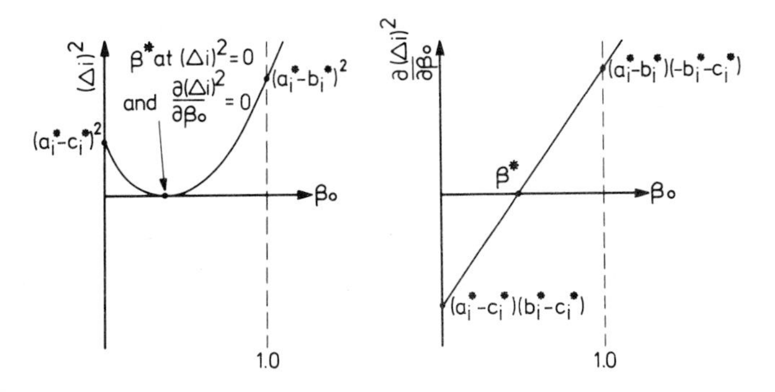

Fig. 7-2. The square and the derivative with respect to β_0 of the mass flow error in component i for Fig. 7-1.

7.2.2 *Flow rate calculation based on multiple-component assays*

The information available in real situations is usually multiple-component experimental data on each stream (a_i, b_i, c_i):

$$a_i \longrightarrow \begin{cases} b_i \\ c_i \end{cases}$$

The estimates of β calculated from eq. 7-7 will rarely be consistent. The problem is to find a "best" value $\bar{\beta}$ of β. The reliability of this best value $\bar{\beta}$ as an estimate of the true value β^* will depend on the accuracy of the data and the extent of separation caused by the process.

There are two basic approaches to calculation of best-fit flow rates or parameters. The usual approach (Deming, 1964; Lees, 1971; Whiten, 1972) is to define a residual in every measured variable; where:

$$\text{residual} = \text{observed} - \text{calculated}$$

$$\Delta a_i = a_i - \bar{a}_i$$

so that the calculated variable values ($\bar{a}_i$, $\bar{b}_i$ . . .) are consistent with the best flow rates ($\bar{\beta}$). The residuals are squared, weighted (multiplied by a factor depending upon their accuracy) and summed. The best-fit flow rates are defined as those which minimise this sum of squares. This approach is very effective because it allows the accuracy of measurement of each variable to be considered. It is also complex and requires a powerful digital computer for general problems. The best-fit flow rates are to a large extent determined by the most accurately measured variables.

The second approach uses two stages. The first stage defines an error in each component mass balance equation between the experimental values and the flow rates. This error is apparent and is due to error in measurement and sampling. This approach is due to Wiegel (1972) and has been generalised by Morrison (1976):

error = mass balance (experimental values)

$$\Delta_i = a_i - \bar{\beta} \cdot b_i - (1 - \bar{\beta}) \cdot c_i$$

These errors are squared, weighted and summed. The best-fit flow rates are defined as those which minimise this sum of squares. The best-fit flow rates in this case are largely determined by the components which are most altered by the process, for example, the concentrations of sulphide mineral in the feed to and product from a sulphide flotation process.

The second stage of this process is to adjust the data to be consistent with the best-fit flow rates by distributing the mass balance errors.

If a large computer and reasonable estimates of sampling and measurement accuracy for each variable are available, the first approach is better. Standard programmes can be readily used in these computers. If only limited computing power is available, the second method is the only possible approach. If accuracy estimates are not available, the better method depends on the choice of weighting factors and, therefore, on the particular case.

Results given by both methods tend towards the true values as experimental accuracy is improved. The second approach is better with highly accurate data because of the simplicity of calculation.

7.2.3. *Calculation of best-fit flow rates based on apparent material balance errors*

Experimental data will generally yield a different estimate β of β^* for each measured component at $\Delta_i = 0$ from:

$$\Delta_i = a_i - \beta \cdot b_i - (1 - \beta) \cdot c_i \qquad (7\text{-}10)$$

The requirement is for a best estimate $\bar{\beta}$ of the true value β^* based on all the component measurements. At this best value of β there will still be mass balance errors ($\Delta_i \neq 0$). The more accurate the data, the smaller these residual errors will be. One approach is to add up these errors ($\sum_i \Delta_i$) and search for a β value which makes this sum equal to zero.

In this case:

$$\sum_i \Delta_i = \sum_i [a_i - \beta \cdot b_i - (1 - \beta) \cdot c_i]$$

$$= \Sigma a_i - \Sigma c_i - \beta \cdot (\Sigma b_i - \Sigma c_i)$$

It will be noted that all components are expressed as mass fractions and that:

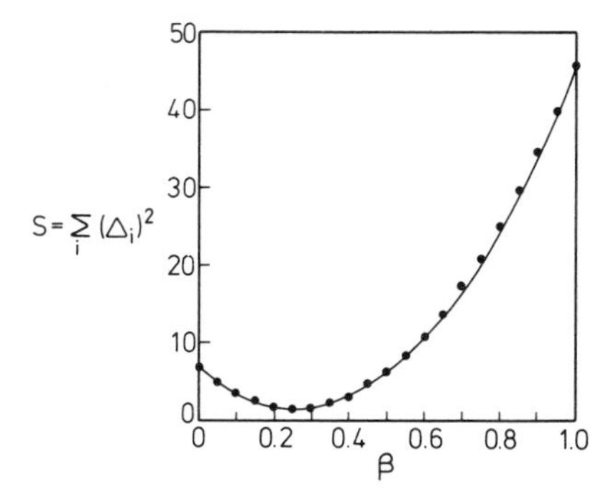

Fig. 7-3. The sum of squares of apparent mass flow errors at different values for the split β for the data of Example 7-1.

$$\Sigma\, a_i \;\equiv\; \Sigma\, b_i \;\equiv\; \Sigma\, c_i \;\equiv\; 1$$

This approach implies that $\Sigma\Delta_i \equiv 0$ for any value of β, and is obviously invalid. The sum can be non-zero if any component is omitted but each omitted component again gives a different estimate of β.

A second approach based on the sum of squares of the errors may be used with a single component. In this case:

$$\Sigma\, \Delta_i^2 \;=\; \Sigma\, [a_i - \beta \cdot b_i - (1-\beta)\cdot c_i]^2 \tag{7-11}$$

$\Sigma\Delta_i^2$ will not be zero at any value of β unless the data are self consistent. It should have a clearly defined minimum for any reasonable set of data as shown in Fig. 7-3.

The simplest method to find the minimum is to differentiate with respect to β and solve for the point of zero slope:

$$\frac{\partial}{\partial\beta}\,\Sigma\, \Delta_i^2 \;=\; 2\,\Sigma\, [a_i - \beta\cdot b_i - (1-\beta)\cdot c_i]\cdot(-\,b_i + c_i)$$

$$= 0 \text{ at the minimum}$$

and

$$\bar{\beta} \;=\; \Sigma\,(a_i - c_i)\cdot(b_i - c_i)/\Sigma\,(b_i - c_i)^2 \tag{7-12}$$

β is now denoted as $\bar{\beta}$ as it is a best-fit estimate of the true value β^* based on the least sum of squares of apparent mass balance errors.

Once $\bar{\beta}$ has been found it is often necessary to adjust the original data to make them self-consistent with this best value.

The accuracy of the data and of the estimate of β can also be estimated The usual measure of experimental error is the estimated variance V.

If an experimental measurement x_i (for example, stream sampling and assay) is repeated n times then the estimated variance is defined by:

$$V_x = \sum_{i=1,n} \frac{(x_i - x_\mu)^2}{n-1} \tag{7-13}$$

where

$$x_\mu = \sum_i \frac{x_i}{n}$$

As the number of repetitions is increased, the estimated variance approaches the true variance which is defined by eq. 7-13 where n approaches infinity.

The accuracy of the data can be assessed by estimating the variance of the residual errors:

$$V_{\Delta i} = \sum_i \Delta_i^2/(n-1) \tag{7-14}$$

It can be shown that the estimated variance of β is given by the equation:

$$V_\beta = \sum_i \Delta_i^2/(n-1) \cdot \sum_i (b_i - c_i)^2 \tag{7-15}$$

In general the accuracy of measurements of the components will not be equal. If one component is measured with absolute precision (that is, with zero variance), the value of β estimated from that component should be exact. One way to ensure that this β coincides with $\overline{\beta}$ is to weight the squared component error with the inverse of the variance*. The weighted sum of squares can be minimised as before:

$$\overline{\beta} = \frac{\sum_i (a_i - c_i) \cdot (b_i - c_i)/V_i}{\sum_i (b_i - c_i)^2/V_i} \tag{7-16}$$

or considered as simple linear regression in $(a_i - c_i)/\sqrt{V_i}$ versus $(b_i - c_i)/\sqrt{V_i}$.

---

* Note: If a number of samples have been taken, the mean and variance of each measurement can be estimated from eq. 7-13. The variance of a function is found by taking its derivative:

$$V_{f(x)} = \sum_i \left(\frac{\partial f}{\partial x_i}\right)^2 \cdot V_{x_i}$$

or in this case $V_i = V_{ai} + \beta^2 \cdot V_{bi} + (1 - \beta)^2 \cdot V_{ci}$ and $\overline{\beta}$ can be found iteratively.

The type of calculation outlined in this section is straightforward. The technique is well suited to a desk or pocket calculator (particularly if it is programmable) or a mini-computer. It is also suitable for a small on-line process control computer when redundant circuit information is available. The flow rates in any circuit can be expressed as the solution to a set of linear simultaneous equations. Application to a more general circuit, such as a closed grinding circuit or a bank of flotation cells, is also possible.

7.2.4 *Data adjustment based on calculated flow rates*

After best fit flow rates have been calculated it is often necessary to adjust the experimental data to be consistent with the calculated flow rates.

All of the adjustment techniques are ways of distributing the mass balance errors Δ_i between the various measured values to give corrected or adjusted values ($\bar{a}_i$, $\bar{b}_i$, $\bar{c}_i$) which are numerically consistent at the calculated flow rates. That is:

$$\Delta_i = a_i - \bar{\beta} \cdot b_i - (1 - \bar{\beta}) \cdot c_i \tag{7-17}$$

and

$$0 = \bar{a}_i - \bar{\beta} \cdot \bar{b}_i - (1 - \bar{\beta}) \cdot \bar{c}_i \tag{7-18}$$

The simplest adjustment is to assume measurement errors are proportional to component flow rates in each stream. Transposing the error equation yields:

$$a_i - \frac{\Delta_i}{2} = \bar{\beta} \cdot b_i + \bar{\beta} \cdot \frac{\Delta_i}{2} + (1 - \bar{\beta})c_i + (1 - \bar{\beta}) \cdot \frac{\Delta_i}{2}$$

or:

$$\bar{a}_i = a_i - \frac{\Delta_i}{2}; \; \bar{b}_i = b_i + \frac{\Delta_i}{2}; \; \bar{c}_i = c_i + \frac{\Delta_i}{2}$$

This procedure does not ensure that the adjusted components add up to unity. If one of the components is measured by difference (for example, insolubles in chemical analysis) minor discrepancies can be absorbed into that component. If the components are completely determined (for example, screen size analysis) it is possible to correct the flow rates and normalise these flows. This method is quite arbitrary.

The least squares method can also be used to distribute the errors to minimise the sum of squares of adjustments of the measured values at the best fit flows. Alternatively the experimental flows, that is, measured assays by best fit flows, can be adjusted and the assays recalculated. Eq. 7-18 can be written:

$$0 = (a_i - \Delta a_i) - \bar{\beta}\cdot(b_i - \Delta b_i) - (1-\bar{\beta})\cdot(c_i - \Delta c_i)$$

and subtracted from eq. 7-17 to yield:

$$\Delta_i = +\Delta a_i - \bar{\beta}\cdot\Delta b_i - (1-\bar{\beta})\cdot\Delta c_i \tag{7-19}$$

where: $\bar{a}_i = a_i - \Delta a_i$; $\bar{b}_i = b_i - \Delta b_i$; $\bar{c}_i = c_i - \Delta c_i$

Now the sum of squares to be minimised for each component is:

$$S_i = \Delta a_i^2 + \Delta b_i^2 + \Delta c_i^2 \tag{7-20}$$

subject to the constraint in eq. 7-19. Eqs. 7-19 and 7-20 can be combined, Δa_i eliminated and S_i minimised by taking the derivative with respect to each of the unknowns (Δb_i and Δc_i) and setting the result to zero. It may be shown that:

$$\Delta a_i = +\Delta_i/k; \ \Delta b_i = -\Delta_i\cdot\bar{\beta}/k; \quad \Delta c_i = -\Delta_i(1-\bar{\beta})/k$$

where:

$$k = [1+\bar{\beta}^2 + (1-\bar{\beta})^2] \tag{7-21}$$

Eq. 7-21 shows that the calculation of all three residuals depends on only one number k once the best fit flow rate is known. This reduction in calculation was generalised by the French mathematician Lagrange.

7.2.5 *The method of Lagrange multipliers*

The method is used to simplify minimisation or maximisation problems which are subject to conditions or constraints. The constraints are expressed in such a way that they equal zero. In this case, eq. 7-19 can be written:

$$0 = +\Delta_i - \Delta a_i + \bar{\beta}\cdot\Delta b_i + (1-\bar{\beta})\cdot\Delta c_i$$

The sum of the squares to be minimised is modified (S_m) by adding each of these constraint equations multiplied by a Lagrange multiplier, that is:

$$S_m = S + \sum_j \lambda_j\cdot\text{constraint } j$$

It will be noted that this approach is valid even if a component has more than one constraint upon it, as in a more complex circuit.

In this case:

$$S_m = \Delta a_i^2 + \Delta b_i^2 + \Delta c_i^2 + 2\cdot\lambda_i\cdot[+\Delta_i - \Delta a_i + \bar{\beta}\cdot\Delta b_i + (1-\bar{\beta})\cdot\Delta c_i]$$

(the + 2 is for convenience).

The modified sum is then differentiated with respect to each of the unknowns (residuals and multipliers) and the Lagrange multipliers are used to substitute for the residuals thus reducing the required calculation:

$$\frac{\partial S_m}{\partial \Delta a_i} = 2\Delta a_i - 2\lambda_i = 0 \quad \text{or} \quad \Delta a_i = \lambda_i$$

Similarly, $\Delta b_i = -\bar{\beta} \cdot \lambda_i$ and $\Delta c_i = -(1 - \bar{\beta}) \cdot \lambda_i$

$$\frac{\partial S_m}{\partial \lambda_i} = 2[+\Delta_i - \Delta a_i + \bar{\beta} \cdot \Delta b_i + (1 - \bar{\beta}) \cdot \Delta c_i] = 0$$

Substituting for Δa_i, Δb_i and Δc_i:

$$\Delta_i = +\lambda_i [+1 + \beta^2 + (1 - \beta)^2]$$

and the solution is seen to be eq. 7-21. As the minimisation becomes more complex the reduction in calculation becomes more useful.

If the variances V_{ai} etc. are known, then:

$$S = \frac{\Delta a_i^2}{V_{ai}} + \frac{\Delta b_i^2}{V_{bi}} + \frac{\Delta c_i^2}{V_{ci}}$$

can be treated in similar fashion. The data adjustments are no longer arbitrary and are justified statistically if $\bar{\beta}$ was calculated from eq. 7-16.

The application of this "curve-fitting" approach to general problems is mathematically complex and is described in Appendix 7. An understanding of Appendix 7 is not essential to the rest of this chapter. However it is essential to the student who wishes to write a general curve-fitting program and would be useful to those who wish to use such programs.

7.2.6 *Comments on material balancing techniques*

The techniques discussed are illustrated in Examples 7-1 to 7-3. With regard to the adjustment of data both the simple proportional adjustment and the least squares adjustment are demonstrated in Example 7-1, the latter method generally being better than the former. If one section is more accurately measured than the others the adjustments can be weighted and the concept of weighting will be discussed briefly in sub-section 7.2.7.

The problem of material balancing and estimation of accuracies of calculated flows and compositions is often more complex than has been indicated so far. However, this discussion gives an indication of the nature of the problems and the methods of solution.

These techniques are more useful in some areas than in others. If accurate sampling can easily be carried out and the streams sampled have a considerable composition difference then the techniques are only useful for producing self-consistent data. The calculated flow rates will not vary significantly between the simple and the more sophisticated techniques.

If sampling is difficult and composition differences are small, the more sophisticated techniques yield more reliable estimates of flow rates and "true" data values. Some typical examples of these areas are:

(1) cleaning and scavenging operations;
(2) any stream where representative sampling is difficult;
(3) where several stages of calculation are required for flow rates;
(4) operations with high circulating loads;
(5) operations where different types of analysis are more suitable for different sections of the process.

7.2.7 *Comments on weighting*

The concept of weighting the sum of squares is no more than the inclusion of previous experience in the calculation procedure. The precise mathematical expression of previous experience is the estimate of variance. The weight of a measurement is the reciprocal of its variance. Knowledge of the process and qualitative experience provide an alternative approach. For example, classifier products outside a certain size range must often be zero. This zero can be strongly weighted (for example, with the inverse of the smallest component measurement possible) to ensure that the residual at that point also approximates zero. Some streams can be more conveniently sampled and can be assigned a higher weighting than more difficult streams.

The simplest general approach, with no variances measured, is to set all the variances to one. In terms of general circuits this is not a particularly good approach. It is probably better to use the inverse of the component measurement itself (simple) or of the component flow rate (iterative) as the weighting factor.

If the relative errors (that is, percentage) in the measurements are reasonably constant, it is reasonable to use the measurement squared as an estimate of the variance. This means that the sum of squares of the fractional (or percentage) adjustment errors is minimised.

As the accuracy of the data is improved all the calculated values tend to the "true" values and the weighting becomes unimportant. For most real data the weighting can make a substantial difference and consideration should be given to obtaining experimental estimates of variance.

7.3 CALCULATION OF MODEL PARAMETERS

A mathematical model of a process is a set of equations which relate the output to the inputs over a range of specified operating parameters and the model parameters are a set of numbers which characterise a particular unit process within a particular plant. For example, in the case of a hydrocyclone classifier the inputs to the machine are the ore and water feed rates, the volume flow rate of the feed and the feed size distributions and the outputs from the machine are the flow rates of ore and water in each product and the ore size distributions. The relationship between the input and outputs depends on the *operating* parameters of the cyclone such as the vortex finder and the spigot diameters. The *model* parameters are the "constant" terms used in the set of equations which describes this relationship. They are only constant for a particular process.

A specific example is the curvature term (α) in the reduced-efficiency curve for the hydrocyclone:

$$E = (e^{\alpha \cdot x} - 1)/(e^{\alpha \cdot x} + e^{\alpha} - 2)$$

Variation in the curvature of this curve with change in α is shown in Fig. 6-8. For a reduced-efficiency curve from plant data, α can be adjusted to give a mathematical function as much like the experimental one as possible. This is a best fit for model parameter α. This approach can be used to derive all the model parameters by seeking a best fit (usually with a least-squares criterion) between the model output and the experimental output (with the same experimental input and operating variables).

7.3.1 *General least-squares methods*

Two methods can be used for the calculation of model parameters. In the first, it is assumed that all the errors are due to the dependent variables, that is, that the weighted residuals which are to be minimised are those of the model constraint function. This approach is called regression. In the second, it is assumed that all of the variables, both dependent and independent, have measurable variances. The objective of the minimisation procedure is to obtain the weighted adjustment residuals in each variable required to satisfy the model constraint function. This approach is called curve fitting. A summary of these approaches is as follows:

(1) regression: $S = \sum_i \frac{\Delta_i^2}{V_{yi}}$ is minimised with respect to β_k

where $\Delta_i = y_i$ − model (x_{ji}, β_k),

(2) curve fitting: $S = \sum_i \sum_j \frac{\Delta x_{ji}^2}{V_{ji}}$ is minimized with respect to β_k and

satisfying the constraint $0 = (y_i - \Delta y_i) -$ model $[(x_{ji} - \Delta x_{ji}), \beta_k]$.

The technique of solution is outlined in Appendix 7.

In both cases β_k are the best-fit parameters, y_i are the variables the model is required to predict, x_{ji} are the input variables j and data set i with estimated various V_{ji}. Δ_i is the residual in the model for the ith data set. Δx_{ji} are the adjustment residuals such that:

residual = observed − calculated

$$\Delta x_{ji} = x_{ji} - \bar{x}_{ji}$$

It is worth noting that the parameters calculated by regression are equivalent to those calculated by curve fitting with residual weighting such that

for each i, $V_{y\,i} = \sum_{i} \left(\frac{\partial f}{\partial x_{ji}}\right)^2 \cdot V_{x_{ji}}$

If the data are sufficiently accurate the weighting becomes quite arbitrary and the two approaches are equivalent.

In the regression method the parameters β_k can be estimated and the sum of squares of the residuals calculated directly. The sum of squares can then be minimised. The regression method is useful for finding parameter estimates where the error of the dependent variable is much more than those of the independent variables. Where the errors are of approximately the same order curve-fitting techniques should be used.

In the curve-fitting method the accuracy of the initial parameter estimates is very important. As the derivatives are calculated at these estimates, non-linear models can often cause non-convergence or false minima with poor parameter estimates. The regression techniques are not so sensitive and can be used to obtain good initial estimates for curve fitting.

Problems can occur with both approaches if the model parameters interact. This simply means that a particular input to the model may produce the same output for several different sets of model parameters. If the data are not sufficiently accurate to produce a reasonably "sharp" minimum sum of squares, then quite different parameters may produce the same "accuracy" of "prediction". The remedy is the same in both cases: reduce the number of parameters.

7.3.2 *Multiple linear regression*

Many industrial processes operate within a limited range and a linear response to process variables often provides an adequate process description over this limited range.

This means the process model is very simple and may be of the form

$$t = a + b \cdot x + c \cdot y + d \cdot z \ldots \qquad (7.22)$$

where t is the process output and $x, y, z \ldots$ are process inputs. Even if this model is inadequate with the actual variables it can often be made adequate by transforming the variables or combining them. For example:

$$t = a + b \cdot \left(\frac{1}{x}\right) + c \cdot (y^2 + z) + d \cdot z^2$$

For any set of experimental data:

$$t_i = a + b \cdot x_i + c \cdot y_i + d \cdot z_i + \Delta_i \qquad (7\text{-}23)$$

where Δ_i is an error term. If there are at least as many data sets available as parameters (a, b, c, d . . .) then estimates (A, B, C, D . . .) of those parameters can be made by minimising the sum of squares of the errors, $\sum_i \Delta_i^2$ or S.

These techniques are documented in standard texts on statistics, for instance Li (1964). For actual calculation, texts on computing are useful (Ralston and Wilf, 1960).

The chief advantages of this method are:

(1) ease of mathematical solution. If the problem has a solution it can be reached in one step;

(2) no parameter estimates are required;

(3) the accuracy of the prediction can be estimated;

(4) the accuracy of the parameters can be estimated;

(5) whether or not a variable should be included in the model (for the operating range used) can often be decided;

(6) other curve-fitting techniques are quite dependent on parameter estimates. Linear regression can often provide reliable first estimates for these other techniques.

Two types of difficulties arise:

(1) if enough of the experimental data sets are similar or are combinations or ratios of each other (to within experimental errors), not enough information is available to estimate all parameters; and

(2) if a linear relationship exists between two of the dependent variables, for example, celcius and fahrenheit thermometers or two measurements of the same thing in different units, the solution will not be available. This problem can be overcome by removing one of the linearly dependent variables.

Standard multiple linear regression programmes are available for most large computers. An example of the use of multiple linear regression is the cyclone model for throughput, water in overflow and d_{50}(c) in terms of operating pressure, vortex finder and spigot dimensions.

7.3.3 *Estimation of accuracy*

If the model is not strongly non-linear the methods of sub-section 7.3.2 for estimating model and parameter variance are quite reasonable and can be extended approximately to the curve-fitting case (Appendix 7).

If the problem is strongly non-linear, the accuracy of the model and its parameters can be estimated by using as inputs a normally distributed set of random variations (over the span of experimental error) in each variable. The distribution of the output errors is then measured. If this distribution of output errors lies within the measured error output distribution, the model provides an adequate description of the process, although it may still contain some bias. This approach is called a Monte Carlo technique.

The same approach can measure the accuracy of the parameters. That is, by making small alterations in the parameters and iterating until the same order of output variation as the experiment is produced. In many cases, especially where there are experimental difficulties, the model will provide more accurate results than the data. The curve-fitting techniques can produce an adjusted set of input data.

If the model is realistic this adjusted set will be a better estimate of the true experimental situation than the experimental data. The criterion is that the pooled standard error is less than unity where this is defined as the square root of the sum of squares of the weighted residuals divided by the number of degrees of freedom. If the object of the model is a process evaluation criterion (such as recovery) care should be taken to ensure that correlation of variables does not reduce the information available within that criterion. Small experimental errors in certain variables may lead to an "optimum" which is related to the errors instead of the process. For example, an "optimum" recovery can easily occur with a high estimate of feed grade. A more reliable approach is to first simulate the process accurately and then to optimise the model inputs subject to a variety of process criteria. The simulation may contain some bias but it is free of random errors in input.

7.4 CIRCUIT SIMULATION

When the mathematical models which represent the processing units in a circuit have been developed, and the dependence of the model parameters upon operating conditions have been determined for the particular ore type, it is possible to calculate the behaviour of the circuit for changes in the operating variables using simulation. This can be done most conveniently with a digital computer and a flow diagram representing the mathematical simulation of a ball-mill—cyclone circuit as given in Fig. 7-4.

A typical simulation procedure uses the standard FORTRAN computer language. A model representing a mineral processing machine may be programmed as a FORTRAN subroutine. Arguments to the subroutine communicate the feed characteristics and the model parameters. For example, the simulation subroutine for a ball mill requires information on the feed rate and sizing analysis of the ore entering the mill as well as values for the breakage and selection functions, and the mill constant. The product sizing analysis is calculated in the program and returned as an argument to the main program. The main program is used to describe the manner in which the machines are connected together to form the circuit, such as whether the new circuit feed goes directly into the ball mill or, alternatively, into the cyclone. A process of iteration and convergence is used to obtain the equilibrium circulating load. Stability of convergence is rarely a problem,

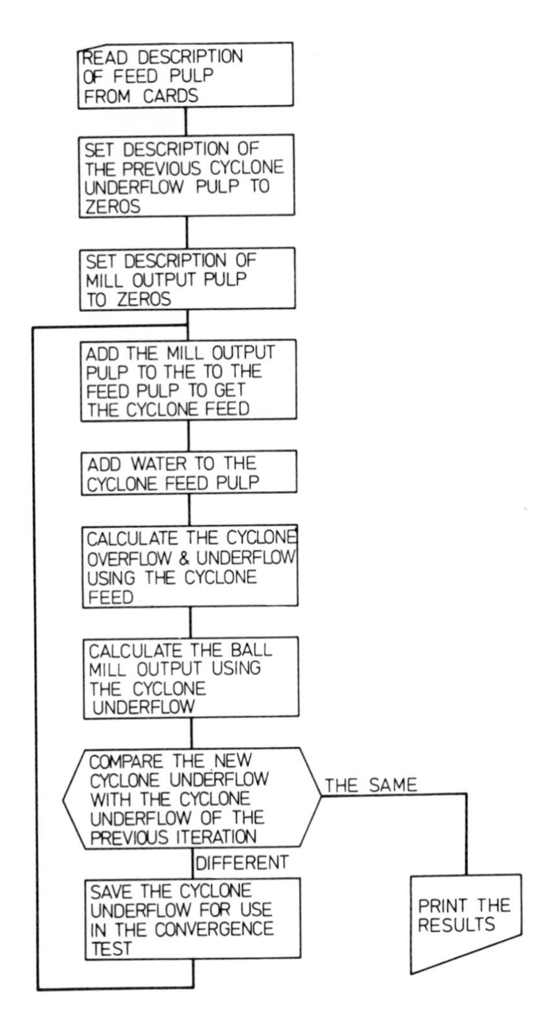

Fig. 7-4. A flow chart for the simulation of a ball-mill—hydrocyclone closed grinding circuit.

but the step size can be decreased by an appropriate amount if the simulation is unstable.

When carrying out simulation on a computer, it is of considerable importance to relate simulation results to practical operating behaviour. Operating equipment has certain physical limitations which define the limits of performance. Pumps are limited on capacity and feed density. Ball mills are restricted to a certain maximum throughput, beyond which the mill discharges part of its ball load if it is an overflow mill, or fills up if it is a grate mill. Cyclones are limited in capacity and in tonnage throughput and density of the spigot discharge. Overloading leads to blocked spigots or inefficient operation through roping or production of tramp oversize.

All these factors must be considered when interpreting the results of simulation procedures. If the predicted flow rates, for example, are beyond the practical physical limitations of the equipment then the simulation is invalid. Additional cyclones may need to be added in the simulation model when the results of a simulation show that feed rates to the cyclones became too high for practical operation. Predictions of solids contents, particularly of classifier sands, must also be checked to ensure that realistic limits are maintained.

An example of the use of simulation to find the optimum operating conditions for a 110 t/h circuit containing three ball mills with cyclones was described by Lynch et al. (1967). The problem was to determine the arrangement of the ball mills which would give most efficient operating performance, the criterion of efficiency being the minimisation of the + 100-mesh particles in the circuit product. Possible alternative arrangements are:

(A) 3 primary mills in parallel;
(B) 2 primary mills in parallel and 1 secondary mill;
(C) 1 primary mill and 2 secondary mills in parallel;
(D) 1 primary mill, 1 secondary mill and 1 tertiary mill.

The best result for each arrangement was as follows:

Circuit arrangement	% + 100-mesh	% − 200-mesh
A	6.2	72.5
B	3.0	75.2
C	6.3	72.1
D	3.3	74.4

Case B was marginally better than case D and appreciably better than cases A and C. In practice, C was the original method of circuit operation and when it was changed to B the predicted improvement was realised.

The possibility of improving circuit performance by double cycloning is another example of the type of problem which can be investigated by simulation. This type of circuit appears to offer three main advantages over single cycloning. These are: (1) reduced mill load; (2) reduced overgrinding; and (3) finer product or increased throughput.

The optimum operating conditions require: (1) all water added to the second cyclone feed; (2) minimum water content in both cyclone underflows; and (3) the $d_{50}(c)$ of the second cyclone within 20% of that of the first.

In practice, (2) and (3) can only be satisfied as closely as the physical limitations allow. However, the advantages of this type of circuit may be reduced to some extent by a considerable increase in pumping requirements. An additional pump is required and the required pump capacities are higher than for the normal circuit.

Thus simulation can be seen to be a very useful aid in circuit optimisation.

It is discussed in detail in Chapter 8 with specific reference to circuit design but the procedures involved in circuit optimisation are the same as those involved in circuit design.

NUMERICAL EXAMPLES

Example 7-1: Mass balance and data adjustment for a simple two-product concentration process. Size analyses of feed and products of an industrial high-tension roll separator shown in Fig. 7-5 are given:

+ screen (μm)	212	180	150	125	106	90	75	63	53
$a_i(\%)$	1.23	4.12	7.70	16.81	32.19	25.22	9.86	2.60	0.28
$b_i(\%)$	0.33	1.96	5.80	15.92	29.42	28.21	13.89	4.34	0.13
$c_i(\%)$	1.74	5.29	7.68	16.24	33.58	25.08	8.35	1.84	0.21

(a) Calculate the fraction of the total feed going into product B on the basis of each size component, and a best-fit fraction based on all the size components. (This type of separation is based on mineral conductivity. Therefore, this is a difficult calculation as the sizing effect is secondary).

(b) Adjust the experimental data to be consistent with the best-fit calculated flow rates by distributing the component errors in proportion to their flow rates and then by the least squares of adjustments.

(c) Plot the data as a linear regression $(b_i - c_i)$ versus $(a_i - c_i)$.

Calculate the standard error in the y estimate and the standard error in the slope or weight split. Plot the error regions about the centroid of the data.

Solution:
(a) Flows $\beta = (a_i - c_i)/(b_i - c_i)$ on each component:

β	0.362	0.351	-0.011	-1.78	0.334	0.045	0.273	0.304	-0.87

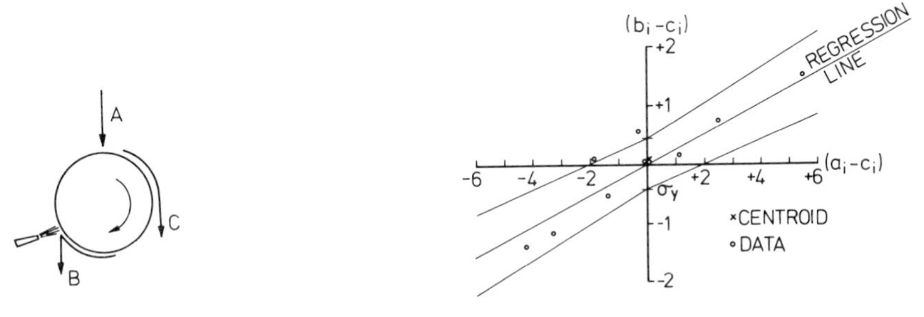

Fig. 7-5. Industrial high-tension roll separator discussed in Example 7-1.

Fig. 7-6. Linear regression approach to Example 7-1.

$$\bar{\beta} = \sum_i (a_i - c_i)\cdot(b_i - c_i) \Big/ \sum_i (b_i - c_i)^2 \text{ on all components}$$

$$= 20.95/80.76 = 0.26$$

(b) Adjustment of data: mass flow errors $\Delta_i = a_i - \bar{\beta}\cdot b_i - (1-\bar{\beta})\cdot c_i$

Error Δ_i	-0.143	-0.304	0.509	0.653	-0.309	-0.674	0.070	0.110	0.091

Simple proportional adjustment: $\bar{a}_i = a_i - \Delta_i/2$; $\bar{b}_i = b_i + \Delta_i/2$; $\bar{c}_i = c_i + \Delta_i/2$

$\bar{a}_i$	1.30	4.27	7.45	16.48	32.34	25.56	9.82	2.55	0.24
$\bar{b}_i$	0.26	1.81	6.05	16.25	29.27	27.87	13.93	4.43	0.18
$\bar{c}_i$	1.67	5.14	7.93	16.57	33.43	24.74	8.39	1.90	0.26

Least-squares adjustment: $\bar{a}_i = a_i - \Delta_i/k$; $\bar{b}_i = b_i + \beta\cdot\Delta_i/k$; $\bar{c}_i = c_i + (1-\beta)\cdot\Delta_i/k$ where $k = 1 + \beta^2 + (1-\beta)^2$

$\bar{a}_i$	1.30	4.28	7.44	16.47	33.35	25.37	9.82	2.54	0.23
$\bar{b}_i$	0.31	1.92	5.87	16.01	29.38	28.12	13.90	4.35	0.14
$\bar{c}_i$	1.69	5.17	7.87	16.49	33.46	24.82	8.38	1.88	0.25

The differences in the computed sizings using the two methods of data adjustment will be noted.

(c) Residual error: $V_{\Delta_i} = \sum_i \Delta_i^2/(n-1) = 1.3734/(9-1) = 0.172$

Standard error: $\sigma = \sqrt{V}$; $\sigma\Delta_i = 0.41$ or σ_y on the graph about the centroid (x_μ, y_μ)
Slope error:

$$V_\beta = \frac{\sum_i \Delta_i^2}{(n-1)\cdot\sum_i (b_i - c_i)^2} = \frac{0.172}{80.76} = 0.00213;\ \sigma_\beta = 0.046$$

The error regions (67% confidence) are shown in Fig. 7-6.

Example 7-2: Mass balance and data adjustment for a ball-mill—hydrocyclone circuit. Size analyses of all streams of the ball-mill—cyclone circuit shown in Fig. 7-7 are given in Table 7-I.

(a) Calculate the circuit flow rates based on a throughput of 100 t/h.

(b) Adjust the data to be consistent with the calculated flow rates. Where the adjusted data is less in magnitude than the smallest component measurement (0.1), set the adjusted measurement to zero.

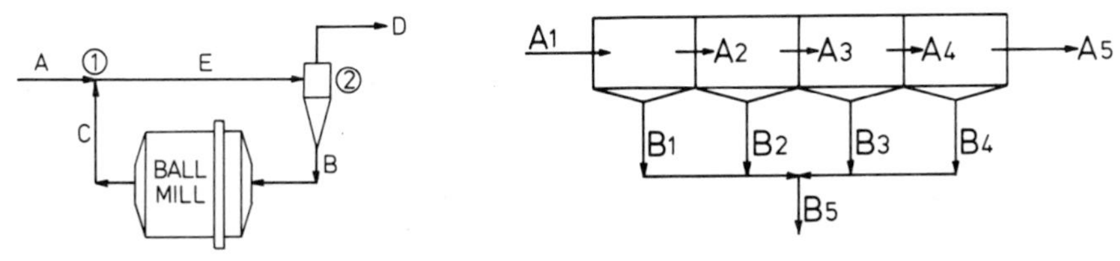

Fig. 7-7. Ball-mill—cyclone circuit discussed in Example 7-2.

Fig. 7-8. Industrial flotation bank discussed in Example 7-3.

TABLE 7-I

Sizing distributions of streams in a ball-mill—hydrocyclone circuit

Tyler mesh	Circuit feed	Cyclone feed	Cyclone OF	Cyclone sands	Mill discharge
+ 8	0.1			nil	
+ 10	0.4			0.3	
+ 14	1.0	nil		0.2	nil
+ 20	1.2	0.4		0.2	0.1
+ 28	1.6	0.3		0.3	0.1
+ 35	2.2	0.3		0.6	0.2
+ 48	2.9	0.9	nil	1.2	0.7
+ 65	4.7	1.7	0.1	2.1	1.5
+ 100	8.1	4.7	0.3	5.7	4.9
+ 150	9.3	8.9	0.8	9.9	9.3
+ 200	12.8	21.6	2.6	25.4	24.6
+ 325	14.1	30.9	13.8	33.5	32.0
− 325	41.6	30.3	82.4	20.6	26.6

Solution:

Flow rates: one parameter completely describes the flow rates based on $A = 1$; $A = D = 1$; $E = \alpha$; $B = C = \alpha - 1$

Calculation of the best-fit flow rates may proceed based on the errors in each component:

$$\Delta_{(1)i} = \alpha \cdot (e_i - c_i) + (c_i - a_i)$$

$$\Delta_{(2)i} = \alpha \cdot (e_i - b_i) + (b_i - d_i)$$

$$S = \sum_i (\Delta^2_{(1)i} + \Delta^2_{(2)i})$$

When $\partial S/\partial\alpha$ equals zero the following equation may be derived:

$$\alpha = \frac{-\sum_i (e_i - c_i)\cdot(c_i - a_i) + (e_i - b_i)\cdot(b_i - d_i)}{\sum_i (e_i - c_i)^2 + (e_i - b_i)^2}$$

and for the present data:

$$\alpha = \frac{-(-111.86 - 753.25)}{24.32 + 117.80} = 6.0872$$

TABLE 7-II

Calculation of Lagrange multipliers

Tyler mesh	Flow residuals at $\alpha = 6.0872$		Lagrange multipliers	
	Δ_1	Δ_2	λ_1	λ_2
+ 8	− 0.10	0.0	0.0024	− 0.0014
+ 10	− 0.40	− 1.53	− 0.0114	0.0305
+ 14	− 1.00	− 1.02	0.0007	0.0103
+ 20	0.73	1.42	0.0023	− 0.0235
+ 28	− 0.28	0.30	− 0.0108	− 0.0109
+ 35	− 1.39	− 1.23	0.0160	0.0099
+ 48	− 0.98	− 0.63	0.0146	0.0013
+ 65	− 1.98	− 0.43	0.0408	− 0.0168
+ 100	− 4.42	− 0.69	0.0947	0.0441
+ 150	− 2.43	3.01	0.0985	− 0.1042
+ 200	− 6.46	− 0.33	0.1477	− 0.0804
+ 325	11.20	3.87	− 0.2110	0.0617
− 325	7.52	− 2.75	− 0.2148	0.1676

TABLE 7-III

Sizing distributions adjusted to allow for experimental error

Tyler mesh	Circuit feed	Cyclone			Ball-mill discharge
		feed	OF	sands	
+ 8	0.1	0.0	0.0	0.0	0.0
+ 10	0.4	0.1	0.0	0.2	0.1
+ 14	1.0	0.1	0.0	0.2	0.1
+ 20	1.2	0.3	0.0	0.3	0.1
+ 28	1.6	0.3	0.0	0.4	0.1
+ 35	2.2	0.5	0.0	0.6	0.1
+ 48	2.9	1.0	0.0	1.2	0.6
+ 65	4.7	1.9	0.1	2.2	1.3
+ 100	8.0	5.0	0.4	5.9	4.4
+ 150	9.2	8.9	0.9	10.4	8.8
+ 200	12.7	22.0	2.7	25.8	23.9
+ 325	14.3	30.0	13.7	33.2	33.1
− 325	41.8	30.0	82.2	19.8	27.7

Data adjustment: minimise: $S_i = \Delta a_i^2 + \Delta b_i^2 + \Delta c_i^2 + \Delta d_i^2 + \Delta e_i^2$

Subject to (1): $\alpha \cdot (e_i - \Delta e_i) - (a_i - \Delta a_i) - (\alpha - 1) \cdot (c_i - \Delta c_i) = 0$

and (2): $\alpha \cdot (e_i - \Delta e_i) - (d_i - \Delta d_i) - (\alpha - 1) \cdot (b_i - \Delta b_i) = 0$

$$Sm_i = S_i - 2\lambda_1 \cdot [\Delta_{(1)i} - \alpha \cdot \Delta e_i + \Delta a_i + (\alpha - 1) \cdot \Delta c_i] - 2\lambda_2 \cdot [\Delta_{(2)i} - \alpha \cdot \Delta e_i + \Delta d_i + (\alpha - 1) \cdot \Delta b_i]$$

where $\Delta_{(1)i} = \alpha \cdot e_i - a_i - (\alpha - 1) \cdot c_i$

Derivatives with respect to residuals yield for each i:

$$\Delta a = \lambda_1;\ \Delta b = \lambda_2 \cdot (\alpha - 1);\ \Delta c = \lambda_1 \cdot (\alpha - 1);\ \Delta d = \lambda_2;\ \text{and}\ \Delta e = -\alpha \cdot (\lambda_1 + \lambda_2)$$

Taking the derivatives with respect to λ and substituting yields for each i:

$$\Delta_{(1)} = \lambda_1 \cdot [\alpha^2 + 1 + (\alpha - 1)^2] + \lambda_2 \cdot \alpha^2$$

$$\Delta_{(2)} = \lambda_1 \cdot \alpha^2 + \lambda_2 \cdot [\alpha^2 + 1 + (\alpha - 1)^2]$$

The errors at each junction in each size range (Δ_{1i} and Δ_{2i}) and then the Lagrange multipliers may be calculated as in Table 7-II.

Data adjustment residuals may be calculated for each i; calc = obs − res, where $\Delta a = \lambda_1$; $\Delta b = \lambda_2 \cdot (\alpha - 1)$; $\Delta c = \lambda_1 \cdot (\alpha - 1)$; $\Delta d = \lambda_2$; $\Delta e = -\alpha \cdot (\lambda_1 + \lambda_2)$.

Adjusted data are given in Table 7-III.

Example 7-3. Mass balance and data adjustment for a flotation bank. An industrial flotation bank is sampled at the points shown in Fig. 7-8 (see p. 157). Assay results, expressed as percent by weight, are given in Table 7-IV.

Calculate the best-fit flow rates on the basis of least sum of squares of apparent mass flow errors.

TABLE 7-IV

Assay data from an industrial flotation bank

	A1	A2	A3	A4	A5	B1	B2	B3	B4	B5
Copper	3.03	1.0	1.15	0.75	0.52	21.2	22.5	17.1	9.7	15.5
Iron	8.2	7.1	6.4	6.0	5.7	23.8	25.9	23.4	20.2	21.4
Sulphur	6.5	5.3	4.4	3.8	3.3	26.5	29.2	25.9	20.8	23.0
Insolubles	46.6	49.0	51.0	51.8	52.8	14.4	11.4	18.4	27.6	21.6

Solution:

Divide all stream flow rates by A_1 to yield α_1 to α_5 where $\alpha_1 \equiv 1$. The sum of squares S_i of errors in one component is:

$$S_i = \sum_{j=1,4} \langle \alpha_j \cdot (a_{ji} - b_{ji}) - \alpha_{j+1} \cdot (a_{j+1i} - b_{ji}) \rangle^2 + \langle \alpha_1 \cdot (a_{1i} - b_{5i}) - \alpha_5 \cdot (a_{5i} - b_{5i}) \rangle^2$$

To compact the working let:

$$x_j = (a_{ji} - b_{ji})\ \text{for}\ j = 1, 5; \quad y_j = (a_{ji} - b_{j-1i})\ \text{for}\ j = 2, 5; \quad \text{and}\ y_6 = (a_{1i} - b_{5i})$$

and use Gauss brackets [ ] to denote summation over all the components. Taking derivatives, setting to zero and shifting to matrix notion gives Table 7-V.

This can be solved as a set of simultaneous equations for α or the symmetric cōefficient matrix can be inverted.

For the present data the equations are:

TABLE 7-V

Determination of α in mass balance calculations

$$\begin{bmatrix} [_iy_2^2+x_2^2]-[_iy_3x_2] & 0 & 0 \\ -[_ix_2y_3] & [_iy_3^2+x_3^2]-[_ix_3y_4] & 0 \\ 0 & [_ix_3y_4] & [_iy_4^2+x_4^2]-[_iy_5x_4] \\ 0 & 0 & -[_ix_4y_5] & [_iy_5^2+x_5^2] \end{bmatrix} \cdot \begin{bmatrix} \alpha_2 \\ \alpha_3 \\ \alpha_4 \\ \alpha_5 \end{bmatrix} = \begin{bmatrix} [_ix_1y_1] \\ 0 \\ 0 \\ [_ix_5y_6] \end{bmatrix}$$

$$5060.75\cdot\alpha_2 - 2888.09\cdot\alpha_3 + 0\cdot\alpha_4 + 0\cdot\alpha_5 = 2149.32$$

$$-2888.09\cdot\alpha_2 + 5087.69\cdot\alpha_3 - 2120.47\cdot\alpha_4 + 0\cdot\alpha_5 = 0$$

$$0\cdot\alpha_2 - 2120.47\cdot\alpha_3 + 3330.44\alpha_4 - 1195.4\cdot\alpha_5 = 0$$

$$0\cdot\alpha_2 + 0\cdot\alpha_3 - 1195.4\cdot\alpha_4 + 3068.23\cdot\alpha_5 = 1499.09$$

The solution is: $\alpha_1 = 1; \alpha_2 = 0.93099; \alpha_3 = 0.88715; \alpha_4 = 0.86056; \alpha_5 = 0.82386.$
Note: the coefficient for α_2 is derived as follows:

$$_iy_2^2 + x_2^2 = (1.9-21.2)^2 + (1.9-22.5)^2 + (7.1-23.8)^2$$
$$+ (7.1-25.9)^2 + (5.3-26.5)^2 + (5.3-29.2)^2$$
$$+ (49.0-14.4)^2 + (49.0-11.4)^2$$

Other coefficients may be derived following the same procedure.
Data adjustment: Following the same approach as Example 7-2 for each component and omitting the i subscript:

$$S_m = \sum_j (\Delta a_j^2 + b_j^2) + 2\sum_{j=1,4} \lambda_j\cdot(\Delta_j + \alpha_j\cdot\Delta a_j - \alpha_{j+1}\cdot\Delta_{j+1} - \beta_j\cdot\Delta b_j)$$
$$+ 2\lambda_5\cdot\left[\Delta_5 + \sum_{j=1,4}(\beta_j\cdot\Delta b_j) - \beta_5\cdot\Delta b_5\right]$$

Changing to matrix notation gives the procedure shown in Table 7-VI.

TABLE 7-VI

Procedure for the calculation of data adjustments

$$\begin{bmatrix} (\alpha_1^2+\alpha_2^2+\beta_1^2) & -\alpha_2^2 & 0 & 0 & -\beta_1^2 \\ -\alpha_2^2 & (\alpha_2^2+\alpha_3^2+\beta_2^2) & -\alpha_3^2 & 0 & -\beta_2^2 \\ 0 & -\alpha_3^2 & (\alpha_3^2+\alpha_4^2+\beta_3^2) & -\alpha_4^2 & -\beta_3^2 \\ 0 & 0 & -\alpha_4^2 & (\alpha_4^2+\alpha_5^2+\beta_4^2) & -\beta_4^2 \\ -\beta_1^2 & -\beta_2^2 & -\beta_3^2 & -\beta_4^2 & (\sum_{j=1,4}\beta_j^2-\beta_5^2) \end{bmatrix} \cdot \begin{bmatrix} \lambda_1 \\ \lambda_2 \\ \lambda_3 \\ \lambda_4 \\ \lambda_5 \end{bmatrix} = \begin{bmatrix} \Delta_1 \\ \Delta_2 \\ \Delta_3 \\ \Delta_4 \\ \Delta_5 \end{bmatrix}$$

As in Example 2, the data adjustments can be calculated directly from the Lagrange multipliers. Note that the matrix only need be inverted once for *all* the sets of components. The multipliers for each component set are the product of this inverse and the vector of errors at each junction.

Example 7-4. Estimation of best-fit parameters by regression and curve fitting. In this example the regression or non-linear least-square approach is compared with the curve-fitting approach.

Data points on the corrected efficiency curve of an industrial hydrocyclone are reported. Use these data to find the best-fit parameters α and $d_{50}(c)$ in the equation:

$$y = (e^{\alpha \cdot x} - 1)/(e^{\alpha \cdot x} + e^{\alpha} - 2)$$

where y is the fraction of particle size d reporting to the underflow and x is the particle size divided by $d_{50}(c)$ the size at which 50% of the particles report to the underflow.

(a) Find the best-fit parameters using the non-linear least squares of model error approach (that is, assume the sizes are exact and all error is in the efficiency). Then use the sizes and best-fit parameters to calculate efficiency values.

(b) Find the best-fit parameters using the curve-fitting approach (that is, assuming errors of unit variance in both size and efficiency).

Data:

Size (μm)	3330	1973	1389	985	700	496	350
Eff. (%)	100.0	100.0	100.0	100.0	94.7	83.1	71.5
Size (μm)	248	175	124	88			
Eff. (%)	58.3	44.4	31.6	21.8			

Solution:

(a) The non-linear least squares approach requires an error sub-program to calculate the error at each experimental point with successive parameter estimates.

This example is for a FOCAL program which is easily interpreted if S is read as "Set". The data are stored as sets within a vector X(I). For example X(1) and X(2) are the first set of diameter and corrected efficiency and X(21) and X(22) are the eleventh set of data. The number of the set being used is N and the number of variables is NV. Lines 8.20 and 8.30 make the N-th data set available as X and Y. The parameter values are stored in vector P(I) where, in this case: P(1) is α, and P(2) is d_{50}.

Lines 8.40 and 8.50 calculate the difference between the model value and the experimental value:

```
08.10 CALCULATE THE ERROR E AT A POINT
08.20 S X = X ((N − 1)∗NV+1)
08.30 S Y = X ((N − 1)∗NV+2)
08.40 S X = X/P(2); S T1 = FEXP (P(1)∗X); S T2 = FEXP (P(1))
08.50 S E = Y − 100∗(T1 − 1)/(T1 + T2 − 2)
08.99 R
```

The results are:

Parameter:	Value:
1	0.100159 E + 01
2	0.209826 E + 03

Parameter:	Error:
1	0.577735 E − 01
2	0.305387 E + 01

Residual standard error: 0.226369 E + 01

(b) The curve fitting approach also requires an error routine. In this case the error routine specifies the error in the constraint. The best-fit parameters are based on least

TABLE 7-VII

Residuals in the parameter estimation for the hydrocyclone model

Point	Residual	$y_{exp.}$	$y_{calc.}$
1	0.0001	100.000	100.000
2	0.0135	100.000	99.9865
3	0.2207	100.000	99.7793
4	1.5228	100.000	98.4772
5	0.5610	94.7000	94.1390
6	−1.9260	83.1000	85.0260
7	−0.1532	71.5000	71.6532
8	1.2995	58.3000	57.0005
9	1.1293	44.4000	43.2708
10	−0.4377	31.6000	32.0377
11	−1.5518	21.8000	23.3518

squares of adjustments to *all* data values required to satisfy the constraint. This example is suitable for use with CURFIT.

```
      SUBROUTINE CURV (NE, IS, X, E)
C  FIT ALPHA AND D50 FOR CYC EFF CURVE
C  R D MORRISON
C  3 12 75
C
      REAL X(20), E(10)
C
      SIZE = X(1)
      EFF = X(2)
      ALPHA = X(3)
      D50 = X(4)
C  FIT EFF CURVE
      T1 = EXP (ALPHA*SIZE/D50)
      T2 = EXP (ALPHA)
      E(1) = EFF - 100* (T1 - 1.)/(T1 + T2 - 2.)
C
      RETURN
      END
```

The best-fit parameters are: α = 1.000; standard error 0.0579; $d_{50}(c)$ = 209.8 μm; standard error 3.08 μm

Adjusted values:

Size (μm)	3330	1973	1389	985	700	496	350
Eff (%)	100.0	100.0	99.8	98.5	94.1	85.0	71.6
Size (μm)	248	175	124	88			
Eff (%)	57.0	43.3	32.0	23.3			

CHAPTER 8

Circuit design by simulation

8.1 INTRODUCTION

Simulation is an important aid in circuit design but the results must be assessed in terms of possible problems, such as ease of circuit expansion. No attempt is made to discuss all problems in this chapter. This would not be possible because the requirements of a circuit are frequently dictated, at least in part, by the locality. Many of the problems discussed refer to circuits being designed for new concentrators; the range of problems is much more limited if existing circuits are being modified, extended or replaced. The results of any attempts to produce an optimum circuit design, whether based on simulation or on other mathematical techniques, must be considered in relation to particular problems and requirements.

The design of new crushing or grinding circuits, or the expansion or modification of an existing circuit, involves three types of problems:

(1) choosing the flow diagram and selecting the equipment sizes such that a satisfactory standard of metallurgical efficiency may be obtained at the required throughput rate;

(2) ensuring that future expansion or modification can be carried out without undue difficulty, and that no serious mechanical problems will arise during the expected life of the circuit; and

(3) planning for minimum capital and operating costs of the circuit commensurate with efficient metallurgical and mechanical performance.

These problems may be broadly classed as metallurgical, mechanical and economic but they are closely related and should not be considered individually. For instance, for each alternative flow diagram or different item of equipment considered, the different capital and operating costs, and the possible mechanical problems in circuit operation or expansion, should also be considered.

8.1.1 *Crushing circuits*

Possible objectives of mineral crushing plants are;

(1) preparation of a product which will be as fine as possible consistent with maintaining the required throughput; and

(2) preparation of a product which will have a maximum proportion of material in certain size ranges.

The first objective is common when ores are being crushed prior to grinding for mineral concentration or leaching, such as in the flotation of metal sulphides or in the leaching of gold or metal oxides. The second objective is common in the production of aggregate and gravel since material wastage and decrease in plant capacity occurs when excess fines are produced which cannot be sold. The market demand for various size fractions of aggregate fluctuates and in a crushing operation the emphasis will be on the production of maximum amounts of those size fractions for which there is a ready market. The crushing and screening operations involved in the recovery of iron ore from the deposits of Northwest Australia are also examples of the type of crushing operation in which a considerable penalty is often incurred for excess fines production.

The design of the crushing circuit involves the determination of:

(1) the number of stages of crushing;

(2) the optimum number and sizes of the crushing and screening units within each stage;

(3) which of the crushing stages are to be preceded by screening and which are to be operated in open or closed circuit;

(4) the number of parallel circuits to be used;

(5) the optimum crusher closed side settings;

(6) the screen apertures; and

(7) the capacities of various surge bins and conveyors.

Some technical, economic and maintenance aspects of crushing operation which must be considered are:

(1) Crushing plants usually require multi-stage operation. It is usual to close the final crushing stage with a screen to eliminate coarse particles from the plant product but close-circuiting of earlier stages generally does not result in the most efficient plant design.

(2) In the selection of crusher types and sizes it is important to ensure that the largest size fractions in the feed to any stage can freely enter the crusher and do not cause bridging in chutes or feeders. The crusher must also be selected on the basis of capacity.

(3) The capital cost per tonne of product discharged at the same closed-side setting does not decrease significantly with crusher size. This is in marked contrast to the capacity/cost relationship for grinding mills. Also the power per tonne of product discharged at the same closed-side setting does not decrease significantly as crusher size is increased. These facts serve to emphasise the importance of capacity and expected feed size in the selection of a crusher. Approximate data illustrating these effects are given in Table 8-I.

(4) The dependence of the power draw of a particular crusher upon feed tonnage is approximately linear but tends towards higher power draws as the tonnage is increased. However, the relationship is dependent upon both feed and product size distributions. Coarser feed and/or finer product cause a

TABLE 8-I

Summary of costs, available kW and open-circuit capacity for various sizes of cone crushers (1976 costs, by courtesy of Allis-Chalmers Australia, Limited)

Crusher size (mm)	Cost ($A)	kW	Capacity (t/h)	$/t	kW/t
914	55,000	93	134	344	0.58
1143	79,000	149	212	336	0.64
1524	139,000	224	315	397	0.64
2134	295,000	373	630	421	0.53

Capacities are given for operation at the same closed side setting.

decrease in throughput for a given power draw. These conditions favour multi-stage crushing rather than single-stage with a very high reduction ratio.

(5) The presence of significant amounts of fine material in the feed to a crusher decreases the tonnage of coarser material which can be processed. This occurs because the presence of fines increases the energy expended by the crusher in the generation of more fine material. The flow of material through the crusher can also be impeded with the accompanying risk of choking the crusher.

(6) For optimum performance a steady continuous supply of ore to the crusher is important. This can be achieved by the installation of a surge bin from which the crusher feed is drawn. Very often this is not possible with the primary crusher where large intermittent tonnages of feed are discharged from wagons or dump trucks. Thus, the primary crusher should have a capacity considerably in excess of the average tonnage required for the remainder of the plant.

(7) It is probable that large changes in the plant feed characteristics will occur. This will result in considerable variation in the tonnages of material which must be handled by conveyor belts. Under some conditions the capacities of one or more of these belts may limit plant throughput, whilst for other conditions the power available for a particular stage of crushing may limit production. During the design of a crushing plant reasonable excess capacity of conveyors and adequate power should be provided.

(8) Better machine availability and a greater degree of flexibility can be obtained by locating the screens, particularly those which are part of closed circuit crushing stage, in a separate screening station. This arrangement will require surge bins before both the crushers and screens and so there is some additional cost due to duplication of surge capacity. However, an assured steady supply of feed to both crushers and screens justifies this cost.

(9) The efficiency of vibrating screens is less than 100%. This must be taken into account when determining anticipated circulating loads. It is also affected by loading and variation in feed size distribution. For the accurate

determination of circulating loads under various conditions it is desirable to take these effect into account.

8.1.2 *Grinding circuits, rod and ball mills*

In the design of grinding circuits which include rod and ball mills some matters for consideration are:

(1) The capital cost of a mill per tonne of finished product decreases rapidly as the mill size and throughput increases. For single-stage overflow ball mills the relationship between mill size, horsepower and throughput is given in Table 8-II and it will be noted that the capital cost per tonne of product decreases from $7000 at 25 t/h to $2160 at 500 t/h. This same trend applies to the other items of equipment.

(2) Separate crushing and grinding circuits will be necessary and in most cases several grinding mills will be required. Consequently, decisions must be made on such matters as the required product size from the crushing circuit, the possible use of a rod mill as a primary grinding unit, the split of duty between the rod mill and the ball mills, the number of parallel circuits, and the arrangement of the processing units within the circuits.

(3) In the case of rod mills the operating availability is about 95% because of the ncessity to shut the mill down to charge rods and tighten bolts. The maximum practical length of a rod mill is about 6 m since beyond this, rod flexing occurs leading to premature rod breakage. The length to diameter ratio should not be less than 1.33 : 1 otherwise rod tangling may occur. A

TABLE 8-II

Summary of costs, design and performance details for single-stage overflow ball mills (1976 costs, by courtesy of Allis-Chalmers Australia, Limited)

kW	Size I.Dia. by EGL	Approx. wt. (kg)	Speed RPM of mill	Approx. cost incl. motor ($A)	Approx. t/h prod.	Approx. kW/t
224	2.40 · 3.05	45,500	21.5	174,000	25	9.0
373	3.05 · 3.05	86,500	17.4	241,000	45	8.3
746	3.61 · 4.53	127,000	15.8	390,000	100	7.5
1492	4.21 · 4.82	205,000	13.7	715,000	210	7.1
2238	4.96 · 5.72	273,000	13.0	943,000	350	6.4
2984	4.96 · 7.52	355,000	13.0	1,040,000	470	6.3
3175	5.42 · 6.32	373,000	12.4	1,080,000	500	6.3

Notes: 1. EGL is the effective grinding length in metres; 2. I.Dia. is the diameter inside shell in metres; 3. The mills are assumed to be in closed circuit with hydrocyclones, the average feed size being — 12.5 mm and the average product size 80% — 210 micrometres; 4. Prices include steel liners, air clutches and slow-speed synchronous motors.

Fig. 8-1. A case of rod tangling.

typical case of rod tangling is shown in Fig. 8-1. Under normal conditions rods wear down to an elliptical section and break up when the length of the major axis is about 2.5 cm. Liner wear and breakage are also a problem in very large diameter rod mills. A limitation of rod-mill operation is their ability to transport ore into the rod mass in the mill. Occasionally this may be improved by the addition of a small amount of fine product such as a cyclone overflow to the rod-mill feed.

(4) Rod mills are most efficient as coarse grinding units and it is important to operate them at as high a feed rate as possible so that comminution occurs mainly in a narrow size range of about 16000 — 4000 μm.

(5) In the case of ball mills the operating availability is high, exceeding 99%.

(6) When ball-mill diameter increases above 5 m the power consumption per tonne of ore at constant product size increases. The reason is not fully understood.

(7) The toughness of the grinding balls must increase as the diameters of the mills increase and forged steel balls must be used in mills above 4.2 m diameter. Grinding media density must be taken into account in specifying motor size.

(8) Grate discharge mills have the advantage that they will maintain a higher grinding media change. These mills are essential for pebble milling to prevent undue spillage of the pebbles. For ball mills their use is doubtful because of: (a) longer lining times, (b) grate blockage, and (c) grate breakage.

If rod mills are to be used the minimum number of circuits is governed by the minimum practical number of rod mills. If the minimum number of rod mills is used expansion is expensive because it must commence with installation of a new rod mill with associated bin discharge and mill feeding system. If more than the minimum number of rod mills is used, expansion may readily be achieved by the addition of ball mills but the initial installation has higher capital and operating costs, and greater instrument requirements.

Some of these problems may be illustrated by the consideration of alternative grinding circuit designs for a 1200 t/h concentrator. One possibility is three 400 t/h circuits, each containing a rod mill and two ball mills, and another is four 300 t/h circuits, each containing a rod mill and one ball mill. These flow diagrams are shown in Fig. 8-2. The former arrangement contains nine mills and the latter eight, although the ball mill sizes will not be the same in each case. Not only must the capital and operating costs, and ease of circuit expansion be considered, the possible operating problems within each type of circuit must also be anticipated. In the case of the 3 rod-mill circuits, the choice of operating the ball mills in series or in parallel must be made. If the mills are in parallel a malfunction in any ball-mill—cyclone closed circuit is transmitted into the grinding circuit product immediately but if they are in series the effect of any malfunction may be minimised. In the case of the 4 rod-mill circuit the circulating loads in the ball mill circuits are very high and the pumping duties are very severe. There is no flexibility for the installation of two-stage grinding and flotation circuits if these are found to be metallurgically desirable.

8.1.3 *Grinding circuits, autogenous mills*

An alternative to conventional rod- and ball-mill circuits is some form of autogenous milling and this may include run-of-mine, pebble and partial autogenous milling. Various combinations of autogenous milling and steel milling (rod and ball mills) have been used in different circuits, for instance rod and pebble mills or run-of-mine and ball mills.

In steel milling, crushing is the major mode of size reduction but in autogenous milling, reduction of the finer sizes occurs by crushing and of the larger sizes by abrasion. For successful autogenous grinding, it is essential that the top size range be capable of generating a grinding load able to crush the finer portion of the feed plus its own progeny but it must not be so durable that it cannot be reduced at a rate equal to that at which it is entering the mill.

Since the success of autogenous milling depends on the coarse fraction of

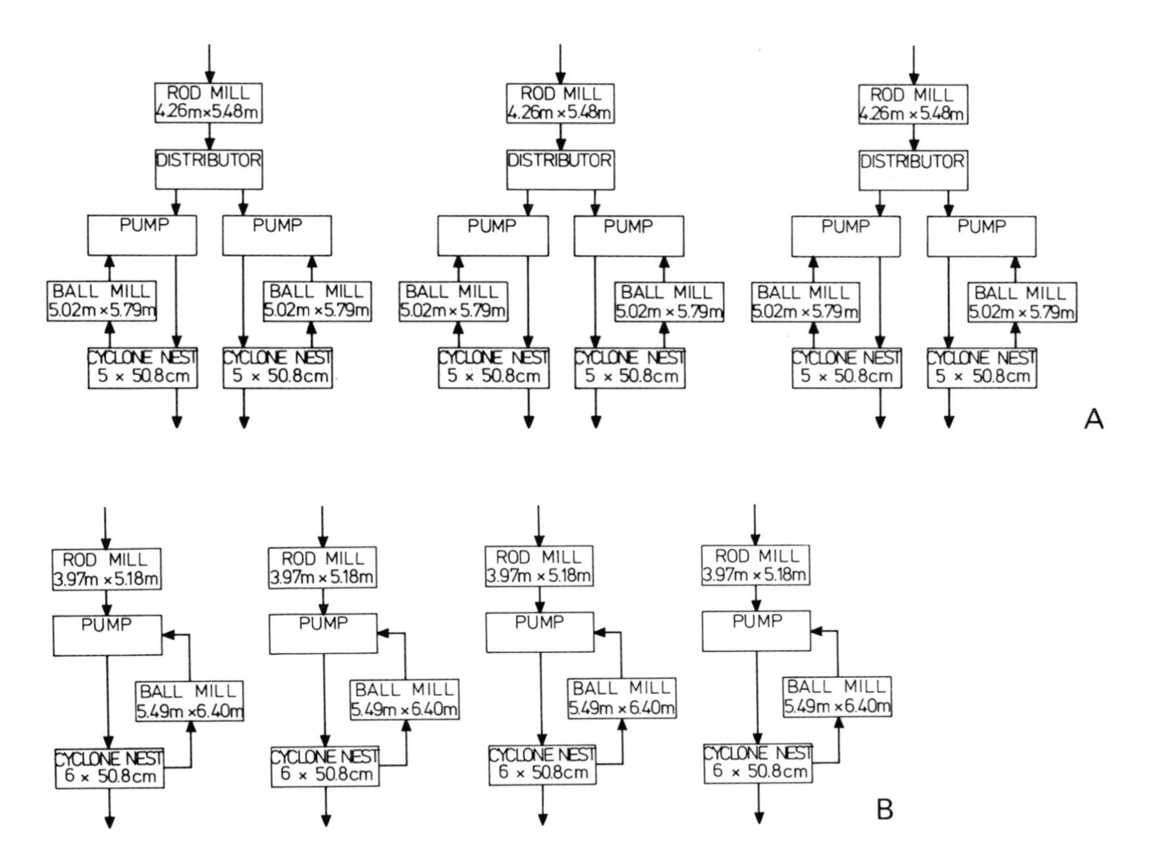

Fig. 8-2. Alternative grinding circuits for a 1200 t/h concentrator: A. 3 rod-mill—6 ball-mill circuit. B. 4 rod-mill—4 ball-mill circuit.

the feed, the mining method adopted must be such as to ensure the adequacy of this fraction. If the ore is hard and blocky, this is not a problem but if conditions are such that the rock tends to break into small fragments consideration must be given to breaking and handling methods to optimise the proportion of coarse material delivered to the mill. Problems encountered in this area may sometimes be overcome by the use of partial autogenous milling, in which a certain proportion of artificial grinding media, such as grinding balls, is added to the mill charge. If there is any doubt as to whether the mine can deliver the necessary proportion of suitable grinding media over the long term the possibility of autogenous milling must be abondoned.

A successful example of pure autogenous grinding is in the Warrego concentrator of Peko Mines Limited at Tennant Creek, Northern Australia. The ore is composed mainly of magnetite with variable admixtures of quartz, and sulphides, predominantly chalcopyrite, occur as veinlets.

The grinding plant consists of one large, "square", 5.1 m diameter by 5.2 m long single-stage autogenous mill in which ore ranging in the size down from 200 mm is reduced to a suitable flotation feed of approximately 80% −200 mesh. The reasons for the choice of autogenous grinding were:

(1) A saving could be made of media per tonne of ore milled. This was important due to the long distances of the Warrego concentrator from its points of supply. Moreover, the Tennant Creek area could become isolated by road during the wet season and reductions of the heavy transport requirements of the mines are to be made where possible.

(2) Fine crushing could be dispensed with and a large saving made on capital, installation, operational and maintenance costs.

A large single unit rather than multiple units was chosen because:

(1) significant savings on capital cost could be made due to the installation of a lesser number of large components;

(2) the layout of the mills and ancilliaries would be simpler and easier to maintain; and

(3) the grinding circuit would be easier to operate especially by unskilled labour.

8.2 GRINDING CIRCUITS

8.2.1 *Introduction*

The optimisation of existing circuits using simulation techniques was discussed in Chapter 7. These simulation techniques may also be used in circuit design although it must be recognised that mill scale-up relationships are a problem when large changes in mill diameter are involved because the relationship cannot yet be defined precisely.

Accurate simulation of existing circuits usually is not difficult because sampling surveys may be carried out on the circuit and parameters for use in the simulation models may be determined. Circuit optimisation may then be investigated with regard to the defined objectives. Expansion or modification of an existing circuit is similar to optimisation because parameters for the existing mills may be determined experimentally and scaling of these parameters can be carried out on the basis of mill power consumption when a change in mill size is considered. This approach may also be used for the design of new circuits which are to replace existing circuits and in which the same type of ore is to be treated.

When a circuit is to be designed using information obtained from laboratory tests carried out on drill cores, or even from pilot-scale continuous grinding tests, the accuracy of simulation techniques is limited because scaling factors which relate small-scale batch or continuous mill behaviour to the behaviour of large mills are not known accurately. In this case, considerable factors of safety are always included in the design of new circuits to allow for local variations in grindabilities through an ore-body and for probable increases in throughput. The optimisation of a design for initial processing requirements for a new circuit is not as important as ensuring that the circuit when installed is: (1) flexible from the point of view of later possible expansion, (2) has an adequate factor of safety in the case of error in the original determination of hardness or grindability, and (3) has no physical limitations such as an incorrectly sized pump.

Simulation may be used in the design of a new circuit for determining the changes in flow rates and size distributions of streams within the circuit for changes in operating conditions. These data may be used to ensure that processing units, pumps and pipe lines are correctly sized.

The use of simulation for design purposes, both in the replacement of existing circuits and in the design of new circuits, will be discussed for crushing, grinding and classification circuits in the remainder of this chapter by considering particular problems.

8.2.2 *Design using data from an existing circuit*

When new circuits are to be designed to replace existing circuits, and rod and ball mills having larger diameters than those installed in the existing circuits are to be used, the assumptions may be made that the selection functions remain unchanged and that scale-up may be carried out by appropriate change in the mill constants. The mill constants may be assumed to vary: (1) with the mill length in a direct ratio, and (2) with the mill diameter to the power of 2.6.

It is known that the effect of change in mill diameter is more complex than this, particularly when the mill diameter exceeds 4 m, but this approximation may be made in the absence of more exact relationships. The error inherent in the assumption should be recognised.

The effect of change in grindability of the ore may be included by scaling down the mill constant with increase in work index. Change in critical speed and ball load in the mill may be included by making changes in the mill constant which is directly proportional to changes in these parameters.

If subscript 1 refers to an existing mill and subscript 2 to a proposed new mill the mill constant of the new mill (MC_2) may be calculated from the equation:

$$MC_2 = \left[\frac{D_2}{D_1}\right]^{2.6} \cdot \frac{L_2}{L_1} \cdot \frac{W_{i1}}{W_{i2}} \cdot MC_1 \qquad (8\text{-}1)$$

The design procedure will be explained by considering the design of the grinding circuit in a 1200 t/h concentrator. Circuits which will be discussed were shown in Fig. 8-2. It will be assumed that the required circuit product size is 78% passing 75 μm at a solids content of 35.0% by weight. Data for design purposes were available from a 110 t/h circuit containing a rod mill (2.74 m by 3.66 m) and three ball mills (3.20 m by 3.05 m) which were operated in closed circuit with 50.8 cm diameter hydrocyclones.

The parameters for the grinding mills and hydrocyclones in the 110 t/h circuit, calculated using the procedure discussed in Chapters 4, 6 and 7 were as follows:

Rod mill:
top size range of particles in feed: 38040—19020 μm; S: ⟨1.0 0.8 0.25 0.2 0.25 0.5 0.5 0.5 0.5⟩; mill constant: 550

Ball mill:
top size range of particles in feed: 9400—4700 μm; S: ⟨1.0 1.0 1.0 0.9 0.5 0.33⟩; mill constant: 170

Hydrocyclone:
water split (eq. 6-6): constant = 0; d_{50}(c): (eq. 6-13); constant = 0; reduced-efficiency curve (eq. 6-20): constant = 2.0

For the new concentrator, rod mills of three different sizes will be considered. Constants for these mills may be calculated from eq. 8-1 assuming that the percent critical speed of the mill speed and the rod load, scaled to the mill size, remained unchanged. The mill sizes and corresponding constants are given in Table 8-III.

TABLE 8-III

Rod-mill sizes considered in the circuit design

Mill	Diameter (m)	Length (m)	Mill constant
A	3.66	4.88	1555
B	3.97	5.18	2040
C	4.26	5.48	2594

For the three rod-mill arrangement a throughput of 400 t/h per line is required and for the four rod-mill arrangement the throughput is 300 t/h. The predicted product size distribution from each rod mill at 300 and 400 t/h for a feed size distribution of ⟨1.0, 2.8, 31.0, 22.6, 14.2, 8.6, 5.4, 4.1, 1.7, (8.6)⟩ are given in Table 8-IV.

TABLE 8-IV

Rod-mill product size distributions for different feed rates and mill sizes

Mill	No. of mills	Feed rate per mill (t/h)	v (calc)	Product size distribution									
A	4	300	3.01	⟨0	0	0	12.8	24.3	17.1	11.0	7.2	5.7	(21.9)⟩
B	4	300	3.61	⟨0	0	0	3.8	17.1	21.7	15.1	9.7	6.5	(26.1)⟩
C	4	300	4.24	⟨0	0	0	0	10.7	21.3	18.4	12.2	7.7	(29.7)⟩
A	3	400	2.48	⟨0	0	5.1	19.2	21.6	14.2	9.1	6.6	5.1	(19.1)⟩
B	3	400	2.98	⟨0	0	0.3	13.3	24.3	16.9	10.8	7.1	5.6	(21.7)⟩
C	3	400	3.50	⟨0	0	0	6.6	19.3	20.3	13.8	8.9	6.2	(24.9)⟩

With the 4 rod-mill—4 ball-mill arrangement, the ball-mill sizes which were considered, and the corresponding mill constants, are given in Table 8-V. The hydrocyclones in the circuit are 50.8 cm diameter.

TABLE 8-V

Ball-mill sizes considered for the four-rod mill—four-ball-mill circuit

Diameter (m)	Length (m)	Mill constant
5.0	5.79	2044
5.02	7.62	1377
5.49	6.40	1464

The steady-state results for the nine cases considered are summarised in Table 8-VI.

With the 3 rod-mill—6 ball-mill arrangement, it was assumed that the ball mills are operated in parallel, two with each rod mill. Ball-mill sizes which were considered, and the corresponding mill constants are given in Table 8-VII. The hydrocyclones in the circuit are 50.8 cm diameter.

The steady-state results for the twelve cases considered are given in Table 8-VIII.

Initial conclusions were that two of the circuits suitable for further investigation were those given in cases 6 (Table 8-VI) and 20 (Table 8-VIII). Change in classification conditions were then studied. Reduction in the

TABLE 8-VI

Steady-state simulation results for a four-rod-mill—four-ball-mill circuit grinding 1200 t/h of ore

Case No.	Rod mill		Ball mill		Cyclone underflow		Cyclone overflow (% minus 75 μm)
	dia. (m)	length (m)	dia. (m)	length (m)	t/h	% solids	
1	3.66	4.88	5.02	5.79	557	81.4	63.1
2	3.66	4.88	5.02	7.62	517	80.3	72.2
3	3.66	4.88	5.49	6.40	509	80.0	74.4
4	3.97	5.18	5.02	5.79	526	80.6	67.7
5	3.97	5.18	5.02	7.6	492	79.4	76.0
6	3.97	5.18	5.49	6.40	484	79.2	77.8
7	4.26	5.48	5.02	5.79	503	79.8	71.1
8	4.26	5.48	5.02	7.62	472	78.7	78.8
9	4.26	5.48	5.49	6.40	465	78.5	80.4

Note. Five cyclones are used with each ball mill, 15.2 cm VF and 12.7 cm Spig.

TABLE 8-VII

Ball-mill sizes considered for the three-rod-mill—six-ball-mill circuit

Diameter (m)	Length (m)	Mill constant
4.27	4.88	571
4.27	6.10	714
5.02	5.79	1044
5.02	7.62	1377

TABLE 8-VIII

Steady-state simulation results for a 3 rod-mill—6 ball-mill circuit grinding 1200 t/h of ore

Case No.	Rod mill		Ball mill		Cyclone underflow		Cyclone overflow (% minus 75 μm)
	dia. (m)	length (m)	dia. (m)	length (m)	t/h	% solids	
10	3.66	4.88	4.27	4.88	469	81.5	51.2
11	3.66	4.88	4.27	6.10	439	80.4	58.7
12	3.66	4.88	5.02	5.79	392	78.6	72.1
13	3.66	4.88	5.02	7.62	362	77.2	81.1
14	3.97	5.18	4.27	4.88	450	80.8	54.8
15	3.97	5.18	4.27	6.10	422	79.8	62.0
16	3.97	5.18	5.02	5.79	380	78.1	74.7
17	3.97	5.18	5.02	7.62	352	76.7	83.0
18	4.26	5.48	4.27	4.88	429	80.1	58.6
19	4.26	5.48	4.27	6.10	405	79.1	65.6
20	4.26	5.48	5.02	5.79	367	77.4	77.5
21	4.26	5.48	5.02	7.62	342	76.2	85.0

Note. Four cyclones are used with each ball mill, 15.2 cm VF and 12.7 cm Spig.

Approximate costs of alternative circuits for grinding 1200 t/h of ore (cases 6 and 20 in Tables 8-VI and 8-VIII)

	Size	Quant.	Cost ($A · 10^3)	Size	Quant.	Cost ($A · 10^3)
Rod mills	3.97 m · 5.18 m	4	1680	4.26 m · 5.48 m	3	1500
Motors	1120 kW	4	588	1490 kW	3	525
Mill charge	136 t	4	196	214 t	3	231
Ball mill	5.49 m · 6.40 m	4	3448	5.02 m · 5.79 m	6	3960
Motors	3360 kW	4	1208	2240 kW	6	1380
Mill charge	261 t	4	355	227 t	6	463
Cyclones	500 mm	24	99	500 mm	30	124
Tramp iron magnets		2	25		2	25
Slot feeders	12 fixed, 8 variable		160	6 fixed, 6 variable		160
Rod charger		1	5		1	5
Liner handler		1	131		1	131
Crane	90/15 t	1	220	90/15 t	1	220
Crane	15 t	1	93	15 t	1	93
Pumps	D.C. motor	4	161	D.C. motor	6	210
Pumps	A.C. motor	4	145	A.C. motor	6	186
Conveyors			59			65
Monorail hoists		8	54		8	54
Installation (labour 70%, materials 30%):						
General			300			350
Civil works			700			770
Structural			1400			1600
Platework, equipment			1300			2000
Piping			800			1000
Electrical			3036			3795
Cladding			446			500
Architectural			380			390
Ventilation			84			93
Painting			687			750
Instrumentation			388			426
Sub total			18158			21006
Contingencies	20%		3632	20%		4201
Total			21790			25207

vortex finder diameter from 15.2 to 12.0 cm diameter showed that the percent minus 75 μm could be increased in each case by 2.5% but at the expense of a considerable increase in operating pressure. These investigations can be extended to include as wide a range of variables as may be considered necessary. In this manner comprehensive information can be built up about a circuit at the design stage and a quantitative basis for choice of a circuit for installation can be established.

The simulation technique gives information about the flow rates, pressures and sizing distributions of solids in the pulp streams and this is valuable for circuit design purposes. The limitations of the technique should be recognised but these limitations will become progressively less important as better data are collected.

To complete this discussion reference will be made to the capital costs of cases 6 and 20. Approximate breakdown of these costs is given in Table 8-IX and it will be seen that the total costs are $A22,000,000 for case 6 and $A25,000,000 for case 20 (based on 1976 costs).

8.2.3 *Design using data from laboratory and pilot-plant tests*

The simulation procedures used for the design of circuits when data are available only from laboratory and pilot-plant tests are the same as those discussed in sub-section 8.2.2. The parameters for use in the hydrocyclone model may be determined from tests on small units. Estimation of parameters for large mills from data on small mills gives reasonable results when the increase in diameter is not too large, approximately double, but if the diameter scale-up is larger than this the errors may become significant. No doubt data which are suitable for the development of valid scale-up relationships will be collected in due course but until this is done the design of new circuits based on data collected from laboratory or pilot-plant tests must be handled with caution and suitable factors of safety must be included.

Despite this precautionary note, simulation may be regarded as a valuable aid in new circuit design even when data available are limited. This will be illustrated by discussion of a typical problem.

A concentrator is to be built to treat a sulphide ore. The grinding circuit is to contain single-stage, closed-circuit ball mills only. The basic data available for the design of the grinding circuit are from pilot-plant tests in a 1.22 m by 1.22 m ball mill operated in closed circuit with a spiral classifier. These data are then used for the calculation of the values of the parameters in the ball mill model which describe the ore.

The parameters that apply to the ore, calculated from the pilot-plant data, are as follows:

top size of particles broken = 13408 to 9362 micrometres; size ratio = 0.707

$S = \langle 0.946\ 0.909\ 0.873\ 1.000\ 0.917\ 0.843\ 0.773\ 0.712\ 0.653\ 0.599\ 0.549\ 0.504\ 0.464\ 0.426\ 0.390 \rangle$

$MC = 4.6$
kW required for actual grinding = 14.9
$W_i = 11.6$

Operating conditions in the full-size circuit are to be as follows:
feed rate = 47 t/h
work index = 12.2 to 12.9
estimated grinding power = 720 to 760 kW
MC (scaled according to power) = 200 for W_i of 12.2, 211 for W_i of 12.9

Circuit feed size:

mm	13.4	6.7	3.35	1.68	0.85	0.42
wt. % passing	100	65.0	41.7	30.2	23.1	18.2

Circuit product size:

mm	0.25	0.15	0.10	0.075	0.065	0.050
wt. % passing	100	97.5	93.0	85.0	80.0	63.0

Hydrocyclone: one 50.8-cm diameter to be used
VF = 16.5 cm
$Spig$ = 10.2 cm

No data are available about the hydrocyclone characteristics of the ore, but because it is a siliceous ore it will be assumed that these will be the same as other siliceous ores.

A preliminary set of simulations was run to investigate the behaviour of the harder and softer ores at the two power consumptions. Results are given in Table 8-X (40% by weight solids in the overflow).

TABLE 8-X

Comminution of ores at different hardness at two power consumptions: preliminary simulations

W_i	Power (kW)	% − 200 mesh (in OF*)	Circ. load (t/h)	% Solids (in UF)
12.9	760	82.4	93.6	75.5
12.2	760	84.1	89.7	74.7
12.9	720	80.5	97.3	76.2
12.2	720	82.4	93.6	75.5

* The required value is 85.0%

A further set of simulations were then run to investigate the behaviour of the harder ore at both 720 and 760 kW and at increasing water additions to the cyclone feed pump, that is, at decreasing solids contents of the cyclone overflow. The results are given in Table 8-XI.

The conclusion is that even with the harder ore 720 kW available for grinding will be adequate provided that the percent solids by weight in the cyclone

TABLE 8-XI

Comminution of the harder ore at different operating conditions: detailed calculations

W_i	Power (kW)	Cyclone overflow		Cyclone feed (t/h)		Cyclone U/F	Cyclone
		% solids	− 200 mesh	ore	water	(% solids)	(Press. kpa)
12.9	760	40	82.4	140.6	100.9	75.5	60.7
12.9	760	35	87.8	147.3	116.1	77.7	75.7
12.9	760	30	92.9	158.0	136.5	80.5	97.3
12.9	720	40	80.5	144.3	100.9	76.2	62.1
12.9	720	35	86.1	151.1	116.1	78.3	76.6
12.9	720	30	91.6	161.5	136.5	81.0	98.9

overflow does not exceed 35. Both mill and cyclone will handle the circulating load in a satisfactory manner and there is an ample margin of safety.

8.3 CRUSHING CIRCUITS

Procedures for the design of crushing and screening plants using simulation may be based on:

(1) large data banks in which performance characteristics, specifications and costs of equipment are stored;

(2) accurate mathematical models of the unit processes developed from plant data.

The first approach has been adopted by Gurun (1972), and Canalog and Geiger (1973), who used stored performance data supplemented by simple empirical relationships to construct models of unit processes and complete circuits. Generally, the models do not take account of changes in the performance of the machines under different loading and operating conditions. The methods do not allow for design of the plant for the highest possible metallurgical efficiency and do not provide data for consideration of future expansion.

The second approach is limited at present to the expansion of an existing crushing facility or the design of a new plant processing similar material from the same ore body. It is desirable that the parameters in these mathematical models can be estimated from laboratory or pilot-scale data but at present it is not possible for this to be done with sufficient accuracy.

An example of circuit design by simulation will be discussed which is based on models developed using data from full-scale operations.

A crushing plant is to be designed to produce up to 1800 t/h of ore containing not more than 25% + 8 mm material. It is to be a single-stage crushing and screening operation, following the flow sheet given in Fig. 8-3, and

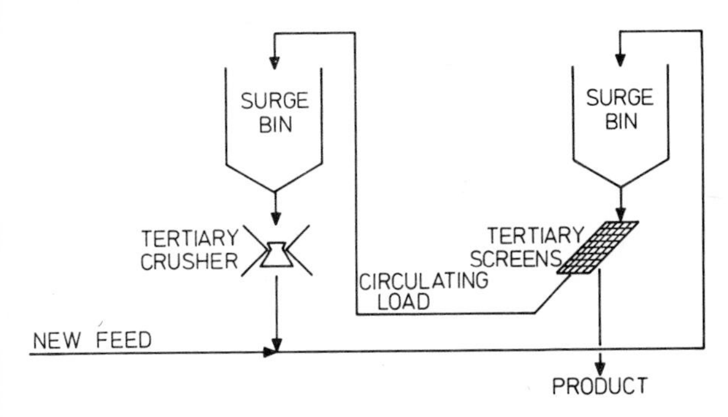

Fig. 8-3. Flow diagram for the proposed crushing plant.

data are available from an existing 200 t/h crusher-screen circuit for the determination of parameters in the models of crushing and screening. For the crusher model a modification of the breakage function discussed in sub-section 4.1.2 has been found to be more accurate in the present case than the original function. The modified breakage function is:

$$B(x, y) = (1 - \alpha' \cdot y^{b-c}) \cdot (x/y)^{u'} + \alpha' \cdot x^b \cdot y^{-c}$$

where α defines the production of relatively coarse particles and b and c are parameters relating to the size-dependent form required to produce fracture. In this model the first term describes the production of relatively coarse particles and the second term describes the production of fines. The classification function used is the same as that described in sub-section 4.1.3. The screen model which has been used incorporates the effects of load which have been mentioned in section 6.2. The values of the breakage and classification parameters used in the simulation are as follows: $u' = 4.6$; $b = 0.64$; $k_1 = 0.423 \cdot g$ (in mm); $k_2 = 22.79 + 1.13 \cdot g$ (in mm). The parameter α' which can vary from 0 to 1.0 was predicted from feed t/h, closed-side setting and feed sizing; c was linearly dependent upon feed tonnage and was generally in the range 0.3—0.5.

In the simulation the discharge from the tertiary crushers is initially set at zero and the units around the closed loop are sequentially simulated following the direction of ore movement. This procedure is continued on an iterative basis until the screen feed sizing converges and mass balance conditions are satisfied. The tonnage of ore in each stream was scaled in accordance with the number of crushers and screens used in the simulation.

The simulation study was carried out on a circuit in which seven short-head 2134-mm hydrocone crushers were used. However, the method can be applied to any desired configuration or number of machines. The example shows the data which can be obtained to allow the determination of: (1) the number of 2.4 m by 6.1 m single deck screens which are necessary to produce the required product tonnage; (2) the screen aperture necessary to achieve the required product sizing; (3) the power requirements and

TABLE 8-XII

Range of new feed size distributions (% wt. passing) expected in the tertiary crushing plant

Sieve size (μm)	% passing	
	distribution (i)	distribution (ii)
64000	100	100
45250	93.0	100
32000	72.8	87.2
22630	46.8	66.9
16000	25.8	45.5
8000	12.8	24.5
4000	9.2	16.5
2000	7.0	11.8
1000	5.6	8.9
125	2.7	3.9

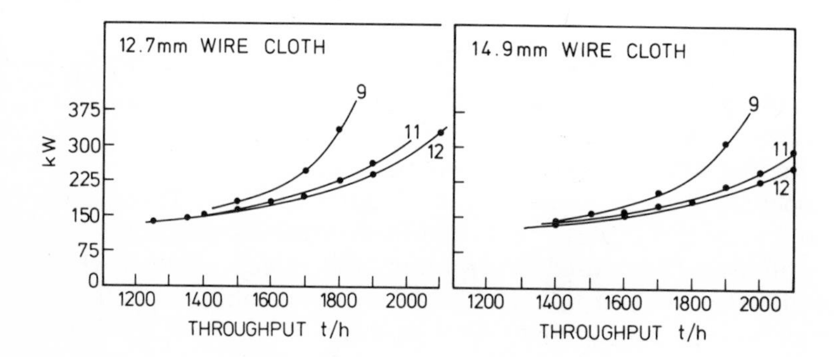

Fig. 8-4. The dependence of power draw upon throughput with 9, 11 and 12 screens—simulated performance.

optimum closed side setting for the crushers; and (4) the conveyor belt capacity required.

Simulation studies were carried out using 9, 11 and 12 tertiary screens fitted with 12.7 or 14.9 mm square aperture wire decks and tertiary crusher closed-side settings of 4.76, 6.35 and 9.53 mm. It was anticipated that the sizing of the new feed to the tertiary circuit would be within the range defined by distributions (i) and (ii) shown in Table 8-XII and the results are given for these two distributions.

The dependence of individual crusher power draw upon throughput of the plant for changes in screen aperture and number of screens is shown in Fig. 8-4. It can be seen that the throughput which can be achieved at a particular power draw increases with the number of screens. However, the additional throughput available per additional screen decreases significantly for 11 or more screens. The number of screens which are required to produce the desired throughput of 1800 t/h is dependent upon both aperture size and the crusher power draw. A power draw up to 260 kW can be achieved in this size of crusher and it is evident that 9 screens will not provide the required throughput at this power draw, the reason being that the crusher is overloaded with fine particles due to screening inefficiency. The installation of 11 screens will give the required throughput provided crusher power level is maintained above 188 kW and there is only a small predicted increase in throughput achieved at this power level by the installation of a twelfth screen. Further simulation results presented are for 11 screens.

The dependence of power draw upon throughput at different closed-side settings is shown in Fig. 8-5. The predicted power required at low throughput decreases as the closed side setting is increased. However, this dependence is reversed at higher throughput and the greatest power is required at the largest closed side setting. This is due to a large increase in circulating load at coarse closed side settings. At a power draw of 260 kW the required throughput cannot be obtained for all of the anticipated conditions using a closed side setting of 9.53 mm. The addition of a twelfth screen when using this closed side setting just fails to provide the additional throughput required.

The dependence of the percentage of + 8-mm material in the circuit product upon throughput is shown in Fig. 8-6. Contours of equal predicted power are also indicated. It can be seen that the requirement for less than 25% + 8-mm material in the product can only be satisfied when using the 14.9-mm aperture screen if the power draw is maintained above 300 kW,

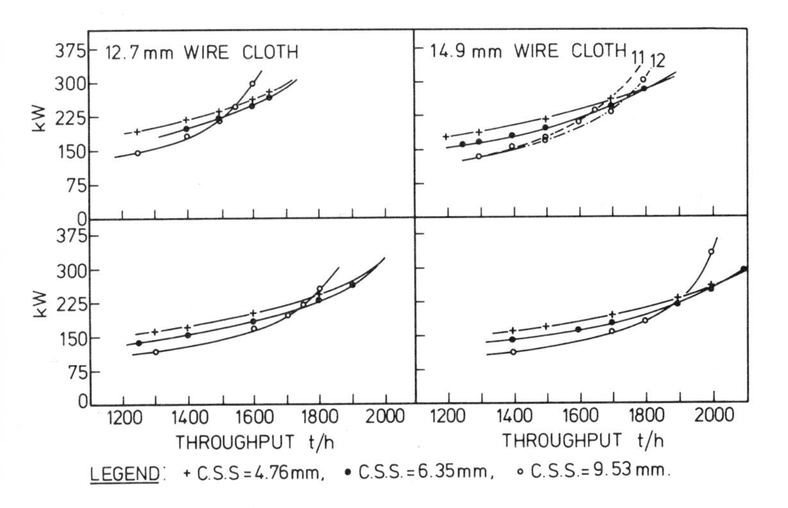

Fig. 8-5. The dependence of power draw upon throughput with change in screen aperture size and closed-side setting of crusher—simulated performance. Upper graphs refer to coarse feed and lower graphs to fine feed.

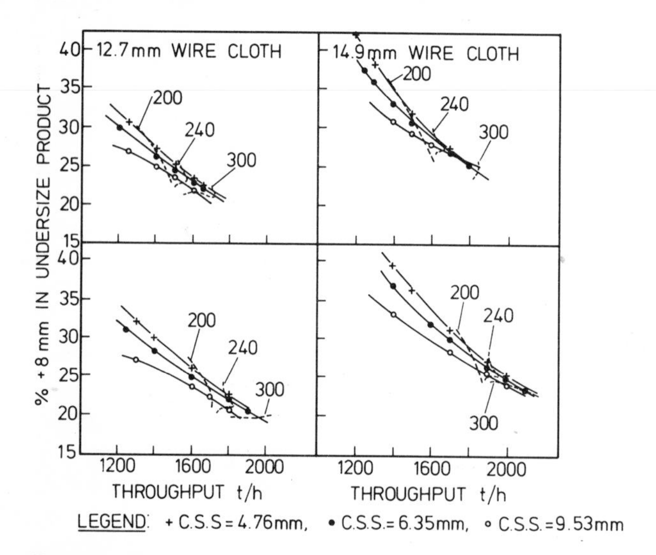

Fig. 8-6. The dependence of the percentage of + 8 mm-material in the circuit product on throughput with change in screen aperture size and closed-side setting of crusher—simulated performance. Upper graphs refer to coarse feed and lower graphs to fine feed. Dotted lines show contours of equal power.

whereas with the 12.7-mm aperture the condition is satisfied at power draws greater than about 224 kW. The decrease in the amount of + 8-mm material with throughput is due to decreased screening efficiency at higher screen feed rates. This means that proportionately more of the coarser fractions in the potential screen undersize report to the screen oversize. For both screen aperture sizes the amount of + 8-mm material increases as closed side setting is decreased at the same throughput. This occurs largely because circulating loads and screen feed rates are lower at smaller closed side settings. The effect of closed side setting upon the amount of 8-mm material is not as large as that of throughput (screen feed rate) and the choice of a closed side setting of 6.35 mm is satisfactory as regards product sizing.

The dependence of circulating load upon throughput is shown in Fig. 8-7. It can be seen that at a closed side setting of 9.53 mm the circulating load increases much more rapidly with throughput than it does at the smaller closed side settings. The co-ordinates of circulating load and new feed tonnage at which a limiting power draw of 260 kW is reached are shown in this figure. It is evident that maximum throughput is not achieved at maximum circulating load. For each combination of screen aperture and feed sizing which was studied maximum throughput at this power limit was achieved using a closed-side setting of 6.35 mm.

In summary, the simulation studies have shown that the production requirements of the plant are satisfied if the following equipment and operating conditions are provided:

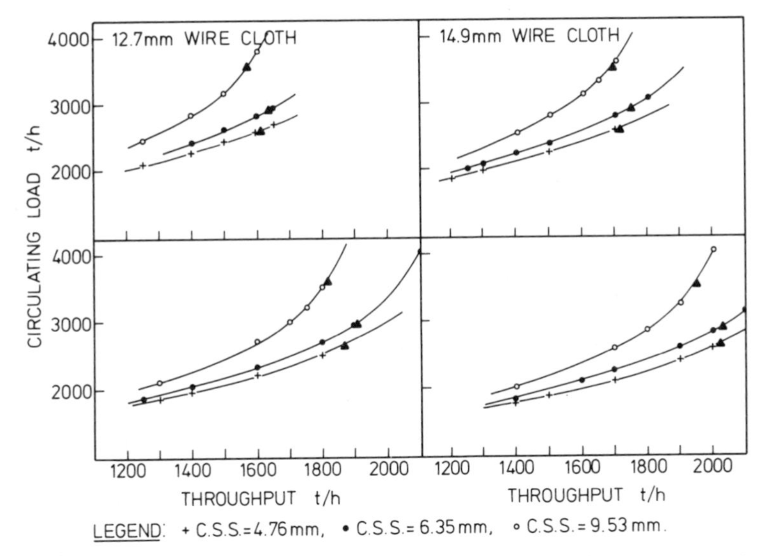

Fig. 8-7. The dependence of circulating load upon throughput with change in screen aperture size and closed side setting of crusher—simulated performance. Upper graphs refer to coarse feed and lower graphs to fine feed.

(1) the installation of eleven 12.7-mm aperture screens for the following reasons: (a) 12.7-mm aperture is required to achieve the desired product quality, and (b) less than eleven screens does not provide sufficient throughput and additional screens provide only a marginal increase in throughput;

(2) the optimum closed side setting is 6.35 mm as: (a) this provides the required throughput at the minimum circulating load and at acceptable power requirements (that is, about 224 kW) and (b) product sizing is satisfactory;

(3) available crusher power should not be less than 260 kW as this power level provides the required throughput at the desired sizing.

The mass flow conditions and individual machine feed rates are given for both screen apertures at different feed sizings in Table 8-XIII for a power draw of 260 kW at a closed side setting of 6.35 mm.

8.4 CLASSIFICATION CIRCUITS

In the grinding circuits discussed in sub-section 8.2.2 and 8.2.3, classification was by single-stage hydrocyclones. Entrainment of fine particles in the coarse product always occurs with hydrocyclones and this may cause problems in later concentration and solid—liquid separation stages due to further breakage of these fine particles.

If hydrocyclones are used in open circuit, for instance in the production of hydraulic underground fill from mill tailings, it is important that the

TABLE 8-XIII

Operating conditions for crusher circuit at maximum power draw and a fixed closed-side setting of the crusher: simulated results

Screen aperture	Feed sizing	New feed (t/h)	Circ. load (t/h)	Tonnes/ crusher	New + circ. load	Tonnes/ screen
12.7	(i)	1640	2925	418	4565	415
	(ii)	1905	2975	425	4880	444
14.9	(i)	1750	2925	418	4675	425
	(ii)	2030	2950	421	4980	453

proportion of the fine particles in the coarse product which is discharged into the worked-out areas should not exceed some critical value. Otherwise, the percolation rate of the water through the fill will be slow and the hold up of water in the stop will be excessive with very serious consequences.

Attempts to minimise the entrainment of fine particles in the coarse product in single-stage units frequently result in coarse particles entering the fine product and this also causes inefficiency. The problem may be reduced by using more than one cyclone and the possible benefits of series cycloning can be investigated by simulation. Two cases will be discussed: (1) series cycloning within a grinding circuit, and (2) series cycloning for the production of a coarse product free of fine particles.

It is important to note, however, that each additional hydrocyclone requires an additional pump and increases both the capital and operating costs of the circuit. In case (1), there are four possible ways of using two

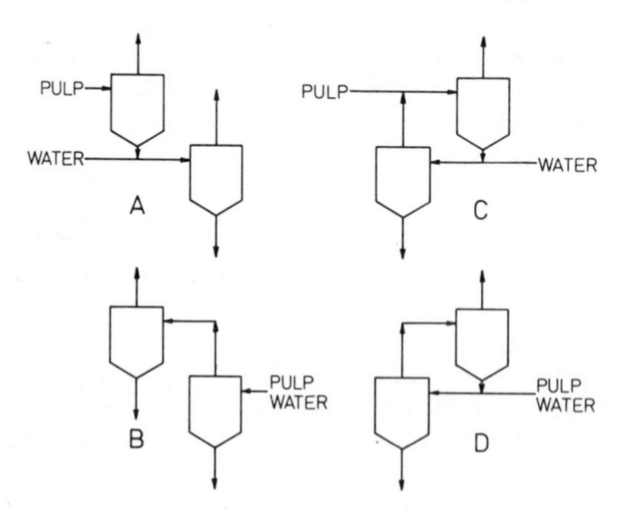

Fig. 8-8. Arrangements of hydrocyclones operated in series.

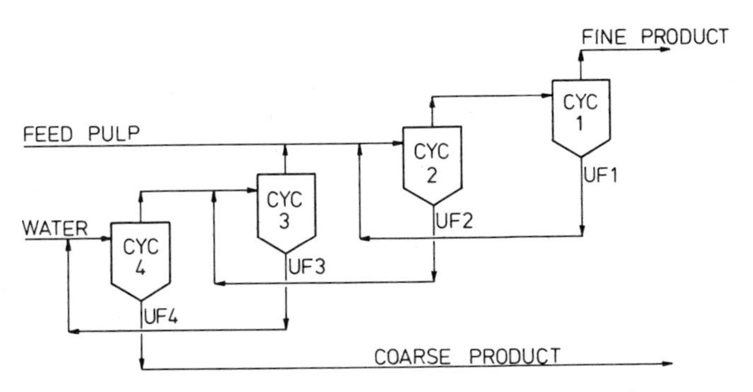

Fig. 8-9. A counter current flow system for hydrocyclones.

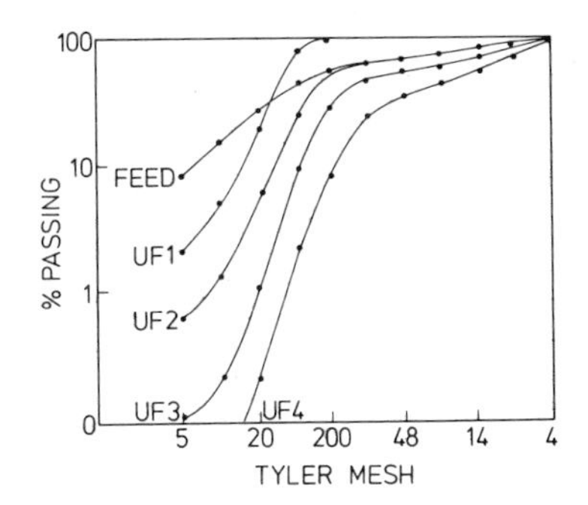

Fig. 8-10. Performance of a counter current hydrocyclone system—simulated.

cyclones in series in a closed grinding circuit and these are shown in Fig. 8-8. Simulations indicated that hydrocyclones used in the Fig. 8-8C formation in closed circuit with a ball mill give a reduction in circulating load and a decrease in the + 150 μ in the grinding circuit product compared with single-stage hydrocyclones.

In case (2), the effect of counter current flow in hydrocyclones in series as shown in Fig. 8-9 may be investigated by simulation. The performance of this system is shown in Fig. 8-10.

The simulation technique may readily be extended to investigate any classification system for which the process models and parameters are known.

CHAPTER 9

Mathematical model of mineral liberation

9.1 INTRODUCTION

One of the most important applications of tumbling mill-size reduction in the minerals industry is in preparing crude ore for concentration into its valuable and waste components. In this case the material being ground does not correspond to either a single homogeneous material or to a mixture of several homogeneous materials. What is usually observed is:

(1) the very coarse particles of ore may be represented as a reasonably homogeneous material which contains inclusions of valuable mineral in a matrix of waste minerals;

(2) the very fine particles of ore may approach a mixture of free valuable mineral and a variety of free waste minerals;

(3) the intermediate sized particles of ore have extreme variations in valuable mineral composition from piece to piece.

It is in these intermediate sizes that size reduction gradually achieves a liberation of the valuable mineral from the waste as particle size becomes smaller.

There is a problem in quantitatively joining the size-reduction and mineral-concentration models because although change in particle size may be adequately described by a size-reduction model (Fig. 9-1), the concentration process depends to a large extent on particle composition which is not included directly in size-reduction models. The usual approach to modelling processes involving the concentration of broken particles has been to arbitrarily assume complete liberation of valuable minerals from waste minerals and to simulate locked middling particles as mixtures of liberated valuable and waste minerals (Fig. 9-2). This has been found to be useful in modelling some flotation processes but where a high proportion of the particles are composite it may give inaccurate interpretations of mechanisms. It does not always provide acceptable quantitative descriptions of the process.

A process which is appropriate for an experimental study of the relationship between size reduction and liberation is the magnetite-taconite process (see Fig. 1-4) because the efficiency of the mineral concentration step is very high even at coarse particle size and all particles containing magnetite can be concentrated into one stream. The usual processing technique is to reject liberated gangue from the process as coarse as possible and save the energy

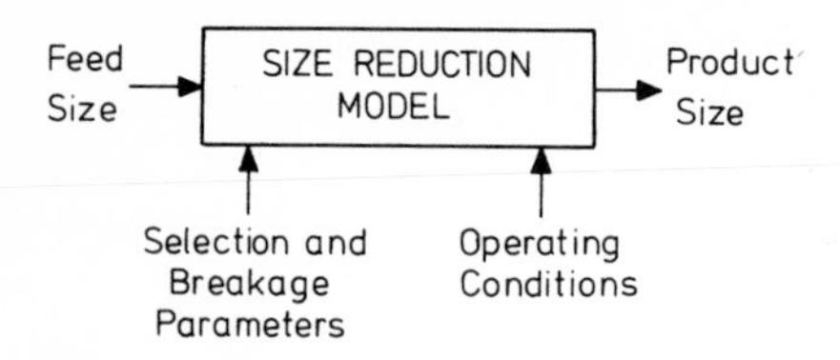

Fig. 9-1. Size-reduction model.

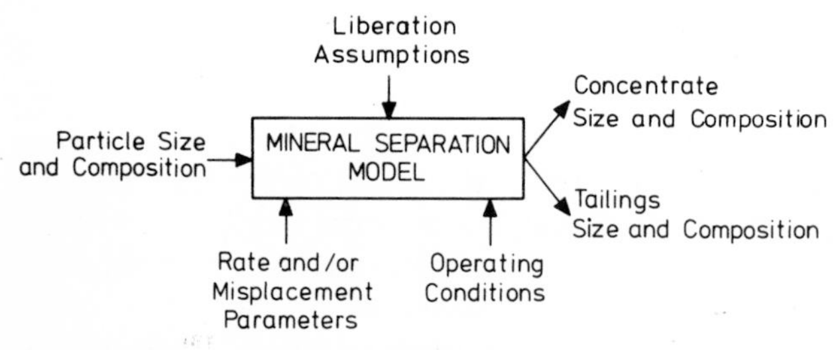

Fig. 9-2. Mineral-separation model.

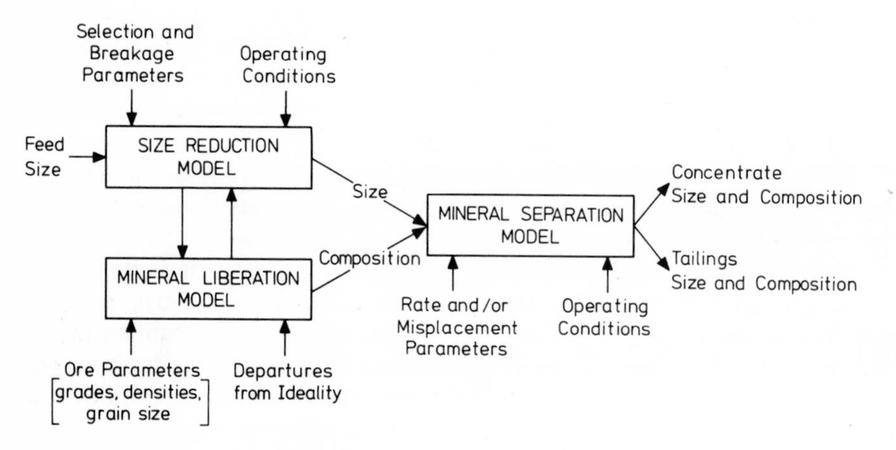

Fig. 9-3. Integration of mineral-liberation model into a size-reduction model to permit effective use of mineral-separation model.

required to reduce its size further. This results in separations taking place on incompletely liberated process streams and concentrate grades of the order of 40—50% valuable mineral (by volume) are the best which could be attained even by perfect separation. Quantitative liberation data should be included in models which attempt to describe processes where mineral liberation is incomplete (Fig. 9-3).

Liberation can be caused by detachment which results from fractures occurring at the mineral grain boundaries, or by shattering of particles which restricts the occurrence of dissimilar mineral locking to a portion of the resulting particles. The detachment effect is dependent on the relative strengths of the mineral grains and the bonds between them, and must be treated empirically for each mineral system. The shattering effect can be treated analytically if certain assumptions are made concerning the shape and geometrical arrangement of the mineral grains.

Gaudin (1939) proposed a model for the fracture of an ideal binary mineral system to explain the shattering effect, and derived equations relating the fraction of each mineral liberated to the ratio of mineral grain size, to particle size, and to the ratio of the volumetric abundance of the two mineral species. The usefulness of that model is restricted by geometric limitations imposed by an assumption regarding mineral-grain arrangement, which results in discrete equations for integer values of the volumetric abundance ratio. The equations also are valid only for grain-size : particle-size ratios greater than unity, and give no information about the distribution of the mineral species in the locked particles.

The model for an ideal binary mineral system proposed by Wiegel and Li (1967) differs from Gaudin's model only in the arrangement of the mineral grains. The model is based on a completely random arrangement of mineral grains, while Gaudin chose to place the grains of the least abundant mineral as far apart as possible. From a practical standpoint, the random model describes more realistically the haphazard appearance of a real mineral system, and it implies the existence of a mineral grain size distribution due to clusters of single grains. From a mathematical standpoint, the random model gives equations which are continuous with respect to volumetric concentration and valid for all values of the size ratio. It also yields information concerning the distribution of mineral species in the locked particles.

The random model will be discussed in section 9.3 and its use illustrated by several numerical examples.

More sophisticated mathematical approaches have been taken by Bodziony (1965) and Andrews and Mika (1975) to the description of mineral liberation, but neither has yet provided a sufficiently practical result to be useful in the prediction or analysis of liberation results. Bodziony's approach makes use of fundamental concepts of integral geometry and attempts to remove the restrictions of grain and particle size and shape imposed in the Gaudin and Wiegel-Li approaches, but requires considerable microscopic measurements as a basis for predictions. The work of Andrews and Mika is aimed at describing liberation in conjunction with size reduction during batch milling, and does not yet suggest a model for liberation itself. The derivation of King (1975) is based on a one-dimensional approach to the probability of liberation based on microscopic measurements in crude ore and does not allow for the three-dimensional complications. Although the unidimensional measurements may be the only ones obtainable from optical microscopic measurements, it is necessary to include three-dimensional effects in the derivation of relationships based upon these measurements.

Several attempts have been made to use release analysis separation data to obtain liberation related parameters. The release analysis technique (for instance, Dell et al., 1972) is based on the hypothesis that high-grade particles tend to float more quickly than low-grade particles, and that successive cleaning stages and selective remixing of concentrates can be used to minimise misplacement of particles based on probability considerations. A standard separation technique is then followed for all samples being compared. In evaluating the results of this procedure it is normal to calculate valuable mineral or metal recovery in the concentrates and concentrate weight. The plot of concentrate recovery versus concentrate weight is then termed the release curve.

The slope of this release curve at a particular point indicates the grade of an infinitesimally small increment of additional concentrate. The grade of this incremental concentrate is referred to as the cut-off grade. Dell (1969) has shown that for several ores this cut-off grade is reasonably well approximated by a linear function of concentrate recovery. It is then suggested that the two parameters defining this linear relationship might be used as measures of recoverable valuable mineral in the ore (intercept) and degree of liberation (slope) for a specific ore and grind.

Only if it can be shown that the release analysis technique does provide a perfect separation, based entirely on the composition of each particle, would the slope of the cut-off—grade-recovery relationship be a measure of liberation. However, since the separation is made with a spectrum of both particle sizes and particle compositions it seems unlikely that the technique could prevent the confounding of size and composition effects in the separation. Although assays on size fractions of the concentrates would indicate if the misplacement due to particle size is important in release analysis separations, these are not available in the literature.

Hall (1971) has also suggested that liberation information can be obtained from release analysis tests, and derived equations for grade-recovery curves and recovery-weight curves as functions of ore grade, grade of the pure mineral recovered and a "liberation coefficient". These equations are derived by forcing a smooth best-fit curve to the data through the point where concentrate grade is equal to the grade of the pure mineral recovered at zero recovery and the point where concentrate grade is equal to crude ore grade at 100% recovery. Although these points are on the grade-recovery relationship it is not necessary that a smooth curve should connect them with the rest of the data. In fact, if there is a significant liberation of either valuable mineral or gangue the curve for a perfect separation would be discontinuous as shown in Fig. 9-4, where the curve segments B, AB and A refer to the recovery of liberated valuable mineral, locked particles and liberated gangue, respectively.

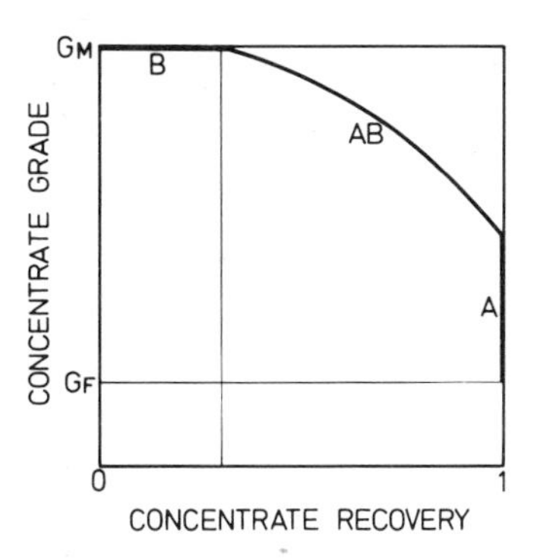

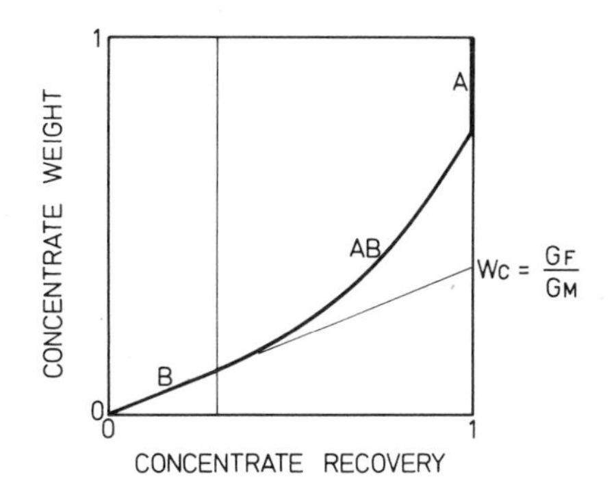

Fig. 9-4. Relationships between concentrate grade, weight and recovery.

9.2 SEPARATION CALCULATIONS USING LIBERATION DATA

Consider a perfect mineral separation taking place on a particular size fraction of a binary ore, in which all particles containing valuable mineral (B) would appear in the concentrate, and all particles of liberated gangue would appear in the tailing. The following relationships for concentrate grade and weight can be derived based on the assumption that the crude ore is a true physical mixture of the two components. The weight fraction of valuable mineral in the crude ore fed to the separation is:

$$G_F = \frac{V_B \cdot \rho_B}{V_B \cdot \rho_B + V_A \cdot \rho_A} = \frac{V_B \cdot \rho_B}{V_B \cdot \rho_B + (1 - V_B) \cdot \rho_A} \qquad (9\text{-}1)$$

The density of the crude ore is:

$$\rho_F = V_B \cdot \rho_B + V_A \cdot \rho_A = V_B \cdot \rho_B + (1 - V_B) \cdot \rho_A \qquad (9\text{-}2)$$

The volume fraction of the crude ore which enters the concentrate is:

$$V = P_B + P_{AB} = 1 - P_A \qquad (9\text{-}3)$$

The volume fraction of the valuable mineral in the concentrate is:

$$C = V_B / V = V_B / (1 - P_A) \qquad (9\text{-}4)$$

The weight fraction of the valuable mineral in the concentrate is:

$$G_C = \frac{C \cdot \rho_B}{C \cdot \rho_B + (1 - C) \cdot \rho_A} = \frac{1}{1 + [(1 - P_A - V_B)/V_B](\rho_A/\rho_B)} \qquad (9\text{-}5)$$

The density of the concentrate is:

$$\rho_C = C \cdot \rho_B + (1 - C) \cdot \rho_A = \frac{V_B \cdot \rho_B + (1 - P_A - V_B) \cdot \rho_A}{1 - P_A} \qquad (9\text{-}6)$$

The weight fraction of the concentrate is:

$$\frac{W - V \cdot \rho_C}{\rho_F} = \frac{V_B \cdot \rho_B + (1 - P_A - V_B) \cdot \rho_A}{V_B \cdot \rho_B + (1 - V_B) \cdot \rho_A} = \frac{P_A \cdot \rho_A}{V_B \cdot \rho_B + (1 - V_B) \cdot \rho_A} \quad (9\text{-}7)$$

The average composition of the locked particles is:

$$C_{AB} = (V_B - P_B)/P_{AB} \quad (9\text{-}8)$$

These relationships are independent of the liberation model used and could also be reversed in order to calculate liberation parameters from concentration data.

Any model of mineral liberation could be used with these relationships, where V_B is simply a volumetric measure of crude ore head grade and P_A, the fraction of particles which are liberated gangue mineral A, is a function of V_B and the ratio of particle size to mineral grain size, β/α.

This same approach can then be applied to each size fraction i resulting from the breakage of a crude ore. If there is no selective breakage of either component, the head grade $V_{B(i)}$ should be constant for each size fraction of the broken crude ore, and the fraction of liberated gangue $P_{A(i)}$ will change as a function of the size ratio for that fraction $\beta(i)/\alpha$.

9.3 RANDOM LIBERATION MODEL

This discussion is taken from Wiegel and Li (1967). The random model is based on the following assumptions:

(1) the grains of both mineral species are cubic and are of the same uniform size α (Fig. 9-5);

(2) the grains are aligned in the mineral aggregate in a lattice-like arrangement so that grain surfaces form continuous planes;

(3) the grains of the two mineral species are randomly arranged throughout the aggregate;

(4) by size reduction the aggregate is broken into particles of uniform size β by a cubic fracture lattice which is superimposed randomly on the aggregate parallel to the grain lattice (Figs. 9-6 and 9-7).

9.3.1 *Probability of liberated and locked particles — size ratio greater than unity*

Consider an aggregate of cubic grains of the same size α of mineral species A and B, which are aligned and arranged as described by the random model. When the aggregate is subjected to a random fracture-lattice parallel to the grain lattice, producing smaller cubic particles of size β, the average number of particles from each grain will be k^3, where k is the size ratio defined as the ratio of grain size to particle size (α/β). Of this number there are, on the

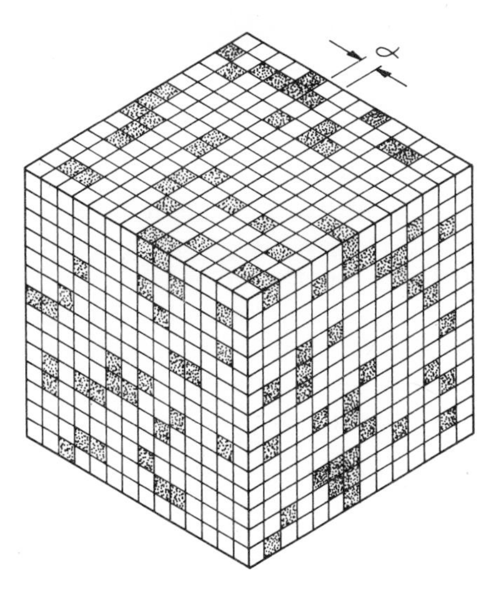

Fig. 9-5. Geometric model of binary mineral aggregate.

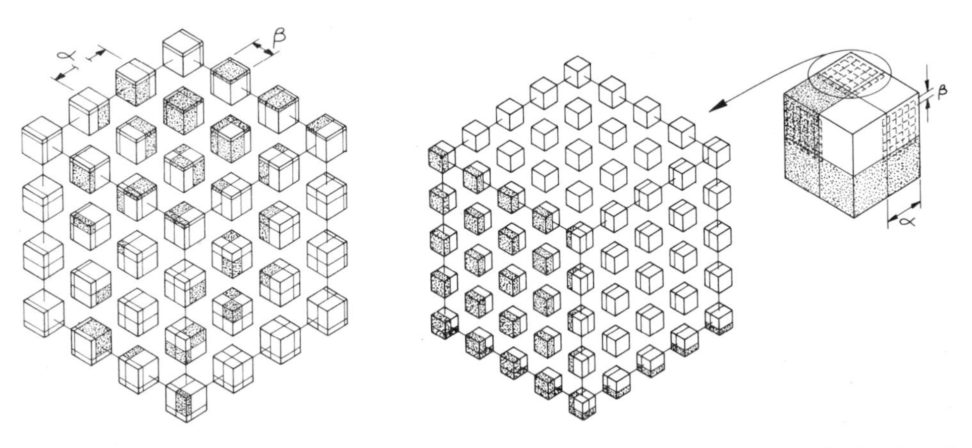

Fig. 9-6. Breakage of binary mineral aggregate into particles of approximately the size of the mineral grains.

Fig. 9-7. Breakage of binary mineral aggregate into particles much smaller than the size of mineral grains.

average, $(k-1)^3$ particles which were contained entirely within a single grain. The remaining particles encompass grain boundaries and are made up of the fragments of a number of grains. The former particles necessarily contain only one mineral species and are termed "liberated". Depending on the surrounding grains being of the same or different mineral species, the latter particles may contain only one mineral species or they may contain both mineral species in which case they are termed "locked".

It is possible to obtain the number of liberated and locked particles resulting from any one grain by considering the various possible arrangements of the two mineral species in the seven adjacent grains in a unit of 2 by 2 by 2 grain cube. On the assumptions that the grains of A and B are randomly arranged, the frequency of each possible arrangement or permutation is given by the binomial probability law. Defining the volumetric abundance ratio n as the ratio of the volume of A to that of B in the aggregate, the probability of each possible arrangement can be expressed as a function of n.

By a straightforward counting procedure, the number of each type of particle was obtained for every possible permutation of the eight grain positions. These numbers were combined with the appropriate probability for each permutation and summed for each type of particle to give the following equations:

$$P_A = \frac{(k-1)^3}{k^3} \cdot \left(\frac{n}{n+1}\right) + \frac{3 \cdot (k-1)^2}{k^3} \cdot \left(\frac{n}{n+1}\right)^2 + \frac{3 \cdot (k-1)}{k^3} \cdot \left(\frac{n}{n+1}\right)^4 + \frac{1}{k^3} \cdot \left(\frac{n}{n+1}\right)^8 \qquad (9\text{-}9)$$

$$P_B = \frac{(k-1)^3}{k^3} \cdot \left(\frac{1}{n+1}\right) + \frac{3 \cdot (k-1)^2}{k^3} \cdot \left(\frac{1}{n+1}\right)^2 + \frac{3 \cdot (k-1)}{k^3} \cdot \left(\frac{1}{n+1}\right)^4 + \frac{1}{k^3} \cdot \left(\frac{1}{n+1}\right)^8 \qquad (9\text{-}10)$$

where P_A is the probability of occurrence of a liberated particle of A, and P_B that of a liberated particle of B. The terms in each equation represent, in order, the probability of a liberated particle being composed of the fragments of one, two, four and eight different grains of the same mineral species, respectively.

The probability of occurrence of a locked particle containing both A and B is given by:

$$\begin{aligned} P_{AB} &= 1 - (P_A + P_B) \\ &= \frac{3 \cdot (k-1)^2}{k^3} \cdot \left[\frac{(n+1)^2 - (n^2+1)}{(n+1)^2}\right] \\ &\quad + \frac{3 \cdot (k-1)}{k^3} \cdot \left[\frac{(n+1)^4 - (n^4+1)}{(n+1)^4}\right] \\ &\quad + \frac{1}{k^3} \cdot \left[\frac{(n+1)^8 - (n^8+1)}{(n+1)^8}\right] \end{aligned} \qquad (9\text{-}11)$$

There is no restriction in these equations concerning the value of n. As n becomes large, the probability of occurrence of a liberated particle of A

increases; as n becomes small, the probability of a liberated particle of B increases. These equations are also applicable to the case of k equal to unity.

9.3.2 *Probability of liberated and locked particles—size ratio less than unity*

When the particle size β is greater than the grain size α, ($k < 1$), a liberated particle must be made up of grains of the same mineral species. Combination of the probability (ϵ terms) for a given number (t terms) of grains being contained in a particle with the binomial probability (n terms) for all grains being of the same mineral species leads to:

$$P_A = (1-\epsilon)^3 \cdot \left(\frac{n}{n+1}\right)^{(t+1)^3} + 3 \cdot \epsilon \cdot (1-\epsilon)^2 \cdot \left(\frac{n}{n+1}\right)^{(t+2)\cdot(t+1)^2} + 3 \cdot \epsilon^2 \cdot (1-\epsilon) \cdot \left(\frac{n}{n+1}\right)^{(t+2)^2\cdot(t+1)} + \epsilon^3 \cdot \left(\frac{n}{n+1}\right)^{(t+2)^3} \quad (9\text{-}12)$$

$$P_B = (1-\epsilon)^3 \cdot \left(\frac{1}{n+1}\right)^{(t+1)^3} + 3 \cdot \epsilon \cdot (1-\epsilon)^2 \cdot \left(\frac{1}{n+1}\right)^{(t+2)\cdot(t+1)^2} + 3 \cdot \epsilon^2 \cdot (1-\epsilon) \cdot \left(\frac{1}{n+1}\right)^{(t+2)^2\cdot(t+1)} + \epsilon^3 \cdot \left(\frac{1}{n+1}\right)^{(t+2)^3} \quad (9\text{-}13)$$

$$P_{AB} = 1-(P_A + P_B) \quad (9\text{-}14)$$

where t is defined as the largest integer contained in $1/k$ and ϵ as the fractional remainder so that:

$$\frac{1}{k} = t + \epsilon \quad (9\text{-}15)$$

The terms in eqs. 9-12 and 9-13 represent, in order, the probability of having a liberated particle made up of the fragments $(t+1)$ grains in each of the three dimensions, $(t+1)$ in two dimensions and $(t+2)$ in the third, $(t+1)$ in one dimension and $(t+2)$ in each of the other two, and $(t+2)$ in each of the three dimensions. Eqs. 9-9, 9-10, 9-12 and 9-13 are derived in Wiegel (1966).

It can be shown that eqs. 9-12 and 9-13 reduce to eqs. 9-9 and 9-10 if $1/k$ is less than unit ($k > 1$) in which case t is zero and ϵ is equal to $1/k$. In this sense, eqs. 9-12 to 9-14 can be regarded as general expressions for all values of k and n.

9.4 FURTHER DISCUSSION OF THE RANDOM LIBERATION MODEL

The volume fractions of the particles in a binary mixture which are liberated minerals, and the weight fractions of the minerals in a concentrate assuming

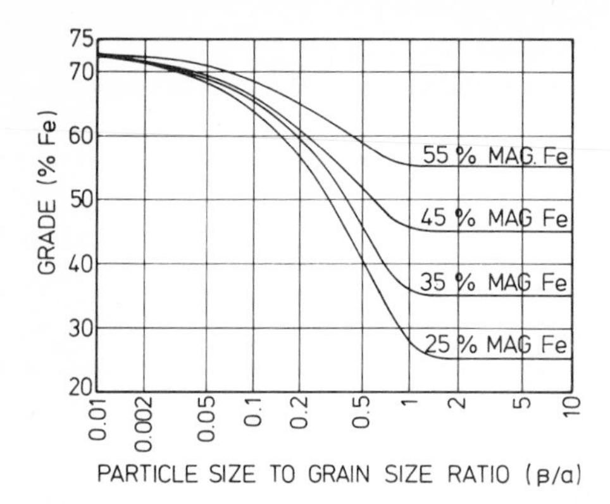

Fig. 9-8. Theoretical liberation curves for magnetite ores of various grades.

perfect separation, can be calculated by the techniques given in sections 9.2 and 9.3. The argument in these sections has been extended to derive a series of theoretical liberation curves and these are shown in Fig. 9-8. It is a property of magnetite that a very efficient laboratory separation can be made on small quantities of size fractions of crude ore. In the Davis tube magnetic separator operating conditions can be chosen so that almost all magnetite is recovered in the concentrate, while essentially all liberated gangue is rejected in the tailing. This result can be verified by making magnetic measurements on Davis tube concentrates and tailings using the SATMAGAN (Saturation Magnetic Analyser).

Davis tube tests were made on size fractions from a group of magnetite samples which were collected from twelve different magnetite concentrators operating in the United States, Canada and Australia (Wiegel, 1975). In this case because there were some variations in head grade with particle size for each sample, a comparison was made between the observed Davis tube concentrate-grade vs. particle-size relationship and that calculated from the random liberation model.

Davis tube concentration data obtained on size fractions of ore can be compared to the theoretical concentration results predicted by the liberation model by plotting Davis tube concentrate grade against log particle size and superimposing the theoretical concentrate-grade vs. log particle-size to grain-size ratio relationship for the appropriate ore grade. The abscissa of the theoretical curve can be shifted to the right or left (because of the log abscissa) to obtain the best visual fit of the data. The particle size corresponding to the location of unity particle size to grain size ratio then represents the "effective mineral grain size" of the ore. This approach has been used in the plots shown in Fig. 9-9 for all twelve commercially ground ore samples. Since the ratio of gangue density to magnetite density for the twelve ores tested is 0.58 ± 0.05, the average value of 0.58 was used to generate the liberation curves. The maximum effect of this deviation in density ratio is less than 1% in concentrate grade.

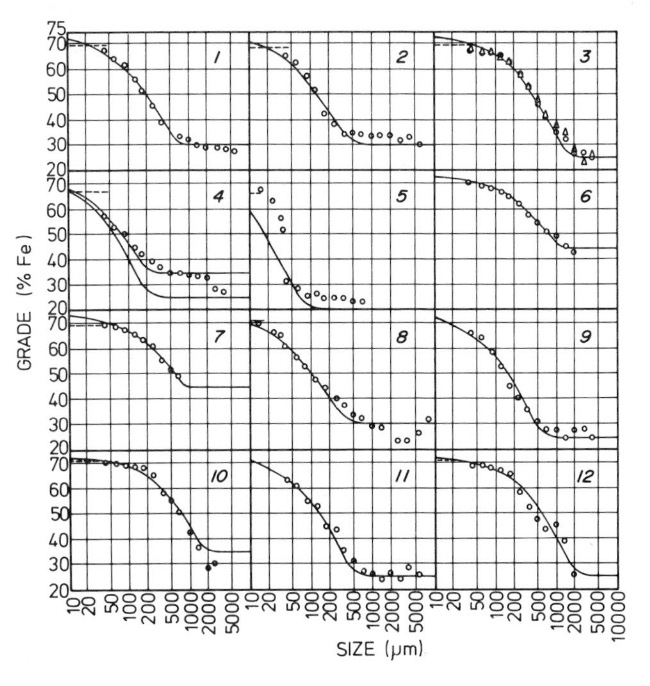

Fig. 9-9. Comparison of theoretical liberation relationship with Davis Tube data for commercial magnetite crude ores.

Of the twelve samples shown in Fig. 9-9, nine indicate reasonable agreement with the theoretical liberation curves. Of the remaining three samples, No. 4 indicates the possibility of several liberation sizes, No. 12 in general follows the theoretical liberation curve but with a wider variation about the curve, and No. 5 indicates a very pronounced tendency to liberate over a very narrow range of particle size, an indication of grain boundary breakage. It is most interesting to note that samples 5 and 12 were the only autogenously ground products tested, and one of these (5) gave very definite indications of preferential grain boundary breakage. All other samples were commercially ground in the more conventional rod or ball mills and only one of these samples did not fit the theoretical liberation curves.

Two of the twelve samples were subjected to laboratory size-reduction tests, with one ore chosen from the coarse grain-size group and the other ore from the medium grain-size group. The four laboratory size-reduction techniques were jaw crushing, roll crushing, dry rod milling and dry pebble milling. These results are presented in Fig. 9-10. In the data for sample 11, grain size 400 μm, jaw crushing, roll crushing and rod milling gave comparable liberation results with pebble milling giving somewhat poorer liberation. Sample-3 data showed that pebble milling gave significantly better liberation than the other three techniques which again gave results comparable within experimental error. The only conclusion which can be reached from these

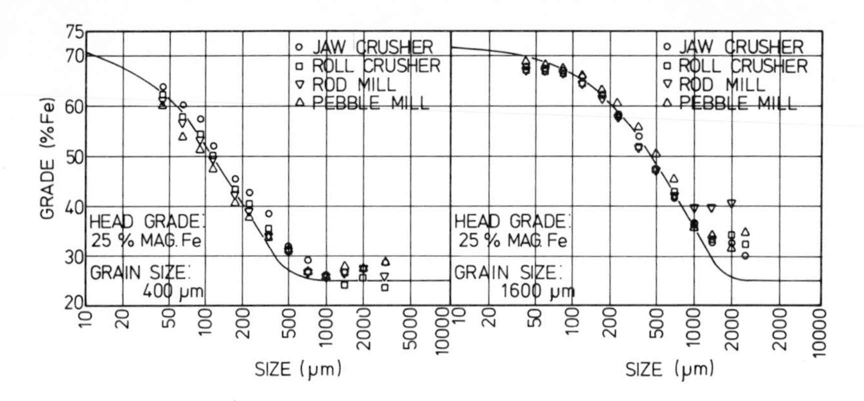

Fig. 9-10. Effect of laboratory size-reduction technique on liberation of magnetite crude ores.

data is that the size-reduction techniques exclusive of pebble milling do not have a significant effect on liberation, while pebble milling may have either a beneficial or detrimental effect depending on the ore treated.

9.5 COMBINATION OF THE LIBERATION AND SIZE-REDUCTION MODELS

In describing the size reduction of a binary mineral system which is undergoing liberation, it is necessary to account for the fact that three types of particles are being broken, the two liberated mineral species and the locked species. When liberated particles break to a finer size fraction, they remain liberated and of the same mineral species. When locked particles break they produce liberated valuable mineral, liberated waste mineral and some particles which remain locked. These progeny locked particles, however, in general have a different average composition from the parent locked particles. Size reduction therefore results in the type of particle identity change shown in Fig. 9-11 as liberation proceeds.

The proportions of locked particles of size j, breaking to size i, which end up as liberated and locked particles are given by directional coefficients which can be calculated from the liberation model as follows:

$$P_{A(i)} = P_{A(j)} + Q_{A(ij)} \cdot P_{AB(j)}$$

$$P_{B(i)} = P_{B(j)} + Q_{B(ij)} \cdot P_{AB(j)}$$

$$P_{AB(i)} = Q_{AB(ij)} \cdot P_{AB(j)}$$

Therefore:

$$Q_{A(ij)} = (P_{A(i)} - P_{A(j)})/P_{AB(j)}$$

$$Q_{B(ij)} = (P_{B(i)} - P_{B(j)})/P_{AB(j)}$$

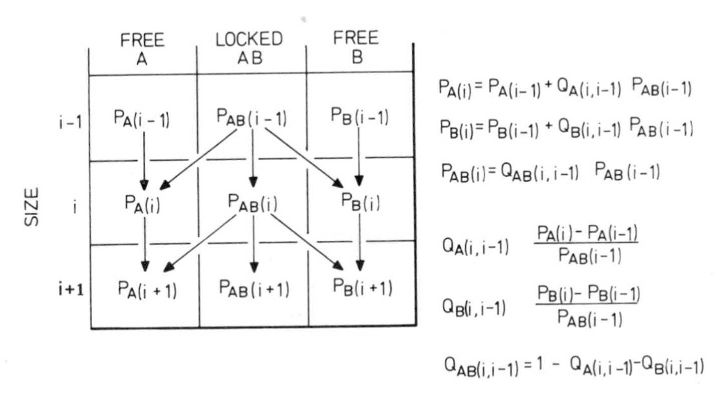

Fig. 9-11. Size-reduction results in particle identity change during liberation.

$$Q_{AB(ij)} = P_{AB(i)}/P_{AB(j)} = 1 - Q_{A(ij)} - Q_{B(ij)}$$

Values of these directional coefficients are therefore dependent on the volumetric grade of the crude ore V_B, the effective mineral grain size α and the particle size from which β_j and into which β_i they are breaking.

If appropriate data can be obtained about an ore it is possible to combine the size-reduction and liberation models and to determine the liberation which occurs as a result of size reduction. This is particularly valuable in determining the effect of size reduction on mineral concentration.

NUMERICAL EXAMPLES

Example 9-1. Liberation as a function of particle size after breakage. An ore containing magnetite (s.g. 5.18) and silica (s.g. 3) is crushed in such a manner that no preferential breakage occurs and every size range in the crusher product contains the same magnetite content. The grain size for both minerals is 1000 μm. What will be the volume fractions of liberated silica and liberated magnetite particles in each size range in the product if the mean particle diameter in the top size range is 8000 μm and the geometric ratio between successive size ranges is 0.5? The iron contents of the crusher feeds are 5, 25, 40 and 60% by weight. Magnetite contains 72 percent by weight of iron.

Solution:
The first step is to calculate the volume fractions of silica and magnetite in the crude ore for each concentration of magnetite, using eq. 9-1.

Eq. 9-12 is then used to calculate the volume fractions of silica and magnetite in each size range in the crusher product. The equation must be solved individually for magnetite and silica in each size range. For each solution t and ϵ may be calculated from β/α, and V_i is known for both minerals from eq. 9-1. The volume fraction of the locked particles in each size range may be obtained by adding the volume fractions of the liberated particles and subtracting from unity. Results are shown in Table 9-I (A refers to silica and B to magnetite).

Example 9-2. Concentrate grade as a function of particle size after breakage and concentration. If it is assumed that the four crusher products from Example 9-1 are concentrated

TABLE 9-I

Volume fractions (calculated) of liberated silica (A) and magnetite (B) particles after breakage of ores with increasing magnetite content

Particle size/ grain size	Iron content of ore (wt. %)							
	5		25		40		60	
	p_A	p_B	p_A	p_B	p_A	p_B	p_A	p_B
8	0	0	0	0	0	0	0	0
4	0.0051	0	0	0	0	0	0	0
2	0.3191	0	0.0007	0	0	0	0	0.0003
1	0.7128	0	0.1167	0	0.0128	0.0010	0	0.0932
0.5	0.8701	0.0058	0.4574	0.0513	0.2428	0.1304	0.0584	0.4262
0.25	0.9213	0.0182	0.6190	0.1231	0.4028	0.2560	0.1367	0.5911
0.125	0.9420	0.0283	0.6942	0.1737	0.4899	0.3332	0.1911	0.6693
0.0625	0.9507	0.0345	0.7300	0.2031	0.5347	0.3754	0.2224	0.7069
0.0313	0.9548	0.0379	0.7474	0.2188	0.3573	0.3974	0.2392	0.7253
0.0156	0.9567	0.0396	0.7560	0.2270	0.5687	0.4086	0.2479	0.7343

Note. Volume fractions can be converted to weight fractions by use of specifc gravities. It will be noted, however, that *volume* fractions must be used in liberation models.

TABLE 9-II

Iron content (calculated) of each size fraction of each concentrate for data given in Example 9-1

Particle size/ grain size	Iron content of ore (wt. %)			
	5	25	40	60
	(% Fe in concentrate from each size fraction)			
8	5.00	25.00	40.00	60.00
4	5.03	25.00	40.00	60.00
2	7.24	25.02	40.00	60.00
1	16.24	27.77	40.40	60.00
0.5	32.19	41.02	49.14	62.37
0.25	47.60	53.02	57.86	65.85
0.125	58.46	61.39	64.04	68.50
0.0625	64.87	66.37	67.76	70.13
0.0313	68.35	69.10	69.81	71.03
0.0156	70.15	70.53	70.89	71.51

in a magnetic separation process and all particles containing magnetite, whether liberated or locked, enter the concentrate, what will be the iron content of each size fraction in each concentrate?

Solution:
Eq. 9-5 may be used in this case, with the weight fraction of the valuable mineral being multiplied by 0.72 to give the weight fraction of the metal. Results are given in Table 9-II.

TABLE 9-III

Calculation of the iron contents of the locked particles in each size fraction for data given in Example 9-1

Fe in ore (wt. %)	: 5		25		40		60	
Magnetite in ore (wt. %)	: 6.94		34.7		55.6		83.3	
Magnetite in ore (wt. %)	: 4.14		23.54		41.99		74.33	
Particle size/ grain size	P_{AB}	C_{AB}	P_{AB}	C_{AB}	P_{AB}	C_{AB}	P_{AB}	C_{AB}
8	1.0	0.0414	1.0	0.2354	1.0	0.4199	1.0	0.7433
4	0.9949	0.0416	1.0	0.2354	1.0	0.4199	1.0	0.7433
2	0.6809	0.0608	0.9993	0.2357	1.0	0.4199	0.9997	0.7432
1	0.2872	0.1442	0.8833	0.2665	0.9862	0.4248	0.9068	0.7170
0.5	0.1241	0.2869	0.4913	0.3747	0.6269	0.4619	0.5154	0.6153
0.25	0.0599	0.3873	0.2579	0.4354	0.3413	0.4805	0.2723	0.5593
0.125	0.0297	0.4411	0.1321	0.4671	0.1769	0.4901	0.1397	0.5300
0.0625	0.0148	0.4662	0.0669	0.4828	0.0899	0.4950	0.0707	0.5151
0.0313	0.0073	0.4795	0.0338	0.4911	0.0453	0.4975	0.0356	0.5076
0.0156	0.0037	0.4865	0.0170	0.4941	0.0228	0.4981	0.0178	0.5038

Note. The locked particles composition tends to 0.50 volume fraction with decreasing size, due to preponderence of two fragment particles.

The calculated weight percents of Fe in the locked particles in each size range are:

Particle size/ grain size	Fe in ore (wt. %): 5 (% Fe in locked particles)	25	40	60
8	5.00	25.00	40.00	60.00
4	5.02	25.00	40.00	60.00
2	7.24	25.02	40.00	60.00
1	16.23	27.77	40.36	58.60
0.5	29.52	36.62	42.99	52.86
0.25	37.61	41.13	44.28	49.44
0.125	41.67	43.37	44.93	47.57
0.0625	43.65	44.49	45.26	46.60
0.0313	44.63	45.04	45.43	46.09
0.0156	45.11	45.32	45.51	45.85

Example 9-3. Composition of locked particles as a function of size. What is the percent by weight of iron in the locked particles in each size fraction in the examples discussed above?

Solution:
The following are calculated in turn:
(1) the volume fraction of locked particles, P_{AB}, in each size fraction where

$P_{AB} = 1 - P_A - P_B$; (2) the average composition of the locked particles C_{AB}, using eq. 9-8; and (3) the percent by weight of iron in the locked particles, using the equation:

$$\% \text{ Fe by weight} = \frac{C_{AB} \cdot \rho_B \cdot \% \text{ Fe in magnetite}}{C_{AB} \cdot \rho_B + (1 - C_{AB}) \cdot \rho_A}$$

The results for the first two stages of the calculation are given in Table 9-III.

CHAPTER 10

Dynamic behaviour of wet grinding circuits

10.1 INTRODUCTION

A study of the dynamic behaviour, or response to a disturbance, of a grinding circuit is important for four reasons:

(1) It gives an insight into the manner in which a change in each variable affects the circuit performance, that is, whether the effect is slow or fast, cumulative over a period of time or smoothed out by a self-stabilising property of the circuit, and into which sections of the circuit are the changes in individual variables felt most.

(2) It enables cause—effect relationships to be identified and a correct selection to be made of the variables which should be used for particular control purposes.

(3) Dynamic data may be used to construct and verify dynamic models of the process which may then be used to design and test control schemes.

(4) Dynamic data which are obtained by studying over a relatively short time interval the response of a process to the deliberate manipulation of a variable give more accurate and reliable data about the performance of a process than do steady-state data. The reason for this is that errors can much more readily be detected with dynamic data.

In an operating circuit the important variables which affect the circuit performance are: (1) ore feed rate; (2) water addition rate at all points of addition; (3) ore feed hardness; and (4) ore feed sizing analysis. These four variables constitute two classes as follows:

(1) ore feed rate and water addition rates are variables which may be deliberately manipulated in order to control the process, and (2) feed hardness and feed sizing analysis are disturbance variables, that is, variables which are not normally controlled and may be regarded as being responsible for random disturbances in the process feed.

For implementation of a satisfactory control system, the former variables may be considered the most important since these are available for manipulation by the control system. Any control scheme must relay on manipulation of these variables to account for disturbances which arise because of a change in the variables of the latter class. It is important to have an accurate knowledge of the manner in which adjustment of these variables affects the behaviour of the process both in the short term (dynamic behaviour) and in the long term (steady-state behaviour). The effects of change in ore and water addition rates are discussed below.

Test work to determine the dynamic effects of ore hardness and feed sizing analysis on circuit performance is difficult to carry out in the industrial environment. A step change in hardness could be achieved by making a transfer from one ore bin to another which had been previously stocked with a different type of ore. However, the bin discharge characteristics would almost certainly produce some unplanned variation in feed sizing analysis, making it difficult to separate the effects of changes in the two variables. Also, it is not convenient to arrange planned changes in feed sizing analysis to investigate the effect of this variable although this would be possible if there were evidence to suggest that this variable had a comparatively significant effect on circuit performance.

Consequently, the discussion below is restricted to the dynamic effects of change in the feed rate and water addition rates. Few detailed dynamic data on industrial grinding circuits are available and extensive reference will be made to two circuits, a rod-mill—ball-mill circuit and a rock-mill—pebble-mill circuit, which have been studied in detail. Other circuits of a similar type may be expected to behave in a similar manner.

10.2 ROD-MILL—BALL-MILL CIRCUIT

The rod-mill—ball-mill circuit which was studied in depth was at Mount Isa Mines Limited, Mount Isa, Qld., and included a rod mill and three overflow ball mills with instrumentation shown in Fig. 10-1. An 8K minicomputer was used for data collection during the dynamic tests.

10.2.1 *Rod mill, response to change in ore feed rate*

The effects on the rod mill discharge of changes in feed rate to the mill are shown in Fig. 10-2.

The response to the increase in feed rate is not a simple reversal of the response to a decrease in feed rate and the classification behaviour of the rod mill produces quite a marked effect in the former that is not apparent in the latter. The increase in flow rate at the rod-mill discharge corresponding to an increase in feed rate consists of a disproportionate increase in water flow rate rather than solids. This indicates the presence of a classification effect in the rod mill whereby the water and fine solids are displaced preferentially after the step. Further evidence for this classification is found in the rod-mill discharge sizings. A graph of the split at one of the finer sizes, say 417 μm (see Fig. 10-2), shows that the percentage of this material initially increases due to the rapid discharge of water and accompanying fine material from the mill. After approximately five minutes, the product becomes coarser due to

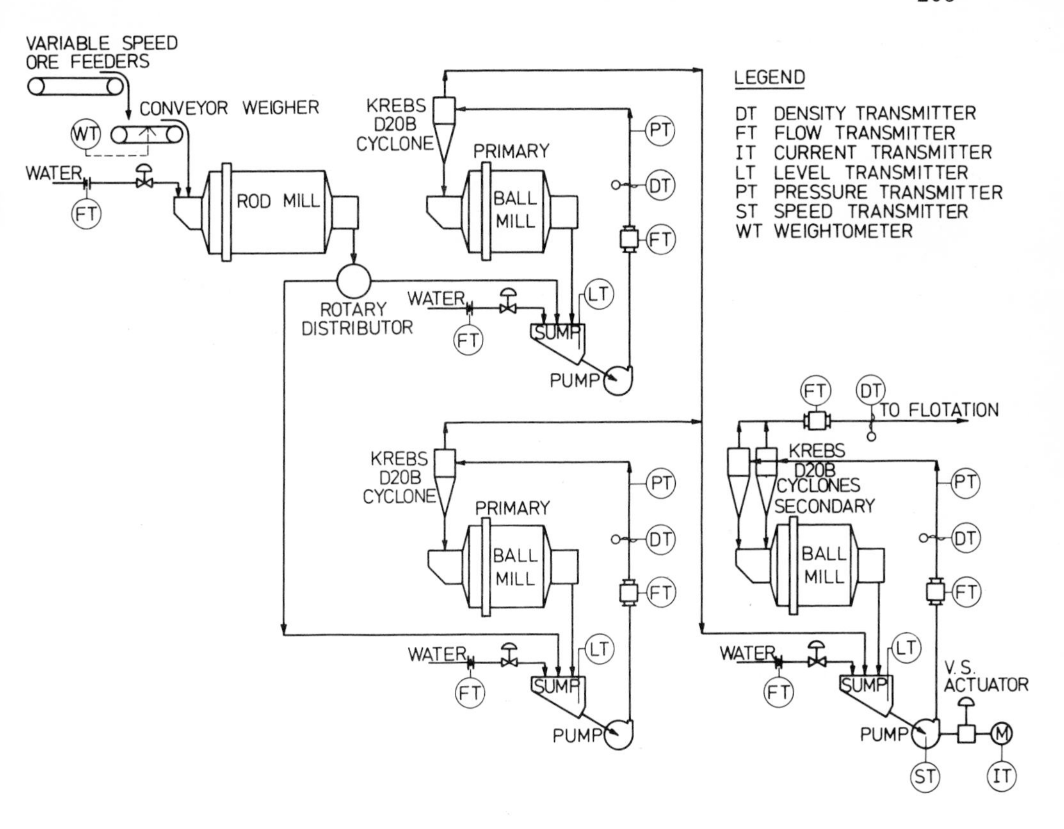

Fig. 10-1. Rod-mill—ball-mill circuit at Mount Isa Mines Limited, Mount Isa, Australia, used for dynamic studies.

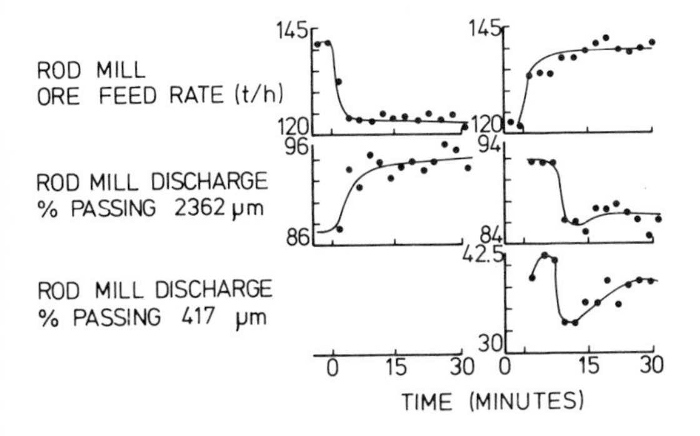

Fig. 10-2. Response of rod-mill discharge to changes in feed rate.

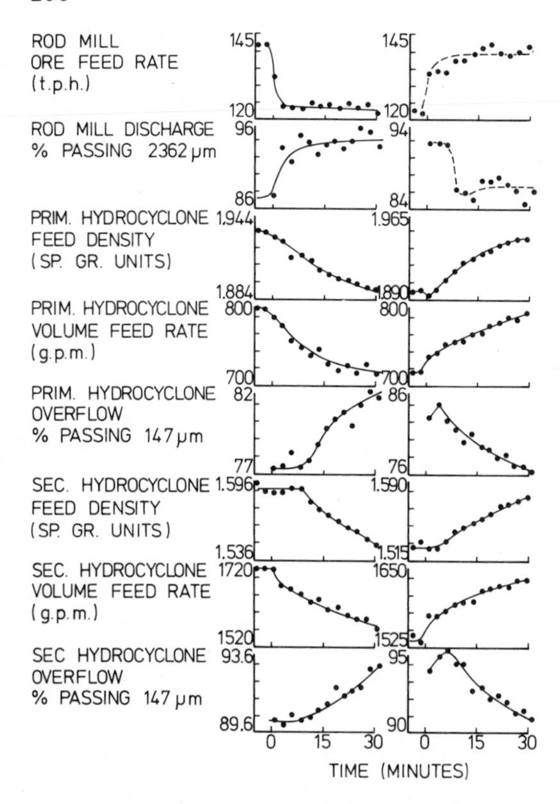

Fig. 10-3. Response of primary and secondary ball-mill circuits to change in feed rate to rod mill.

the depletion of available fines and the transport of the products of breakage at the higher tonnage rate through the mill. This is followed by the steady return of material passing 417 μm to approximately the same value as that observed before the step. By contrast, in the coarse sizes where the effect of the water is negligible, the change in product sizing is permanent.

The effect of the rapid discharge of water from the rod mill is also to produce a distinct step in the mass flow of water to the primary cyclones. This will be discussed further in the next section.

The results of both tests indicate that the rod mill approximates a constant-volume machine. It is also significant that the main change in rod-mill sizing takes of the order of five minutes to appear from the mill indicating that the contents of the mill are not well mixed.

10.2.2 *Overflow ball mill—hydrocyclone, response to change in ore feed rate*

In the circuit which was studied it was not possible to make a sharp change in the feed rate to the ball-mill circuit or to change the feed rate to this circuit without affecting the feed size because the primary ball mill was preceded by the rod mill. However, the change passed through the rod mill fairly quickly and the effect on the ball-mill circuit of a change in feed rate to the rod mill is shown in Fig. 10-3.

It will be seen that at all points in the circuit there is a gradual transition from one operating condition to the other and the maximum effect is obtained at the steady-state conditions. This is in contrast to the effect obtained with water which will be discussed in the next sub-section.

10.2.3 *Overflow ball mill—hydrocyclone, response to change in water addition rate*

The results of the step changes in water addition are shown in Figs. 10-4 and 10-5.

In all tests involving a step in water addition, there was a large immediate response in both cyclone feed density and cyclone overflow sizing analysis. This was followed by a slow return of the density of the circulating load to a point between the steady-state value and the value after the initial response.

The initial response to the step change represents the change in classifier performance with change in feed density and the additional volume flow rate. The continuing response then represents the combined effects of change in mill and classifier performance as the circulating load changes. It is difficult to assess the individual contributions of changes in volume flow and density to the initial step response of the cyclone, but a detailed analysis of the data of Rao (Draper et al., 1969) tends to indicate that the density effect is the more significant of the two.

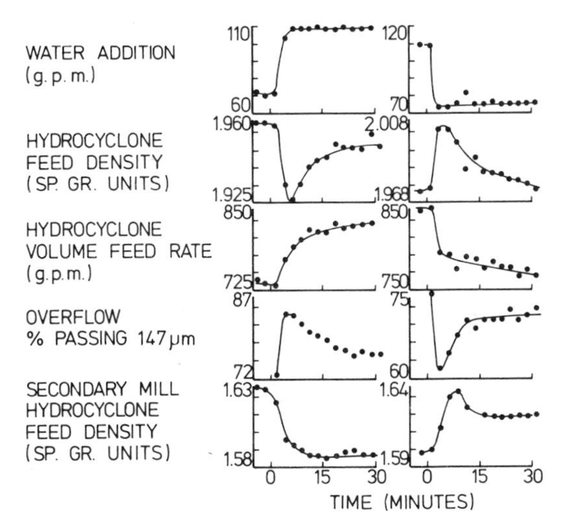

Fig. 10-4. Response of primary circuit to change in water addition.

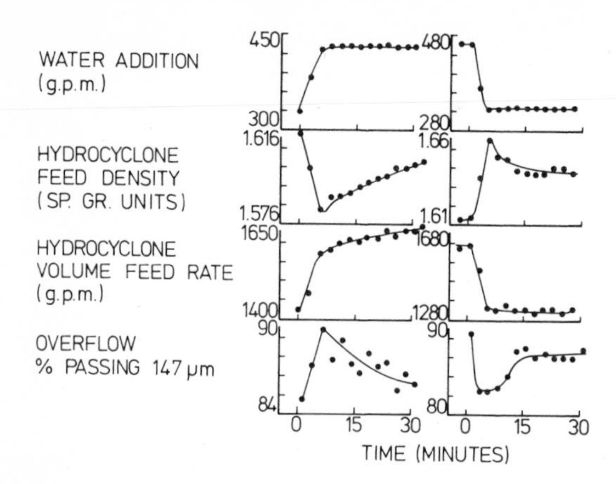

Fig. 10-5. Response of secondary circuit to change in water addition.

The relatively large initial response of the primary cyclone for a unit change in water addition as compared with the secondary cyclone is very important from the aspect of circuit control. The necessity of maintaining careful control of primary-circuit volume feed rate and density is clearly demonstrated. Any change in the sump water addition will produce an effect similar to that shown in Figs. 10-4 and 10-5. Change in volume flow rate of the rod-mill discharge will have a twofold effect. Firstly, the performance will be affected by any change in feed rate, and secondly, provided the circuit sump water addition remains constant, the net rate of dilution at the classifier feed pump will vary with the changes in rod-mill discharge flow rate. The effect of this is to vary both classifier feed density and flow rate, with consequent variation of cyclone performance.

The results indicate that the ball mill provides very little damping of volume flow rate changes in the circuit. This is shown by the fact that the volume flow rate of the circulating load in the secondary circuit changes almost immediately by a considerably greater amount than the size of the step change. The effect is less pronounced in the primary circuit presumably because the pulp has a much coarser size distribution and a higher density, which do not permit it to flow as readily. The fact that the mill does not damp out volume flow rate changes very satisfactorily indicates that the effect of feed rate variation tends to be amplified in the circulating load. A surging fixed speed pump unit has a similar effect to large feed rate variation.

These results confirm the importance of : (1) the use of the variable speed pump unit, and (2) the use of a level control system in which the sump level provides a remote set point for the classifier feed volume flow rate controller.

This system of pump speed control allows the sump level to vary over a considerable range while maintaining the classifier feed rate relatively constant. This effectively eliminates the short-term flow rate variations by use of the capacity of the large sump.

Probably an even more important consideration is the effect on the secondary circuit of the relatively small change in water addition at the primary circuit pump sump. There is a large and almost immediate effect on the density and sizing analysis of the feed to the cyclones in the secondary mill circuit. The magnitude of the change in feed density at the secondary mill for a 160 l/m step in water addition to one primary mill circuit is similar to that obtained by a 570 l/m step in water addition to the secondary mill pump sump. The shapes of the response curves are very similar.

As the changes in primary mill water addition not only produced a large change in the secondary mill cyclone feed density but were also responsible for considerable variation in the feed to these cyclones, it is expected that the product from the secondary mill cyclones would undergo a greater initial change in sizing analysis for water added at the primary mill than for the same amount of water added at the secondary mill. When it is considered that this product passes immediately to flotation, the careful control of the primary mill circuits is very important. From the point of view of flotation feed control, the results indicate that quite rapid adjustment of the circuit product can also be achieved by manipulation of water addition at the primary mill pump sumps.

In the case of overflow ball mills, change in solids content of pulp within the mill in the usual operating range does not have a large effect on the mill load or the rate of breakage of the particles within the mill, although an effect does exist. In the case of grate discharge mills, a major effect exists and this is discussed in the next sub-section.

10.2.4 *Grate ball mill—hydrocyclone, response to change in water addition rate*

In general, these circuits respond in the same manner as the overflow ball-mill—hydrocyclone circuits. However, one effect which grate mills have on circuit dynamics and overall performance and which is present only to a minor extent, if at all, in the case of overflow mills is that the grate presents a resistance to flow of pulp through the mill because the apertures are small. When the pulp has a low viscosity it flows through the grates freely but when it is high the pulp accumulates in the mill until there is sufficient "head" to force it through the grates at the same rate at which it enters the mill. Normally, the variable which controls the viscosity is the solids content and, although quantitative data are lacking, it appears that there is a fairly sharp change from free to hindered flow through the grates. The response of a grate-mill—cyclone circuit to an increase from 450—510 l/m in the ball-mill feed water, with the mill discharge water maintained constant at 2180 l/m, is shown in Fig. 10-6.

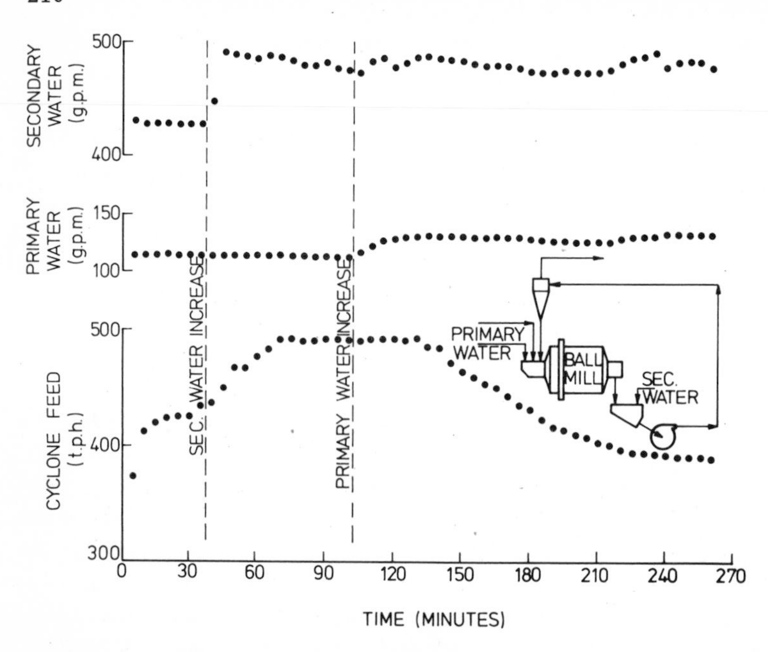

Fig. 10-6. Importance of maintaining solids content of grate ball-mill feed below a critical value—effect of change in primary water.

10.2.5 *Implications for circuit control*

Studies of circuit dynamics emphasise the important difference between the response of the circuit to ore feed rate changes and the response to change in water addition. Changes in ore feed rate initiate a slow progressive change in which the final equilibrium state represents the maximum product response. By contrast, change in classifier water addition gives the maximum response instantaneously, while the equilibrium product response is relatively small.

This leads to the following conclusions about grinding circuit control. (1) Relatively long-term control of ore feed rate will allow satisfactory long-term control of the circuit with respect to change in ore characteristics. (2) Highly effective short-term control of the classifier feed conditions is required in order to maintain the stability of the product sizing analysis.

The studies confirm the importance of using a high-capacity sump and variable-speed drive pump in maintaining effective control of the classifier feed conditions. The contribution to stability of these units on the secondary circuit can be assessed from a comparison of the standard deviations of measurements obtained from the instruments on the primary and secondary cyclone feed streams. Typical measurement values and their standard deviations are shown in Table 10-I.

TABLE 10-I

Comparison of primary and secondary cyclone feed conditions

	Primary circuit		Secondary circuit	
	variable typical value	standard deviation	variable typical value	standard deviation
Density s.g.	1.94	$6.2 \cdot 10^{-3}$	1.61	$2.9 \cdot 10^{-3}$
Flow rate (l/m)	3620	136	7900	45

10.3 ROCK-MILL—PEBBLE-MILL CIRCUIT

The rock-mill—pebble-mill circuit which was studied in depth was at Kambalda Nickel Operations, Kambalda, and is shown in Fig. 10-7. There were three sources of new ore entering the circuit: (1) − 9.5-mm ore to the rock mill, (2) grinding media, 127—203-mm ore, to the rock mill, and (3) grinding media, 76—127-mm ore, to the pebble mill. Data were recorded on charts during the dynamic tests.

The media addition to each mill was controlled by an on—off controller (with a differential gap of 20 kW) according to the mill power draft. The media feeders were activated when the power draft was 10 kW below the set

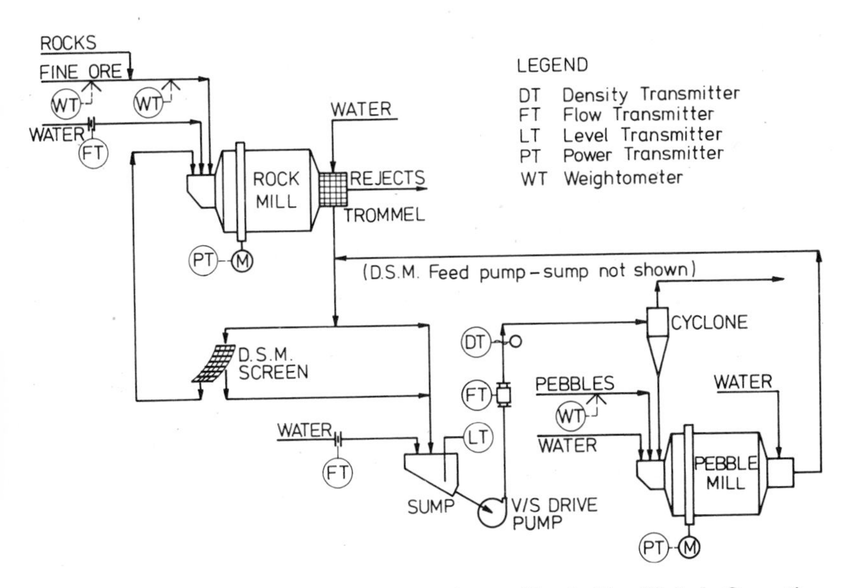

Fig. 10-7. Rock-mill—pebble-mill circuit at Kambalda Nickel Operations, Kambalda, Australia, used for dynamic studies.

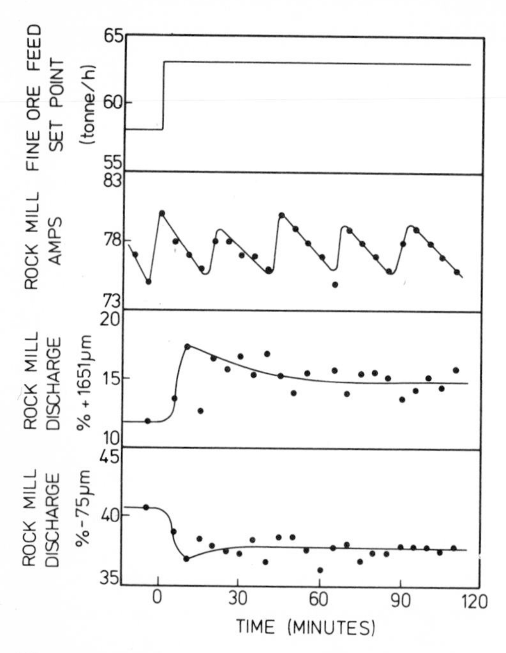

Fig. 10-8. Response of rock mill to change in fine-ore feed rate.

point and deactivated at 10 kW above the set point. Consequently, recordings of the rock- and pebble-mill power drafts had saw-tooth shapes. The relatively sharp power build-up was caused by the addition of the grinding media, while the gradual fall-off in power represented the grinding-out period.

10.3.1 *Rock mill, response to change in fine-ore feed rate*

The results of an increase in fine-ore feed rate are shown in Fig. 10-8. As the rock-mill feed water was under ratio control with the fine-ore feed rate, the step change was effectively a step in volumetric throughput at approximately 74% solids by weight. Taking into account the transportation lags of the feed conveyors, the step change produced an immediate response in the rock-mill discharge sizing, the initial response being greater than the steady-state response. The overall result was that a coarse product was produced. The rock media addition appeared to be unaffected by the disturbance.

10.3.2 *Rock mill, response to change in rock-mill power set point*

The dynamic response of the rock mill to an increase in power set point from 349 to 375 kW is shown in Fig. 10-9. These experimental data contain a considerable degree of scatter. The oscillation in the mill discharge density and the reject rates closely followed the mill stator amp. readings. Consequently,

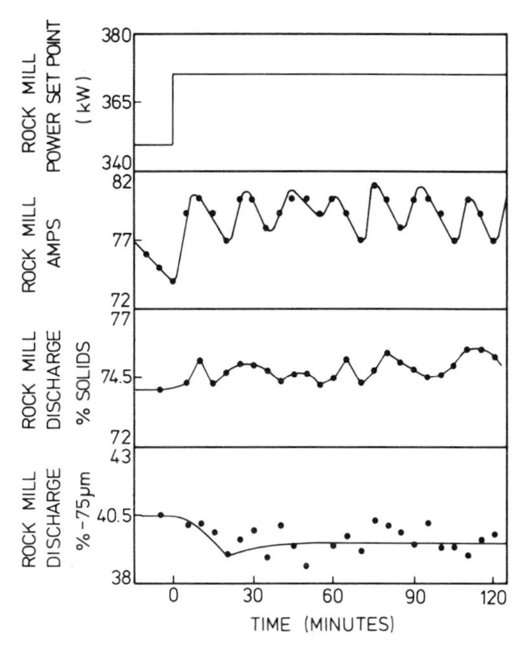

Fig. 10-9. Response of rock mill to change in power set point.

the intermittent feeding of the rock grinding media affected the discharge of solids from the mill. Despite the scatter in the data, the trends are still evident, the mill discharge becoming coarser and the reject rate of pebbles through ports in the discharge grates increasing.

10.3.3 *Pebble mill, response to change in pebble-mill power set point*

The response of the pebble-mill—cyclone system to a step up in the pebble-mill power set point from 480 to 510 kW is shown in Fig. 10-10. The large pebble addition of 3.6 t required to increase the mill power to its new set point caused a decrease in the mill discharge density. The cyclone overflow responded quickly, becoming finer. This large initial reaction then subsided and the overflow gradually became finer. At the new equilibrium conditions, all streams in the circuit showed decreased pulp densities and increased fineness.

10.3.4 *Pebble mill, response to change in water addition rate*

The response of the secondary circuit to an increase in the pebble-mill feed water addition is shown in Fig. 10-11. This sudden increase in water resulted in an immediate rejection of solids from the mill, with the cyclone feed and overflow showing increased densities before becoming more dilute.

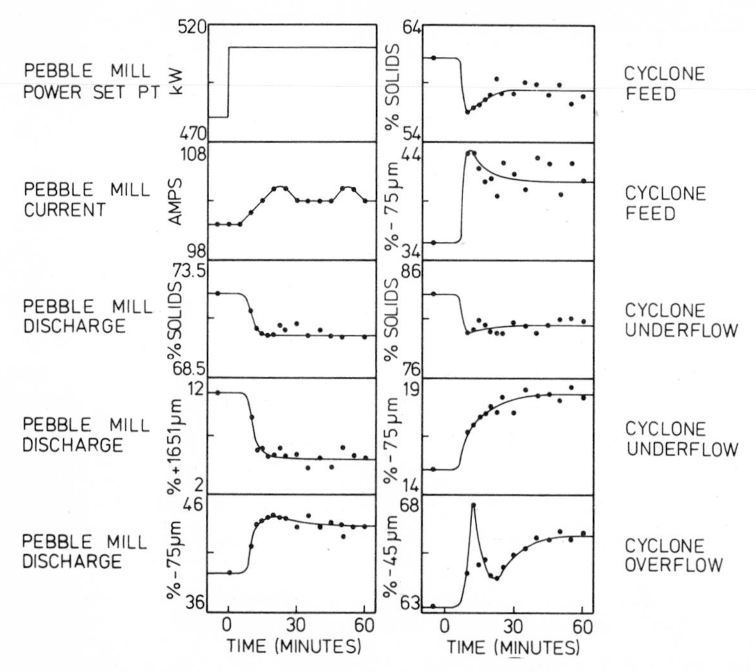

Fig. 10-10. Response of pebble mill to change in pebble mill power set point.

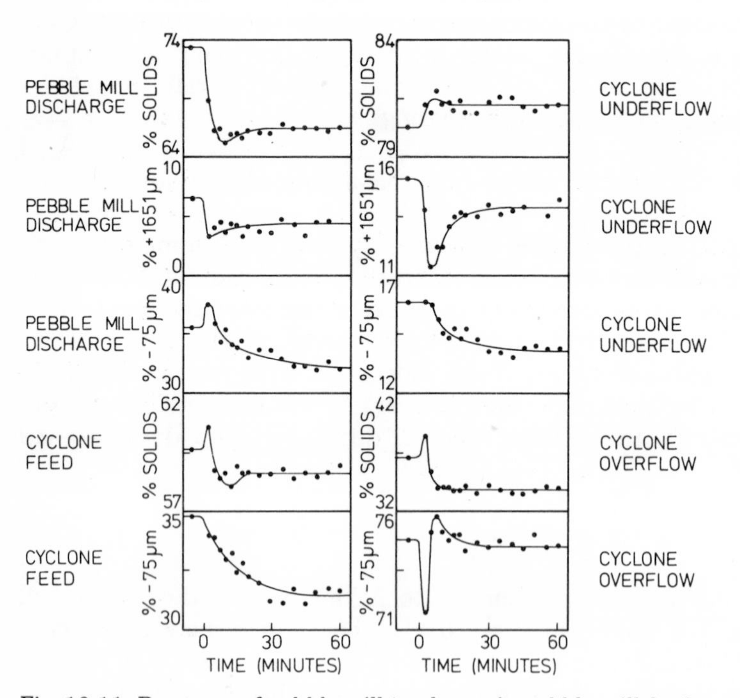

Fig. 10-11. Response of pebble mill to change in pebble mill feed water addition.

This flushing action caused a sharp drop in the pebble-mill power draft, which resulted in a small addition of pebbles to return the power within the operating band of the power controller. The combined effects caused the cyclone overflow to become denser and coarser for the first few minutes and then, as the density settled down, the overflow sizing became quite fine and then progressively coarser. The equilibrium overflow was only marginally coarser than before the step change.

The pebble-mill discharge and cyclone underflow showed large reductions in the percentages of fine material ($-75\,\mu$m) and coarse material ($+1651\,\mu$m). The increased pebble-mill feed water produced a narrower mill discharge sizing distribution and resulted in an increase in the sharpness of the cyclone separation and an increase in the circulating load. The final cyclone overflow sizing was not significantly different from the initial sizing (as measured by the percentage passing $75\,\mu$m), but the cyclosizings indicated a significant reduction in the $-12\,\mu$m (nominal) fraction.

The dynamic response of the secondary circuit to the step change in pebble-mill feed water addition closely followed the response obtained by introducing a similar disturbance in a pilot-scale ball-mill—cyclone circuit (Gault, 1973). In particular, the cyclone overflow responses were identical.

10.3.5 *Implications for circuit control*

Studies of the dynamics of a pebble mill—classifier circuit show that the same type of conclusions apply to this circuit as to a ball-mill—classifier circuit. These are:

(1) relatively long-term control of ore feed rate will allow satisfactory long-term control of the circuit with respect to change in ore characteristics, and

(2) highly effective short-term control of the classifier feed conditions is required in order to maintain the stability of the product sizing analysis.

There is an additional flexibility in control of rock-mill—pebble-mill circuits because the mill power consumptions are controllable over the short term whereas this is not the case with rod-mill—ball-mill circuits.

If the rock- and pebble-mill power set points are kept at constant operating levels circuit control becomes similar to control of a rod-mill—ball-mill circuit. If the requirement is for constant product size at constant fine-ore feed rate for an ore of varying hardness the pebble-mill power set point can be used to a limited extent to control the product size.

CHAPTER 11

Control of wet grinding circuits

General concepts are discussed in this chapter. Applications are discussed in Chapter 13 in which case studies of control systems for several different types of wet grinding circuits are described.

The function of a grinding circuit in a processing plant is to prepare the ore either for concentration by liberating the valuable minerals from the waste minerals or for chemical reaction by exposing the surfaces of the valuable minerals. The term "automatic control" when applied to grinding circuits may have several different meanings depending on the particular process being considered, and it is important that the objective to be achieved by installing an automatic control system should be defined explicitly. The grinding circuit performs one operation only in the sequence of operations involved in processing the ore and the objective must be formulated for the entire system and not for the grinding circuit or any other individual circuit only.

If a circuit is to be controlled automatically, it must be possible to detect changes occurring within the circuit or in the characteristics of the product leaving the circuit, and to compensate for these changes by suitable variation in the controllable variable. The abrasive nature of mineral particles and their tendencies to block sampling lines and form unstable deposits in regions of low fluid velocity has provided serious problems in the development of sensing instruments which are only now being solved. In developing an automatic control system for a grinding circuit the following questions must be considered: (1) what is the objective to be achieved? (2) what are the controllable variables? (3) what sensing instruments may be used to detect changes in circuit operation? and (4) how must the controllable variables be altered to compensate for these changes?

11.1 CONTROL OBJECTIVES

Possible objectives are:

(1) the sizing analysis of the circuit product is to be maintained constant at constant feed rate;

(2) the sizing analysis of the circuit product is to be maintained constant at maximum feed rate; or

(3) both sizing analysis and solids content of the circuit product are to remain constant.

The limitations these different objectives may impose on the manner a control system may operate are apparent; for instance, ore feed rate must be maintained constant in the first case whereas it is the important controllable variable in the second case.

Frequently these objectives must be modified by local conditions. For instance, when the concentration circuit is the limiting circuit in the plant with regard to the mass flow rate of mineral which can efficiently be processed, it may be necessary to adjust the plant feed rate to ensure that the amount of valuable mineral in the feed to the concentration circuit does not exceed this maximum value. In this case, the feed rate to the grinding circuit may be varied independently of the ore hardness yet the control system may still be required to maintain the product size constant or as close to the required value as possible.

It is not uncommon for an objective to be defined which is physically unattainable. For instance, the objective of constant product size and maximum throughput cannot be attained since two constraints are involved and one must be relaxed to achieve the other. Thus if a circuit is operating at maximum capacity and the ore becomes harder, a coarser product size must be accepted until the reduction in feed rate which compensates for the harder ore takes effect. Otherwise, the circuit overloads and spillage occurs. In this case, the definition of the objective should include a statement concerning the extent to which one constraint may be relaxed before relaxing both.

In defining an objective for a control system, the assumption can be made that the more stringent the requirements for the system the more complex and expensive it will be and the larger will be the technical and professional staff required to develop and support it. Control systems may be installed at various levels of complexity, an increase in complexity normally requiring an increase in expenditure and resulting in an improvement in control. If the relationship between incremental expenditure and incremental improvement could be determined, it would be possible to decide the level of control to be installed but this is a difficult problem.

For a wet grinding circuit three levels of control are:

(1) local controllers on all ore and water inputs;

(2) variable speed drives on all classifier feed pumps, these being controlled from sump level detectors; and

(3) complete particle size and circuit control based on a digital computer.

An intermediate level involving particle-size control using an analogue computing relay would now be regarded as out-moded.

The basic equipment requirements for each of the three levels of control applied to a typical 400 t/h grinding circuit consisting of one rod mill and two ball mills are shown in Fig. 11-1. Control system costs, 1976 values, are given in Table 11-I.

TABLE 11-I

Capital costs, 1976 values, for three levels of control for a grinding circuit which includes a rod mill and two ball mills (see Fig. 11-1), installation costs not included

Level 1. Control of ore and water feed rates to set points	
Water control:	$
3 orifice plates	105
3 differential pressure transmitters	2200
3 control valves	3180
3 controllers	2700
Ore control:	
1 weightometer	8000
1 controller	900
1 variable speed drive for conveyor	5000
Miscellaneous:	
1 control panel	2500
Progressive total (approx.)	$25000
Level 2. Additional to level 1—control of cyclone feed pumps to stabilise internal circuit flows:	
2 differential pressure transmitters	2000
2 controllers	1800
2 variable speed drives for cyclone feed pumps	20000
Progressive total (approx.)	$49000
Level 3. Additional to level 2—particle-size and throughput control by digital computer:	
2 radiation density gauges	9000
2 magnetic flow meters	9000
1 digital computer	40000
Progressive total (approx.)	$107000

While it is relatively simple to estimate the costs of various systems, it may be more difficult to evaluate the gains. If the objective of automatic control is to increase the ore treatment rate without deterioration in metallurgical efficiency, this may often be achieved by installation of ore and water controllers, and variable speed drives on the classifier feed pumps. These stabilise the operation of the system and remove the large variations which can be induced by unstable pump operation, and a higher average throughput can be obtained without loss in metallurgical efficiency. It is not difficult to assess economically the value of the control systems in terms of increased throughput. If the objective is to improve the metallurgical performance at the same treatment rate, the problem of economic justification may be more difficult. In the case of a grinding circuit preparing the ore for flotation, this can best be done by considering the flotation results but these are affected by many other variables also. A control system for grinding circuits is difficult to justify on the basis of a statistical improvement in flotation efficiency

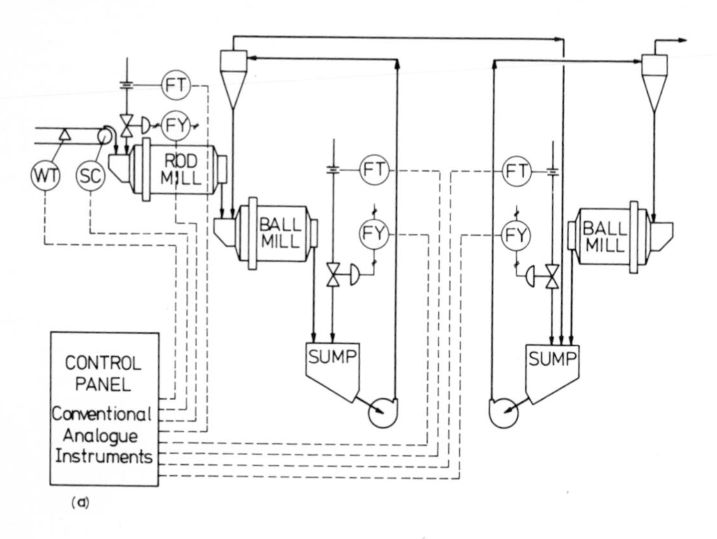

(a)

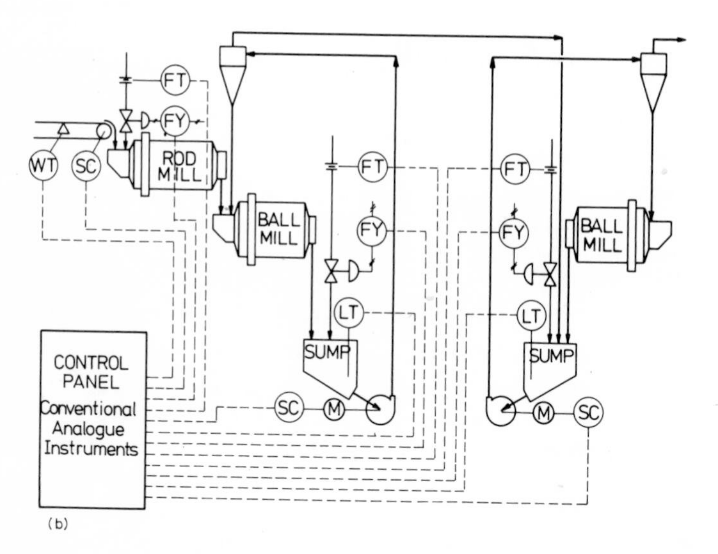

(b)

---

because the data are so difficult to obtain. It is possible that this problem of economic justification will become more common as more control systems are installed. If identical controlled and uncontrolled circuits may be operated in parallel for long periods, the problem may be solved, but this is rarely possible. The question of the cost of evaluating a control system is important as this cost can exceed the cost of the system. There is no simple answer to this.

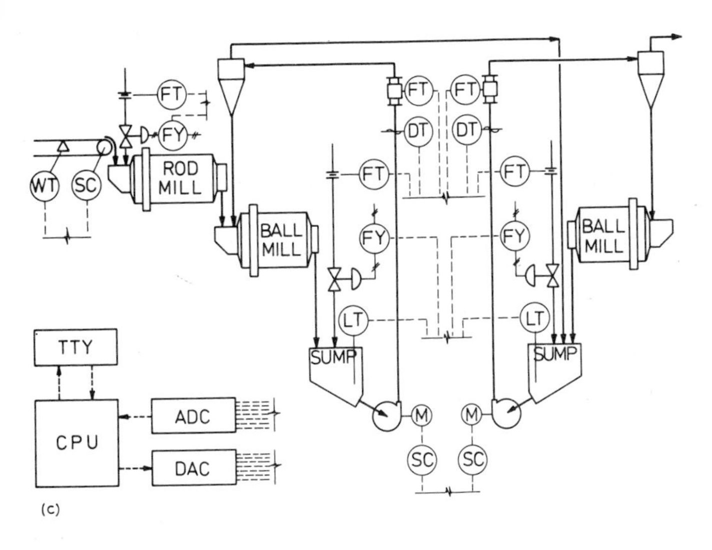

Fig. 11-1. (pp. 220, 221). Instrument requirements for different levels of control for a wet grinding circuit. a. Ore and water control to set points. b. Control of cyclone feed pumps in addition to (a.) c. Total circuit control using a digital computer.

11.2 TYPES OF DISTURBANCES

There are several types of disturbances which may occur in an operating grinding circuit and which may cause irregularities or undesirable changes in the size distributions of the particles in the circuit product. Some of the more important types are:

(1) change in the characteristics of the ore entering the circuit, and in the hardness, size distribution, flow rate or mineral composition;

(2) change in the flow rate of water entering the circuit;

(3) changes in the nature of the cyclone feed pulp, such as intermittent aeration of the pulp due to a surging pump;

(4) changes in the flow rates of pulp within a circuit due to mechanical reasons, such as partial or complete blockage of a launder or a spigot in a hydrocyclone;

(5) changes in the distribution of pulp from a splitter feeding parallel processing units so that the proportion of the total pulp which enters each unit varies in an intermittent and random manner;

(6) long-term changes in circuit performance due to wear in the mill liners, partial blockage of pipe lines by gypsum, etc.

Changes in ore characteristics cause a gradual change in the flow rates and size distributions of the solid particles in all streams in a closed grinding circuit. The full effect of the change is felt when steady state has been achieved.

Change in pulp flow rates, caused by change in water addition or a surging pump or other reasons, causes a rapid and large change in the size distributions of the solid particles in all streams but these are normally dampened out quickly. Continuous disturbances, as in the case of a continuously surging pump, cause continuous large changes.

It is better to prevent disturbances occurring than to compensate for them by a control system once they have occurred, if this is at all possible. Thus, if it is necessary to use pulp splitters these should be of a type which ensures that the split is accurate although these units are much more expensive than fixed splitters. Local control systems designed to maintain ore and water feed rates to the circuit at set points which may be adjusted manually or remotely as required, are most important. Local alarm systems which detect when blockages occur are also valuable.

In theory, an ore blending system of the type shown in Fig. 11-2, to minimise changes in the ore entering the plant, is also advisable. In practice, this system is very expensive and, in the case of sulphides, may cause deterioration in the flotation properties of the minerals due to oxidation.

Fig. 11-2. An ore-blending system designed to smooth out feed fluctuations.

11.3 SENSING TECHNIQUES

A control system has four component groups:

(1) various means of measurement;

(2) a computer to calculate, and infer if necessary by the use of mathematical models, the values of the important variables;

(3) controllers to maintain the desired variables at the required values; and

(4) ancillary controls to maintain desired external conditions, present information to the operators and ensure practical operation.

Mineral processing circuits present a particularly difficult environment for the operation of instruments which are required to give consistent and accurate performance continuously over long periods of time, and this has had a major effect in retarding the development of control systems. It is usually difficult to make measurements on large volume streams, and continuous accurate sampling of streams is a major problem. In continuous plant use, the instruments must be able to resist mechanical shocks, occasional inundation with pulp or water, dusty atmosphere and wide ranges of temperature and humidity since even with the best housekeeping, these conditions will inevitably be encountered. Consequently, suitable instruments are expensive.

Sensing instruments which have been used for the continuous measurement of ore, water or pulp characteristics are listed in Table 11-II.

In any wet grinding circuit control system, the basic requirements are:

(1) measurement and control of all ore and water flow rates to the circuit;

(2) measurement of the pulp level in the sump, so that the sump may be prevented from overflowing or running dry; and

(3) measurement of the circulating load, so that overload may be prevented.

There is a range of instruments which has been successfully used for the first two requirements but only one unit has been found to be generally suitable for the last, namely, a mass flow metering system which incorporates a magnetic flow meter and a gamma density gauge. It may be shown that, if V is the volume flow rate (magnetic flow meter output), D is the specific gravity (gamma density gauge output) of the slurry and S is the specific gravity of the solid particles.

$$M = \frac{K \cdot V \cdot S \cdot (D-1)}{(S-1)} \qquad (11\text{-}1)$$

and:

$$W = \frac{K \cdot V \cdot (S-D)}{(S-1)} \qquad (11\text{-}2)$$

where M and W are the mass flow rates of ore and water in the slurry and K is a scaling factor. The advantage of this system is that it gives a direct measure of the circulating load using instruments which have been found to be accurate and reliable, and the measurement is made on the full stream.

TABLE 11-II

Sensing instruments which have been used in wet-grinding-circuit control systems

Instrument	Purpose
Weighers: mechanical, electrical or nuclear	mass flow rate of dry ore on a conveyor belt
Orifice, venturi meters	volume flow rate of water in a pipe
Magnetic flow meter	volume flow rate of any conducting fluid in a pipe
Gamma nuclear gauge; differential pressure cell; "Halliburton" tube	solids content or specific gravity of a slurry
"Autometrics" particle-size monitor; "RSM-Mintech" particle-size analyser	particle-size monitor
Power meter: autogenous mills	inference of load retained in the mill
Power meter: rake classifiers	inference of mass flow rate of coarse particles
Noise meter: ball mills	inference of load retained in mill
Vacuum gauge: air core of hydrocyclone	inference of solids content of cyclone underflow
Bubble tube	level of fluid in a tank such as pulp in a pump sump

A balance must be kept between instrument accuracy and reliability on one hand and cost on the other. An instrument which is unreliable is of no value in a plant measurement system and is potentially dangerous in a control system. However, a reliable instrument which gives minor inaccuracies over a normal operating range may often be preferable to a very accurate instrument which is much more expensive.

11.4 CONTROL SYSTEMS FOR BALL-MILL—HYDROCYCLONE CIRCUITS

11.4.1 *Physical limitations of circuits*

Any control system for a ball-mill—hydrocyclone circuit must ensure that the circuit is operated at all times within its physical capabilities.

Overloading or blockage in an overflow ball-mill—hydrocyclone circuit may occur:

(1) at the cyclone when the solids content of the cyclone underflow stream exceeds a value which is a function of the ore type and size distribution causing partial or complete blockage at the spigot;

(2) at the ball mill when the mass flow of solids through the mill exceeds a value which is a function of the mill size and speed, solids content of the pulp, ore type and size distribution, causing discharge of the grinding media with the discharging pulp.

In the case of a grate ball mill, overloading in the mill tends to occur when the solids content of pulp entering the mill exceeds a value which is a function of the ore type and particle size, and discharge through the mill grates is restricted.

Many grinding circuits have additional limitations which exist due to local conditions. For instance, the motor on a variable-speed drive pump may not be large enough to handle peaks in the pulp flow entering the sump and the danger of burning out the motor exists if an upper limit of pulp power consumption is not recognised in a control system.

When considering a control system for a circuit, the physical limitations of that circuit must be assessed thoroughly and these must be recognised in the system. The importance of the limiting effect of mill throughput in particular is emphasised by the fact that the efficiency of closed grinding circuits in eliminating "oversize" from the circuit product tends to increase as the load circulating through the mill increases and it is at a maximum just before the mill overload point.

Particular attention must be given to the pump installation. A fixed-speed pump means that the volume of pulp delivered to the cyclone must remain essentially constant irrespective of whether or not the metallurgical requirements require a varying flow. A variable-speed pump means that the pulp flow can be controlled according to metallurgical requirements and a separate control loop can be implemented between the sump level and the pump speed to ensure that the pump can cope with the volume flow requirement. A fixed-speed pump may impose a serious restriction on the efficiency of a control system. Some alternatives which have been proposed to combine the flexibility of a variable-speed system with the lower cost of a fixed-speed system are: (1) use of a control valve on the pump discharge pipe to restrict the flow and maintain the required level in the sump, or (2) recycling portion of the hydrocyclone overflow to maintain the level.

11.4.2 *Control system based on classifier feed density*

One of the early control systems based on the use of a gamma gauge to measure cyclone feed density was installed by Erie Mining Co. at Hoyt Lakes, Minn. U.S.A. and is shown in Fig. 11-3 (Thornte et al., 1964). The principle of operation of the system is that any change in ore feed characteristics causes a change in the circulating load and in the level of pulp in the pump sump. The sump level is continuously measured and controlled by varying the water added to the sump; this causes a change in the cyclone feed density. This is then controlled by a change in the feed rate. A system using a similar principle was installed at Bancroft Mines, Zambia, about the same time (Barlin and Keys, 1963), and systems of this type are now in common use.

These systems respond well in the long term to changes in ore characteristics but have poor short-term response in that over the short term they

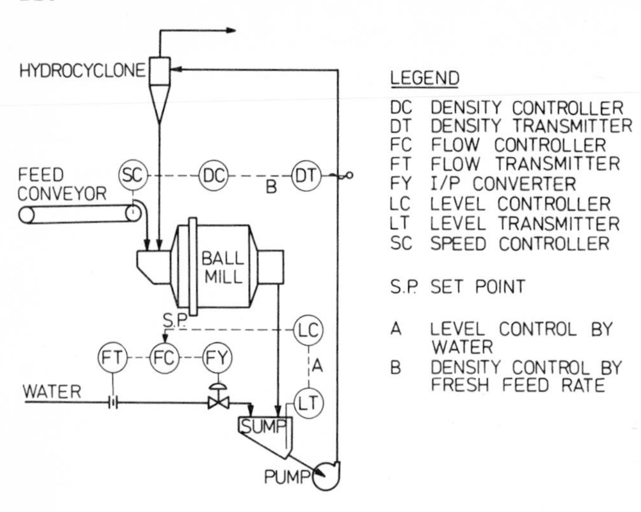

Fig. 11-3. A grinding circuit control system based on maintaining constant cyclone feed density and using a fixed-speed drive pump.

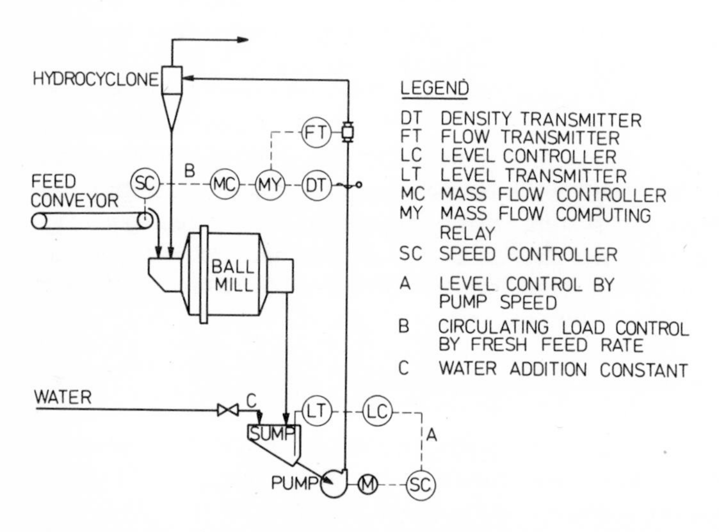

Fig. 11-4. A grinding circuit control system based on maintaining constant circulating load and using a variable-speed drive pump.

tend to magnify in the product any changes which have occurred in the feed. This may be seen by considering the response of the system to an increase in ore hardness. The nature of the response for a fixed-speed pump with the sump level controlled by water addition and a variable-speed pump with the sump level controlled by pump speed, as shown in Fig. 11-4, is shown in Fig. 11-5.

If the ore hardness increases and the water added to the sump remains

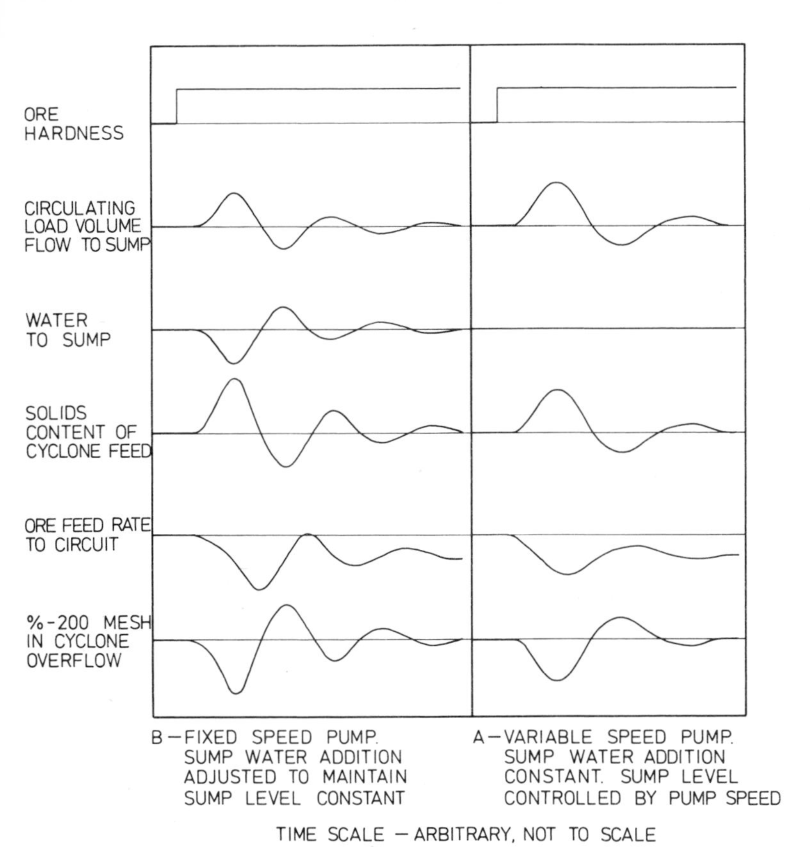

Fig. 11-5. Comparison of dynamic behaviour of control systems based on fixed- and variable-speed pumps.

constant and provided that a variable-speed drive controlled by the level probe in the sump is used on the pump, the circulating load, the cyclone feed density, and the proportion of coarse product in the cyclone overflow increase steadily. The feedback control to the ore feed rate to compensate for the changes will result in fairly steady behaviour of the circuit although a coarser product size will occur for a short period. However, with a fixed-speed drive and control of the sump level by water, an increase in the circulating load results in a decrease in water added until the change in feed rate due to increased cyclone feed density has an effect. Reduction in water as the circulating load increases is the wrong action and increases the rate of discharge of coarse particles from the circuit over the short term. Severe cycling will also occur.

Any control system for a ball-mill—cyclone circuit which includes a fixed-speed drive on the pump will have these undesirable short-term response characteristics.

It should also be noted that when a variable-speed drive is used, the pump does not restrict the volume of the circulating load and this may be maintained at an optimum level without consideration of the pump characteristics. In addition, the problem which arises concerning the variation in flow rate from a fixed-speed pump over the wearing life of the impellor, with the consequent effect on circuit capacity, no longer exists.

11.4.3 *Control systems based on size sensors*

Two types of control systems have been discussed which are based on on-line inference of particle size. Hathaway (1972) has described the operation of a device which relates the attenuation of a sonic signal during its passage through a flowing stream of pulp to the size of the particles in the pulp, and he has shown that this may be used as the primary sensor in controlling the size of the particles.

Lynch and colleagues (Lynch et al., 1967a, b; Lynch and Dredge, 1969), took the view that any change in conditions in a grinding circuit results in a change in the circulating load and in the final product size. Consequently, if the mass flow of ore and water in the cyclone feed stream is monitored continuously and a change is observed, and if sufficient is known about the behaviour of the cyclone, it should be possible to infer the nature and magnitude of the resultant change in product size over the limited range of conditions encountered in normal operation.

One reason for this choice of a size sensor was that control of a grinding circuit involves not only control of the size of the final product but also continuous monitoring of the circuit conditions to ensure that the limiting values of the critical physical variables, in particular the circulating load and the solids content of the cyclone underflow, are not exceeded. A separate size sensor provides no information about the circuit operating conditions other than about particle size, and a mass flow metering system on the cyclone feed line is required to give this additional information. However, a size predictor based on the mass flow metering system provides information about the particle size and also about the circuit condition in terms of mass flow through the mill and the solids content of the classifier sands stream. Ideally, both a size sensor and a mass flow metering system should be used, but economically there is no doubt of the value of using both sets of instruments if one will do the job.

The use of the hydrocyclone as a size sensor was based on the following reasoning (Lynch et al., 1967a):

(1) The calculated $d_{50}(c)$ of the cyclone operation is closely related to the -200 mesh in the cyclone overflow.

(2) For a cyclone in an operating circuit, $d_{50}(c)$ may be calculated on-line from an equation of the form of eq. 6-14, with *VF*, *Spig* and *Inlet* constant, if a mass flow metering system is installed on the cyclone feed line.

(3) The cyclone responds immediately to a change in feed conditions so it will detect a change in conditions immediately.

The equation used in practice to predict particle size was of the form:

$$\log(\% - 200\ \text{mesh}) = K_1 \cdot MFW - K_2 \cdot MFO - K_3 \cdot RMF + K_4$$

where MFO, MFW are the mass flow rates of ore and water to the cyclone, and RMF is the rod-mill feed rate, or the fresh feed rate to the circuit. MFO and MFW are calculated directly from the density gauge and magnetic flow meter signals and the constants K_1 to K_4 can best be found by regression analysis of experimental data.

This approach to on-line particle-size prediction has now been used successfully in several grinding control systems.

11.4.4 *A comprehensive control system*

Control of a grinding circuit involves: (1) control of the size of the final product to some required specification, (2) monitoring of the streams within the circuit to ensure that the limiting values of the critical physical variables are not exceeded.

The particular features of an operating ball-mill—cyclone circuit, which should be recognised in developing a total control system, are:

(1) for steady input conditions, the higher the circulating load the finer will be the circuit product *provided that* increase in load is obtained by rejecting the coarser particles from the circuit product;

(2) this improvement continues up to the overload point, at which point the circuit becomes inoperable;

(3) any disturbance to the process is reflected in changes in cyclone feed conditions, the circulating load, the product size;

(4) the purpose of a control system is to detect when a disturbance occurs and to compensate for this disturbance.

A total control system for a closed grinding circuit must:

(1) stabilise and prevent rapid changes in flows within the circuit since these are transmitted rapidly into the product and are accompanied by changes in product size distribution;

(2) ensure that the limiting conditions of circuit operation are not exceeded and, in many cases, ensure that the circuit is operated as close to these conditions as possible;

(3) maintain the size distribution of the product as close to the set point as possible;

(4) control the fresh feed rate so that the required obective such as constant feed rate or maximum feed rate is attained.

Case studies of control systems which have been designed and installed on different circuits and which include some or all of these loops are given in Chapter 13.

11.4.5 *Control of a variable-speed pump in a ball-mill—hydrocyclone circuit*

The reasons for using a variable-speed pump are:

(1) to cope with the variations in circulating load which accompany any disturbance in the circuit operation;

(2) to remove any limitation which is placed upon circuit performance due to inability of the pump to handle the required flows.

Incorrect implementation of a level-speed control loop may cause continuous abrupt changes in the pump speed, the volume flow rate to the hydrocyclone and in the size distribution of the hydrocyclone overflow. Thus a control system aimed at maintaining an arbitrary level in a sump by a simple level-speed control loop, using proportional plus integral control action to ensure return to set point in a minimum time after a disturbance, will give good long-term and start-up performance but poor short-term performance. The reason is that it responds to high-frequency changes in level, which are induced by the sensing and control system, and to low-frequency changes, which are due to changes in ore type and which are the important changes. In one case (Draper, et al., 1969), it was found that although input disturbances were essentially random, this type of level-control loop reacted in a cyclic fashion responding to changes at a predominant frequency of about 0.5 cycles per minute. Therefore, variations in the cyclone feed pump speed contained both cyclic and random components.

The theory of level control has been explained in some detail by Buckley (1964). In his terms, the control loop should be a "material balance" system which responds only to low-frequency disturbances. With this revised objective, proportional plus integral control action may be replaced by proportional action only. In this situation, the actual operating level is determined by throughput, and input fluctuations are absorbed to some extent by changes in pump box level. A significant reduction in cyclone feed stream "noise" is achieved by operating at the lowest value of proportional gain consistent with avoiding extremes in level under transient conditions.

A further improvement involves filtering of high-frequency components from the loop so that it does not attempt to counter transient changes by adjusting pump speed. Under these circumstances, the pump box absorbed high-frequency input fluctuations by changing level.

An alternative method of controlling a variable-speed drive pump, which is better than the low-gain proportional control system discussed above, is to use a flow-level system. In this system, the set point of the cyclone feed volume flow loop is determined by the sump level and the variable-speed drive is controlled to give this set point. This eliminates problems due to the dynamics of the drive system and reduces high-frequency variations considerably.

11.5 CONTROL OF AUTOGENOUS MILL CIRCUITS

Autogenous mills may be operated in open circuit or in closed circuits with hydrocyclones. Unlike mills which contain steel grinding media and which transmit changes in feed type into the product in the form of a change in size distribution, autogenous mills are to some degree self-stabilising with respect to product size but any change in ore characteristics is reflected as a change in the mill load. An increase in mill load is accompanied by an increase in power draft to a maximum value but this decreases as the load is increased further. Maximum grinding efficiency is obtained at maximum power draft as shown in Fig. 11-6.

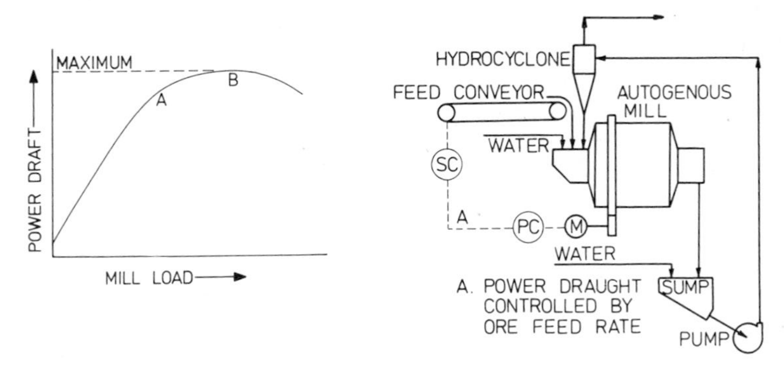

Fig. 11-6. Mill-load—power-consumption relationship for an autogenous mill.

Fig. 11-7. A conventional control system for an autogenous mill.

Some cause—effect relationships in autogenous milling are:

(1) an increase in feed rate at constant size and hardness causes an increase in the mill load and the power draft; variation in the size of the mill product is small;

(2) an increase in ore hardness at constant rate and size has a similar effect to a change in feed rate; and

(3) an increase in feed size at constant rate and hardness has the effect of increasing the pebble wear rate and the breakage of the smaller particles, and giving a coarser product size.

Because of the partially self-stabilising feature of autogenous mill operation and the dependence of mill load on feed characteristics, the approach in developing a control system for an autogenous mill has been to assume that the size distribution of the mill product will look after itself and to use a feed-back loop of the type shown in Fig. 11-7. By reference to Fig. 11-6, it will be seen that if the maximum power which can be drawn for a particular installation is below the peak on the curve, due to motor, shell or gear-box limitations or very large trunnions, and the power is to be controlled at *A*, the problem can be handled by a simple three-term controller. When the power is to be controlled at B a "Williamson" controller has been found to be successful. This works in a 4-min cycle as follows:

(1) during the first two minutes, ore is fed to the mill;

(2) the mill power is then read and compared with the power observed at the same point in the previous cycle;

(3) if the power has decreased, the mill is operating at an excessive load and the feed is turned off;

(4) if the power has increased or remained stationary, the mill has not reached this condition and the feed is left on.

11.6 USE OF A DIGITAL COMPUTER FOR CIRCUIT CONTROL

Digital computers for on-line control have had a significant impact on many of the processing industries, for instance sugar mills, oil refineries and chemical plants. While their use in the mineral processing industry cannot yet be regarded as extensive, sufficient experience has been gained with digital computers in several operating plants to enable informed comments to be made on their advantages and disadvantages in this type of environment. Details of the choice and installation of digital computer systems will not be given here since up-to-date information is available in recent textbooks and in brochures prepared by computer manufacturers.

The advantages of a digital computer are:

(1) The high quality of the data which are available about plant performance. The better the data, the more reliable is the analysis of the plant operation and the better the chance of operating the plant at maximum efficiency. The extraction of data from chart recorders associated with analogue computing relays and instrumentation is time-consuming and difficult, and the accuracy is generally poor because it is limited by the accuracy of the recorder. This is not a problem with a digital computer.

(2) The flexibility which permits control policies of any degree of complexity to be tested. For instance, feed-back loops with large delay times, non-linear control functions and predictive techniques have all been implemented readily.

(3) The ease with which emergency action routines can be established. For instance, the occurrence of a break in a cyclone feed line can be detected immediately and appropriate action taken to prevent possible damage, blockages can be detected, and motors prevented from burning out.

(4) The suitability of the data for off-line and on-line reduction.

(5) The ease with which an on-line data base may be provided for the Company Management Information System.

(6) By appropriate use of redundant instrumentation proper identification of abnormal conditions can be ensured. That is, a correct decision can be made

whether an apparently abnormal condition is real or is due to an instrument error.

In summary, on-line digital computers have changed the understanding of industrial grinding circuits from qualitative to quantitative and this has resulted in substantial improvements in the performances of circuits where the effort has been made to capitalise on this improvement in knowledge.

There are also disadvantages associated with the use of an on-line digital computer. These are:

(1) The computer should be installed in a vibration-free, air-conditioned room. This adds considerably to the cost.

(2) High-quality sensing instruments of proven ability to operate in mineral processing plants should be used in conjunction with the computer, otherwise most of its value will be lost.

(3) Maintenance of instrumentation on the computer must be of a high standard, and this may be difficult to ensure in an isolated area.

(4) There is, and will continue to be, a shortage of metallurgists with the skills necessary to develop, operate and maintain digital-computer-based control systems. Installation of a computer without ensuring that appropriate staff are available is not to be recommended.

11.6.1 *Detection of change in ore specific gravity or hardness*

A digital computer can be used for more purposes than as a highly efficient data logger and process controller. Its use for the on-line determination of particle size by inference has already been described. It can also be used to determine the nature and extent of changes which may occur in ore specific gravity or hardness. The manner in which the computer may be used to detect and compensate for change in ore specific gravity will be discussed in the case study of the control system on ASARCO's Silver Bell (Ariz., U.S.A.) circuit.

Change in ore hardness may be detected by utilising the ball mill in a closed circuit as a hardness sensor. Watson et al. (1970) have indicated that, with a simple regression model of closed-circuit behaviour, it is possible to detect changes in ore hardness by sensing the circulating load in the closed circuit. These authors have in effect produced an empirical model which is a combined size—hardness sensor, and which appears to work reasonably well in the limited application discussed in their work.

The logical extension of the techniques of Lynch and Rao (1965) for on-line size detection, and Watson et al. for on-line hardness detection, is to apply the complete circuit model to the combined task of size and hardness detection. In this role the cyclone model becomes primarily the size sensor, and the ball-mill model may be used as a hardness sensor. This sensing method would utilise the mass flow measurements on the classifier feed as the on-line measurement required for implementing the technique.

For on-line application the steady-state model would be programmed on the control computer and, for the measured feed rate conditions, used to predict the circulating loads of solids and water and the size distribution of the circuit product. In order to do this it would be necessary to assume a feed size distribution and a value for the ore hardness or overall breakage rate. The predicted circulating loads of solids and water would then be compared with the known values obtained from the instruments. Any discrepancy between the two must then be due either to discrepancies in the model or changes in the feed hardness and size distribution, since the model should account for the other changes in ore and water feed rates.

The ball-mill model is known to give a reasonable prediction of both the dynamic and steady-state behaviour of ball-mill circuits and it may be assumed that if the model was applied to these circuits then the discrepancy between the observed and predicted behaviour would be due to changes in feed characteristics. The method of using the model would be to attribute the error in predicted circulating load to changes in the overall breakage rate, thereby adjusting the breakage rate until good agreement between observed and predicted circulating load was obtained. The breakage rate at which this agreement was achieved would be a measure of the ore hardness. This method neglects the effect of changes in feed sizing analysis and attributes all the change to a variation in breakage rate.

This method of ore hardness detection has not yet been tested on-line but simulation studies show that it is quite possible. It is an indication of what can be done with a digital computer on-line with a grinding circuit.

CHAPTER 12

Control of crushing plants

12.1 CONTROL OBJECTIVES

Objectives for a control system for a crushing plant depend upon the purpose for which the plant is operated although for almost all operations maximum throughput consistent with correct product quality will be necessary. The objectives must be formulated with regard to all operating components in the entire processing system. Some possible objectives are: (1) the product sizing is to be maintained constant at constant throughput; (2) the product sizing is to be as fine as possible at maximum throughput; (3) the product sizing is to be maintained constant at maximum throughput.

Modification of objectives is frequently necessary as conditions within the entire processing system alter. For instance, where the product from a crushing plant enters a grinding circuit, the crushing plant throughput may be limited by grinding circuit capacity under some circumstances and it may itself be the limiting circuit at other times. It may not be physically possible to attain some objectives, thus the objective of constant product size at maximum throughput would have to be relaxed under some conditions of fresh feed. In these cases the minimum standard of product quality required must be defined.

Two basic principles must be recognised in a crushing plant control system: (1) the necessity for ensuring continuity of flow of solids within the circuit, and (2) the necessity for operating at optimum conditions because of the effect of change in conditions on circuit performance.

Two levels of control which can be implemented are:

(1) Local and interlocked on/off control to avoid power overload, overfilling of surge bins, etc. This should be coupled with the setting of operating variables at values which give generally good performance. Electrical interlock ensures correct start-up and shut-down sequence; for example, overload of crusher in Fig. 12-1 would cause shut down of feeder and screen.

(2) Complete circuit control using a digital computer for logging, control action, on-line calculation of variable set points and on-line modification of circuit constraints. This control system may be implemented using analogue controllers.

These two levels of control are shown in Fig. 12-1.

In designing a control system it is necessary:

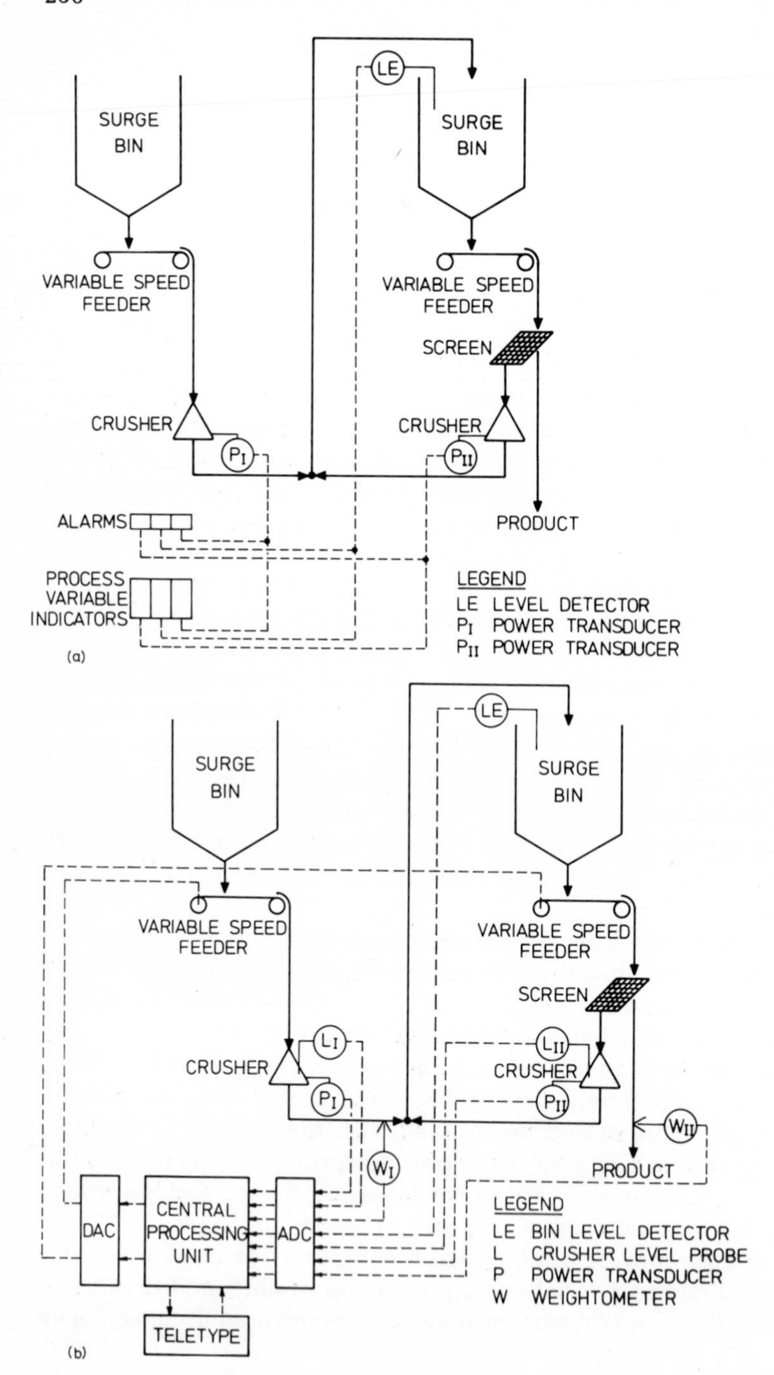

Fig. 12-1. Requirements for different levels of control for a crushing plant: a. local and interlocked on/off control; b. complete control using a digital computer.

(1) to identify those operating variables which are suitable for use for continuous control, such as fresh feed rate, feed rates from internal circuit surge bins, and crusher closed side setting if this is appropriate;

(2) to determine the optimum values of other variables which are not altered during operation;

(3) to determine the circuit limitations at various operating conditions; and

(4) to determine the optimum number of crushing and screening machines for various operating conditions. Thus at times it may be desirable to operate with less than the maximum number.

With regard to the economics of a crushing control system the cost can be estimated relatively simply. The evaluation and economic justification is considerably less ambiguous than for grinding circuit control systems. The factors which are important in the evaluation are throughput, size distribution of the product and circuit availability and these can be measured directly although averaging of the data over sufficiently long periods to minimise any effects of changes in ore hardness may be necessary in some cases.

12.2 TYPES OF DISTURBANCES

Disturbances which affect the throughput and product sizing can originate either from the feed to the plant or from the condition of the plant. Some of the more important types of disturbances are:

(1) Changes in the hardness or size of the ore entering the circuit; these are particularly significant in the secondary and later stages of crushing and have effects on: (a) the size distribution of the feed to various crushing stages, and (b) the relative throughput requirements of the crushing stages as the amount of material which is removed by inter-stage screening varies.

(2) Variation in the flow rate of material from surge bins to vibrating and belt feeders resulting in changes of feed rate at the same nominal feeder speed.

(3) Variation in the number of machines which are operating in parallel in a particular stage of crushing or screening, such as occurs during maintenance periods.

(4) Changes in the performance characteristics of crushers and screens due to liner or screen cloth wear.

In many large mining or quarrying operations ore is supplied to the crushing plant simultaneously from a number of locations and its hardness and size may vary considerably between locations. If the locations supplying ore are altered, changes can occur in the characteristics of ore delivered to a coarse ore stockpile in a relatively short time. Control systems should be designed to utilise as much as possible of the crushing capacity of the plant and this may involve the transfer of crushing work load to different stages as feed characteristics vary.

Changes in the flow rate of ore discharging from surge bins to feeders can cause rapid changes in screen loading and crusher power draw. Changes in screen loading can have a long-term effect upon the metallurgical efficiency of the plant while changes in crusher power draw can result in a power overload condition requiring rapid control action. Short-term changes in the size distribution of the feed to a crusher also occur and can cause significant changes in power draw. Thus, the installation of local control loops designed to maintain power draw at a set point is most important.

Alteration of the number of machines operating in a particular stage or variation in the performance of these machines will cause changes in the amount of material which must be processed by individual machines and crushing or screening stages. For instance, if one tertiary crusher is taken out of a circuit the remaining tertiary crushers must process an increased tonnage if throughput is to be maintained. If this cannot be done additional crushing load must be transferred to the secondary crushers.

12.3 SENSING TECHNIQUES

The basic plant data required as inputs to an automatic control system are:

(1) Tonnages of ore entering and leaving the plant. These are used to control ore flow rates within the plant, to monitor throughput and to avoid overload of conveyor belts.

(2) Levels of ore in all surge bins. These are used for control of bin discharge rates and for short-term prevention of overfilling or emptying of the bins.

(3) All circulating loads to ensure stability of plant operation and prevention of belt overload.

(4) The amperage or power drawn by all crushers in the plant. This is used to avoid power overload and to obtain optimum crusher performance. In some cases the hydraulic support pressure of the mantle/main shaft assembly is used as an indirect measurement of amperage or power.

(5) The closed-side setting of all crushers. This is normally done on an intermittent basis by "leading" the crusher. The measurement is important for maintaining the desired size distribution of the crusher product.

It is important that the measuring instruments selected for crushing plant control should be highly reliable although small inaccuracies in instruments used for process control can be tolerated. The sensitivity of the instrument to the adverse operating environment should also be considered. This is particularly important for weighers and bin level detectors. Sensing instruments which are used in achieving plant control systems are listed in Table 12-I.

TABLE 12-I

Instruments and techniques used in process control systems for crushing plants

Process measurement	Instruments	Comments
Mass flow of ore	electronic load-cell mounted on a weigh-length formed by suspending one or more conveyor idlers independently	output is normally by continous integrators which multiply instantaneous belt speed by belt loading to give instantaneous tonnage rates; for production control purposes, batch totalisers may also be used
Bin level	(1) ultra-sonic instruments	determine level by measuring travel time for sound reflected from the ore; problems can arise from multiple reflections from idling
	(2) gamma-ray instruments	have proved to be reliable in service; may be either on/off using single collimated beam or continuous over a limited range using wide angle source
	(3) weighing of surge bin by mounting on load cell	high initial cost but low maintenance and satisfactory service
Power draw	(1) ammeters measuring current (2) thermal convertors measuring power	care should be taken with the phase difference of current and voltage in crusher operation; generally, power measurement is more reliable when load charges are substantial; totalising kWh meters are also used for production control
Closed-side setting	(1) leading	inconvenient, time-consuming and cannot be done under load
	(2) calibration against mainshaft position	only available in crushers equipped with hydraulic mainshaft support systems

12.4 PHYSICAL LIMITATIONS OF CIRCUITS

An automatic control system for a crushing plant must ensure that the physical limitations of the circuit are not exceeded. These limitations may apply to either the ore transport facility (for example, conveyor belts) or local machine constraints (for example, available power draw).

In many large, modern crushing plants several machines in one crushing stage may discharge product onto the same conveyor belt. In some plants the products from two stages of crushing discharge onto the same conveyor and the number of machines discharging onto a belt can vary considerably when maintenance is carried out. In addition, the tonnage delivered to the belt by individual machines will vary. The problem of avoiding belt overload in such circumstances is not efficiently solved by simply limiting feeder speeds so that when all machines are running belt overload does not occur. This system would limit throughput in direct proportion to the number of machines shut down even though in many cases it may be desirable to maintain throughput by increasing the feed rates to the remaining machines. This is particularly important with regard to conveyors transporting the circulating loads since if this is not maintained at the correct level the power draw of the crushers in the closed circuit will generally decrease with a consequent coarser product. This drop in power draw can be avoided in crushers which provide adjustment of closed-side setting while under load. However, as is discussed in section 12.5, such a control loop is not always desirable.

Overfilling of surge bins must also be considered. This is not a serious problem in small crushing plants using only one surge bin before a crushing stage since a high-level condition can be corrected by decrease of feed rate. The problem is more complex in large plants where several interconnected surge bins are fed from a single tripper discharge conveyor. In this case it is not generally desirable to decrease feed rate to the surge bins if a high-level condition exists in only a small proportion of the surge bins. Commonly, by-pass gates are installed so that material is diverted from the bin to the surge capacity of the previous stage when an individual high bin level is encountered. Computer control of tripper position can also be used to allow feed rate to multiple surge bins to be maintained when some individual high bin levels are encountered. When an averaged high or low bin level condition is detected then control action must be taken to decrease or increase the feed rate, respectively.

Overloading of crushers can occur:

(1) when the feed tonnage exceeds a certain value, which is a function of size distribution, ore type and closed side setting, and causes power overload, and

(2) when the feed tonnage exceeds the maximum physical flow possible through the crusher for a particular closed side setting, that is, the crusher feed chamber is completely filled and the crusher "chokes".

The power draw variations of crushers occur rapidly and are often of large amplitude. To ensure safe operation if a crusher is run using a fixed nominal feed rate and closed-side setting, the feed rate must be considerably less than that which can be obtained at full power draw. This means that the full crushing capacity of the plant is not utilised. Variable-speed feeders and crushers equipped with hydraulically positioned mantle/main shaft assemblies are used to attain safe operating conditions at high power levels.

If it is decided to run a plant using choked feed conditions or if such a condition may occur during normal operations it is necessary to measure the level in the crusher feed chamber and alter the feed rate accordingly.

12.5 EFFECTS OF OPERATING VARIABLES

The performance of a cone crusher is strongly dependent upon: (a) power draw, (b) closed-side setting and (c) size distribution and hardness of the feed. These parameters can either be directly controlled or, in the case of (c), the effect counteracted as far as possible by an automatic control system. Other parameters which can affect performance are mantle and concave configuration, countershaft speed and mainshaft throw. These parameters are normally fixed by machine design and not varied by control action.

It has been found that the feed tonnage of a cone crusher increases with the power draw of the crusher (Kellner and Edmiston, 1975). Simulation studies which have been carried out using accurate mathematical models of crushers developed from data obtained at Bougainville Copper Limited, indicate that the relationship between power draw and feed tonnage tends to higher values of power draw at high feed rates. This effect is shown in Fig. 12-2. The higher energy consumption per tonne of processed material is accompanied by a finer product size. An example of this effect is given in Table 12-II. If the objective of a crushing plant operation is to produce a fine product at high throughput then control loops should be implemented to maintain the highest possible safe power draw.

As the closed-side setting is increased the power draw at a particular feed tonnage decreases considerably as shown in Fig. 12-2 and the product size becomes coarser. Variation of closed-side setting can be used in control loops: (a) to avoid excess power draw at constant feed rate, (b) to maintain a fixed product size at variable feed rate, or (c) to alter the size of the product

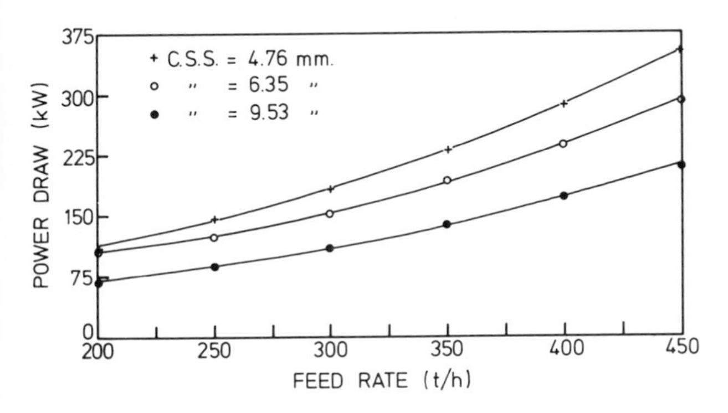

Fig. 12-2. The dependence of power draw upon feed rate for a 2134-mm crusher.

TABLE 12-II

The predicted product-size distribution for a 2134-mm tertiary cone crusher at three power levels (closed-side setting 6.35 mm)

Aperture (mm)	Feed (% ret.)	Product (% ret.)		
		123 kW	153 kW	236 kW
64.00	0.05	0.0	0.0	0.0
37.00	9.22	0.0	0.0	0.0
16.00	40.25	14.54	13.08	12.39
8.00	45.40	54.26	50.40	47.85
4.00	3.35	17.11	17.70	17.80
2.00	0.51	5.48	7.09	8.11
1.00	0.09	2.77	3.88	4.63
0.50	0.08	1.77	2.49	2.98
0.25	0.09	1.17	1.64	1.95
0.125	0.22	0.93	1.22	1.42
Submesh	0.74	1.97	2.50	2.87

with an accompanying change in feed rate. The third control action can be very important in transferring crushing load from one stage of crushing to another. These control loops are only feasible in plants equipped with automatically adjustable closed-side setting if frequent control action is necessary.

Simulation studies which have been carried out on the crushing operations at Bougainville Copper Limited, indicate that a coarser product results from a coarser feed size distribution when power draw and closed-side setting are maintained constant. The main effect of this is to produce a large increase in circulating load. This may mean that new feed tonnage has to be decreased significantly in order to avoid overloading of the circuit. Under these circumstances it is sometimes necessary to transfer crushing load from one stage to another.

Generally, the efficiency of vibrating screens decreases as loading increases. It is desirable to operate the screens at reasonable efficiency otherwise a significant amount of potential undersize product in the screen feed reports in the oversize and increases circulating load without a corresponding increase in plant throughput. The problem of avoiding highly inefficient screen operation can be overcome to some extent by limiting feed tonnage so as not to produce circulating loads (per individual screen) greater than a fixed value.

12.6 CONTROL SYSTEMS FOR CRUSHING AND SCREENING PLANTS

In crushing plants which are followed by grinding circuits the most important objective of a control system is to ensure supply of crushed ore at the rate required by the grinding plant. The fineness of this ore is generally maintained at a minimum standard by the selection of the appropriate aperture for the final closed circuit screen. For a specified screen aperture the most effective means of increasing throughput is to maintain the highest possible crusher power draw. It has been found by Fewings and Whiten (1972) and Kellner and Edmiston (1975) that when increased throughput is achieved in this way the product is also finer. This effect is also evident in the results of studies carried out on crushing operations at Bougainville Copper Limited and is due to increased fines production at high power draw. If the increased throughput provided by the installation of the control system cannot be processed by the grinding plant, then the higher average power draw can be used to produce a finer product. This is done by using a smaller aperture for the closed circuit screens, thus increasing the circulating load relative to throughput.

It is generally found that for those crushing plants there is an optimum closed-side setting for the crushers operating in closed circuit. This setting provides the highest tonnage of finished screen product for a particular power or circulating load limit although the actual feed tonnage to the crusher increases at larger closed-side settings. Thus, control systems which require frequent adjustment of closed-side setting to values other than this optimum may not achieve the full potential increase in throughput available from the maintenance of high power draw.

In crushing plants which produce a final saleable product an important objective of a control system is often to maximise the production of certain size fractions from each tonne of feed. In such plants the maintenance of high power draw is often not as important as control over product size distribution by variation of the setting of the crushers. However, increased throughput in such plants can be achieved by ensuring a continuous supply of ore to all sections of the plant and by utilising surge bin capacity when the feed supply to the plant is intermittent

12.6.1 *Control system based on fixed-crusher closed-side setting*

Automatic control systems designed for use in these plants have generally had the objective of increasing throughput and producing a finer product. To achieve this objective the power draw of the crushers must be maintained at as high a value as possible.

The power draw can be maintained by using belt feeder speed to vary the feed rate to the crusher. In this loop changes in ore hardness or size distribution or sudden surges onto the feeder due to the flow characteristics of

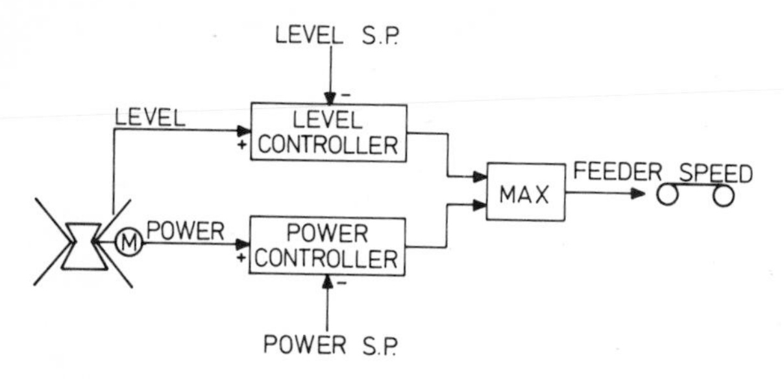

Fig. 12-3. Control loop for maintaining power draw or feed chamber level in a crusher without automatic setting adjustment.

the ore are compensated for by a change in feeder speed. This control loop is shown schematically in Fig. 12-3. Fewings and Whiten (1972) described the use of this loop for crushers operated in both open and closed circuit in Mount Isa Mines. The determination of optimum controller settings may be a problem because of the lag time introduced by screens which discharge oversize directly into the crusher. Whiten (1972b) has reported a method for determining the optimum controller settings and set point. In some plants, such as that of Bougainville Copper Limited, closed-circuit screening is carried out in a separate screening station and the tertiary crushers are fed directly from the belt feeders. Thus, the problem of lag time is eliminated. In plants where choked feed conditions are anticipated the ore level within the crusher feed chamber should be measured and controlled. This can be done by modifying the feed rate controller so that the error signal used is the maximum of the power draw or level errors measured. This loop is also shown in Fig. 12-3.

In many plants the efficiency of screen operation decreases significantly at high screen loading. This change in screening efficiency is greater for particles which are just smaller than the aperture than for finer particles and at high loads the product is finer. This effect provides a means of achieving a finer product during periods of either excess closed-circuit crushing capacity (when there is inability to maintain bin levels) or reduced throughput requirement. If the number of operating screens is decreased during such periods and the loading on the remaining operative screens is increased a finer product can be produced. In addition, circulating load can be increased by less efficient screen operation thus allowing a higher power draw to be maintained. This also leads to a finer crushing plant product. The implementation of this type of control loop requires accurate knowledge of the behaviour of the plant under various operating conditions.

Control loops designed to maintain a continuous supply of ore to all stages of the circuit use the measurement of bin level to control feed rate to the prior crushing or screening stage. In plants without automatic crusher setting adjustment it is not desirable to simply control the feed rate to a

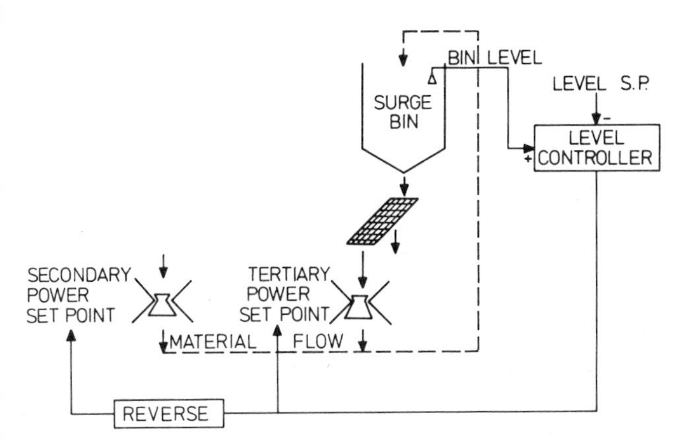

Fig. 12-4. Control loop for maintaining the level in the surge bin. The bin level also provides information on which stage is limiting production.

crusher using the crusher product surge bin level as it is imperative that this feed rate be used for the control of power draw and to avoid overfilling of the crusher chamber. In such cases the bin level controller should be cascaded to the power draw controller of the crushers to provide a remote power set point. A diagram of this type of control loop is shown in Fig. 12-4. The first element of the cascade loop has reverse action since for a high bin level it is necessary to decrease the power set point. The reduction in power draw set point will result in a decrease in feed rate. The measurement of bin level also provides information on which stage of the circuit is limiting production. For instance, if the level of the surge bin shown in Fig. 12-4 is high then the succeeding crushing stage must be limiting, not the preceding stage. Although this system identifies which stage of crushing is limiting production and maintains it at its maximum capacity, it does not transfer additional crushing load to the secondary crushers if they are not operating at maximum capacity. This can be done if the first control action taken when a high tertiary bin level is detected is to decrease the secondary crusher closed-side setting. This will result in the production of a finer secondary crusher product at reduced tonnage. If the high bin level persists when the secondary setting is at its minimum then the secondary power set point is reduced. In plants without hydraulically variable crusher setting this control loop is impracticable if frequent manual adjustment of secondary setting is required.

12.6.2 *Control systems based on variable-crusher closed-side setting*

Some commercial crushers are equipped with an hydraulic mainshaft support system in which crusher setting can be adjusted rapidly while under load. This allows the Svedala Arbra automatic control system to be implemented. In this system a closed-side setting is selected for the operation and an overload condition, registered by high power draw and/or hydraulic pressure, is compensated for by increasing the closed-side setting. The control system is essentially a sequential two-level system in that if an overload condition occurs for greater than a present time the setting is increased by a fixed amount by draining fluid from the support reservoir. Persistence of the overload causes a further incremental increase in setting. The setting is decreased in an incremental manner until the desired setting is regained. It should be pointed out that the system, as described, does not decrease closed-side setting if a low pressure/power draw condition persists. Further increases in throughput would be obtainable if this control action was also included.

Kellner and Edmiston (1975) have reported a control system which uses this type of loop for control of the tertiary crushers and propose to use it for control of the secondary crushers. However, in this system analogue controllers with continuous signal output are used. The position of the mainshaft/mantle is controlled by the operating pressure such that a pressure above the set point causes an increase in closed-side setting and a pressure below the set point decreases the closed-side setting. In conjunction with this control loop to maintain pressure and power draw, it is proposed to maintain a choke-fed condition. This is done by using a nuclear level gauge installed in the crusher feed chamber to control feed rate. If choke-fed conditions are maintained the pressure and power draw remain relatively constant and the mantle position is seldom changed. Thus, large variations in product size do not occur and the performance of the circuit approaches that obtainable at the optimum closed-side setting.

In the control system described by Kellner and Edmiston the level of the tertiary surge bin is used to control the feed rate to the secondary crushers. If the tertiary bin level is falling, indicating that the tertiary crushers have excess crushing capacity, the feed rate to the secondaries is increased. This causes an increase in secondary crusher pressure, resulting in an increase in secondary closed-side setting. Thus, a higher tonnage of coarser material is delivered to the tertiary crushers. Conversely high tertiary bin levels, indicating that these crushers are limiting production, result in a decrease in secondary closed-side setting. In this case a greater fraction of the crushing load is accepted by the secondaries and they produce a lower tonnage of finer material.

In crushing plants which produce a final saleable product there is a very wide range of criteria which define product quality. In the iron ore industry it is often desirable to maximise the fraction of lump product (approximately -33 mm/$+6$ mm) relative to fines (-6 mm). A study of tertiary crushing and screening operations carried out in this industry has shown that this objective can be achieved by increasing the closed-side setting until limiting circulation load is reached. Further, it was found that production was generally not limited by either choke feeding or power draw. Thus, the

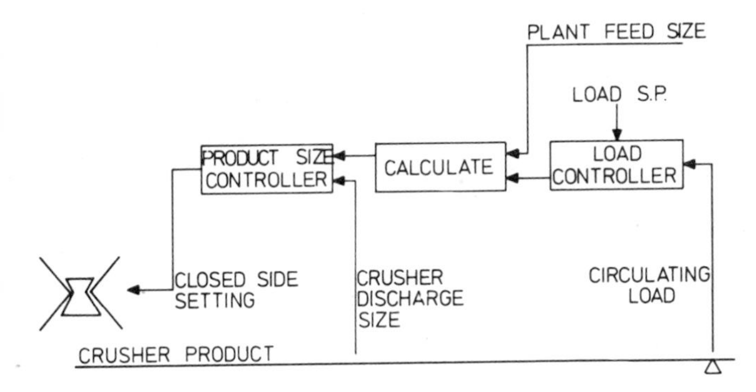

Fig. 12-5. Control loop for maximum lump production in an iron-ore tertiary crushing and screening circuit.

required control action is to maintain circulating load at a set point by alteration of closed-side setting. There is a considerable time delay before the circulating load is affected by changes in the closed-side setting and this direct loop would be slow and possibly unstable. This problem can be overcome by using circulating load to control a product size set point and varying closed-side setting to maintain product sizing of this set point. This control loop is shown schematically in Fig. 12-5. The required change in closed-side setting is calculated using an accurate mathematical model of crusher performance. It is preferable to measure product size directly. However, if circuit behaviour is known it can be inferred from the known closed-side setting, lump product tonnage and circulating load.

12.6.3 *Cascade loop to control feed rate to bins*

In many crushing plants it is desirable to control the level in a surge bin by varying the feed rate to the previous stage. Examples of this are the loop between tertiary bin level and secondary crusher feed rate given in the previous section and the control of tertiary crusher bin level by varying the feed rate to the tertiary screens in plants which have a separated tertiary screening station. In such cases changes in bin level occur slowly in response to changes in the feed rate or the number of machines operating in the previous stage.

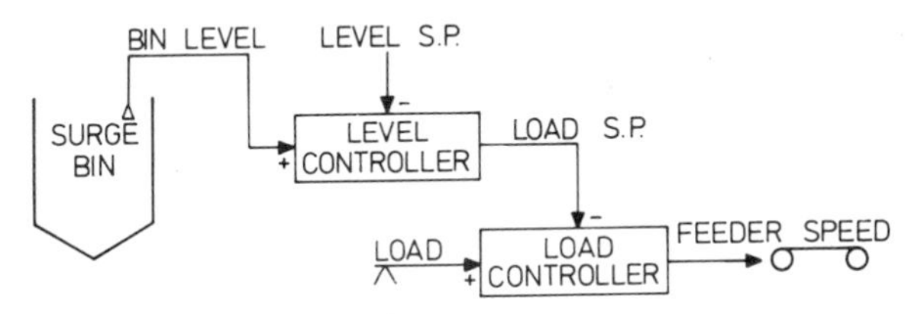

Fig. 12-6. Cascade loop for stable control of bin level.

However, the load on the connecting conveyor belt changes much more rapidly in response to these changes. In these circumstances it is better to use the bin level error in a cascade loop to alter the set point of the load on the conveyor. Limits can be imposed on this set point to avoid overloading of the conveyor. The error in load on the conveyor is then used to vary feeder speed. This cascade loop would provide much more stable operation and, in particular, prevents belt overloading. The loop is also suitable for controlling the level of a secondary surge bin by varying the speed of the apron feeders reclaiming ore from a coarse ore stockpile. A schematic diagram of the loop is shown in Fig. 12-6.

CHAPTER 13

Case studies of automatic control systems for various applications

13.1 INTRODUCTION

A control system for any crushing or grinding circuit must be designed with full knowledge of the characteristics and limitations of the circuit. The case studies which are discussed in this chapter have been chosen to illustrate the different types of problems which are encountered with different ores and different machines. All systems are still undergoing development and refinement and some are further advanced than others, but valuable information can be obtained from each case study. A summary of the case studies and the reasons why they are chosen for discussion are given below.

Case study 1. Ball-mill—wedge-wire screen circuit at Renison Limited (Tasmania). The ore being treated contains cassiterite which is concentrated by gravity methods and "over-grinding" of the high-specific-gravity cassiterite must be avoided. Therefore, wedge-wire screens are used for classification. The response to disturbances of a circuit which includes wedge-wire screens as classifiers is different from a circuit which includes hydrocyclones as classifiers.

Case study 2. Rod-mill, ball-mill—hydrocyclone circuit at Mount Isa Mines Limited, (Queensland). Two circuits are discussed, copper ore being processed in one and lead—zinc ore in the other. Both are conventional circuits.

Case study 3. Ball-mill—hydrocyclone circuit at Bougainville Copper Limited (Papua New Guinea). This is a high-capacity circuit containing a very large ball mill, and problems have been found to exist with large mills which do not exist with small mills. These must be recognised in a control system.

Case study 4. Ball-mill—rake classifier circuit at New Broken Hill Consolidated Limited (New South Wales). An inexpensive control system is described which makes use of particular characteristics of rake classifiers.

Case study 5. Grate ball-mill—hydrocyclone circuit at the Silver Bell Unit of ASARCO Inc. (Arizona). There are constraints on the performance of grate ball mills which are not important with overflow discharge mills and the method of handling these is discussed. In addition this mill must handle ores of varying specific gravity and the procedure for on-line compensation for change in ore specific gravity is discussed.

Case study 6. Semiautogenous-mill, ball-mill—hydrocyclone circuit at Cyprus Pima Mining Company (Arizona). This circuit must treat ores with a wide range of grindabilities and this may lead to mill overloading. A control system to maximise throughput at a stable product size while preventing mill overloading is described.

Case study 7. Crushing control systems at Mount Isa Mines Limited (Queensland). This is a case study of a control system on a two-stage crushing circuit.

An important point which emerges from all studies is that considerable attention must be given to ensuring that the performance of the circuit is not limited by ancillary items such as pumps, conveyors and surge bins, otherwise it is incapable of operating at maximum efficiency.

13.2 CASE STUDY 1. BALL-MILL—WEDGE-WIRE SCREEN CIRCUIT AT RENISON LIMITED*

13.2.1 *Introduction*

Renison Limited is a member of the Consolidated Goldfields Group. The Company operates a mine and concentrator treating complex lode tin ores.

General concentrator operations. Renison ores exhibit a high degree of variability. The major mineral is pyrrhotite which coexists with siliceous and carbonaceous gangue and the tin-bearing mineral is cassiterite. The liberation characteristics of the ore vary widely according to ore type. The pyrrhotite found at Renison is extremely unstable and this fact dictates against the manipulation of large stockpiles. As a result, ore blending is fairly poor and the concentrator feed is quite erratic in terms of its treatment properties.

A simplified concentrator flow sheet is shown in Fig. 13-1. The flow sheet is reasonably conventional for a plant treating this type of ore except that almost all the table tailings are reground. This causes a very high circulating load within the gravity plant which normally sets that section as the rate-limiting stage for the concentrator. The only positive metallurgical control on the gravity plant circulating load derives from manipulation of the primary grinding circuit.

The primary grinding circuit. A schematic diagram of the primary grinding circuit is shown in Fig. 13-2.

The items of principal general interest in the Renison primary grinding circuit are the wedge-wire screens which close the circuit. The screens were installed to avoid the selective overgrinding of cassiterite which can occur in mill circuits closed with cyclones.

---

* By B.R. Spiers, Metallurgist Renison Limited.

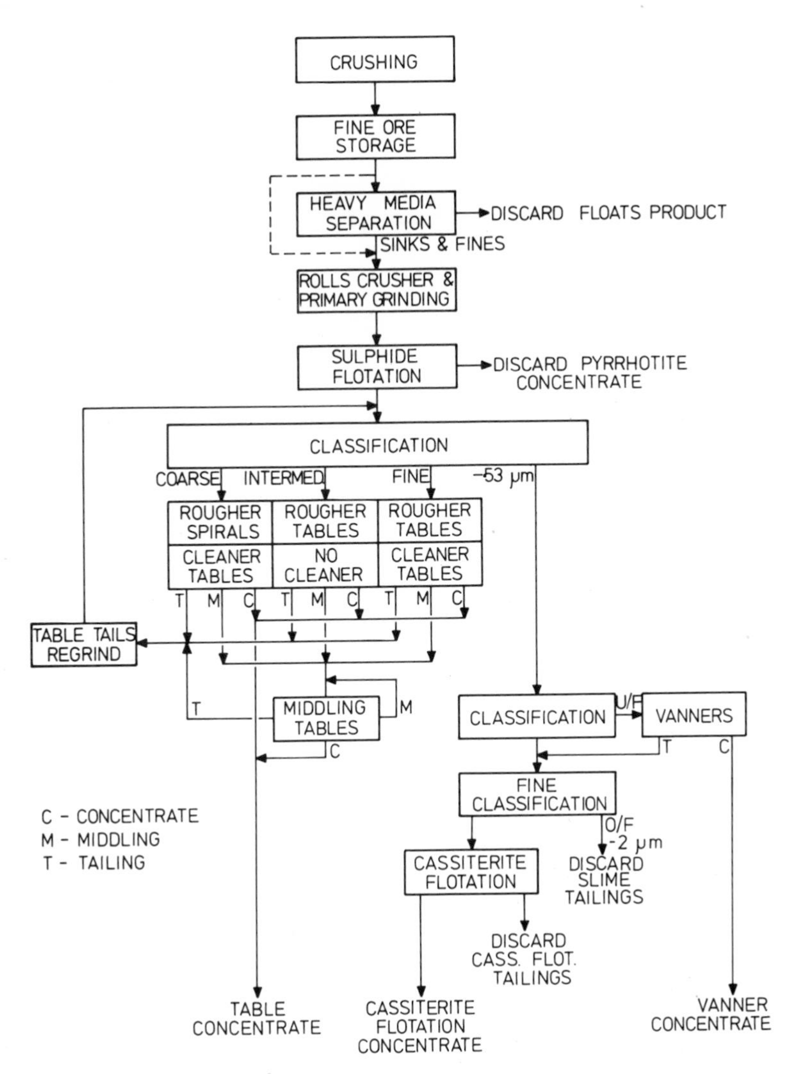

Fig. 13-1. Flowsheet of the cassiterite ore grinding and concentration circuits, Renison Limited.

Wedge-wire screens sizing abrasive feeds suffer a rapid loss of efficiency in all size fractions as the leading edge of the wedge bar wears. This efficiency drop was very pronounced at Renison when using stainless steel screens. Efficiencies fell by 60% in the first eight hours of screen operation. Partial solutions to this problem have been found at many mines by installing screens in cradles which allow frequent turning, and may other devices are available to improve the efficiency of wedge-wire-type screens.

At Renison, simple inclined flat screens are used. They are pneumatically

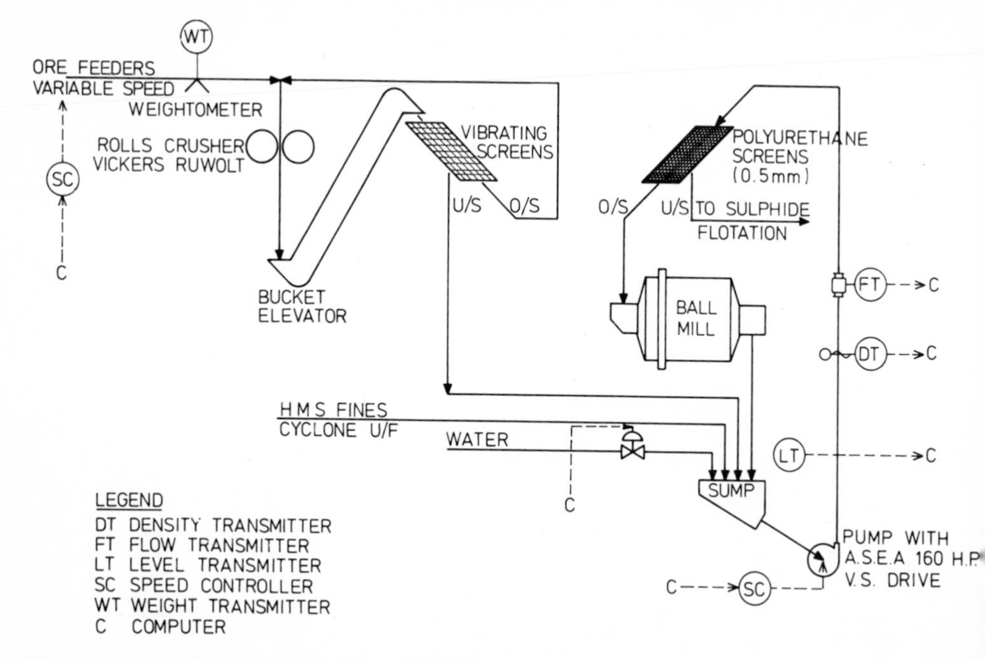

Fig. 13-2. The primary grinding circuit at Renison Limited.

rapped and the local solution to the problem of high wear rates has been to substitute polyurethane for stainless steel. The average efficiency of the polyurethane screens is lower than for stainless steel screens of the same product size. Selection of this material merely followed from a desire to reduce manual handling in an awkward environment. Actually, the screens perform to consistent levels for considerable periods of time which probably assisted in stabilizing the circuit under control. It has become necessary, however, to plan to expand the available screen area to provide more flexibility in the selection of the product size and to allow for increases in throughput.

13.2.2 *Automatic control*

Background considerations. Effective control of the grinding operations in a complex concentrator to stabilise the rate of production and the quality of the grinding circuit product is a very considerable aid to the optimisation of subsequent operations. Improvement of grinding circuit stability may also ensure higher average throughput rates and this type of benefit was important to the operating plant at Renison as availability of the primary grinding circuit was relatively poor. Operators were provided with mill running load meters and an indication of the level in the circuit pump sump. The only modes of control were to vary the numbers of screens and/or rappers to control circulating load, or in an emergency, to reduce the incoming feed rate to the circuit. Control actions were frequently drastic and circuit blockages were a common occurrence.

The expectations of control. This section has deliberately not been titled "the aims of control". Expectation is in this a case a more precise word. In some cases, predictions of the benefits of control amounted to great expectations.

The expectations at the commencement of serious study of a control philosophy and its implementation were:

(1) the maximisation of circuit throughput to a particular selected product sizing;

(2) the control of product sizing, preferably by an automatic system;

(3) the improvement of circuit availability by providing for operational stability within the control system;

(4) the provision, by the above steps, of a constant rate of a selected product to subsequent treatment stages.

In quantitative terms, it was predicted that an increase in throughput rate of some 5% could be expected. Improvements to circuit availability were expected to be of the order of 1.5%.

Planning the control philosophy. A program was laid out which covered: (1) data gathering; (2) circuit simulation; (3) selection and testing of control approaches based on the results of the simulations; (4) selection and procurement of instrument hardware to satisfy the planned control program; (5) installation and commissioning of the control system; (6) system evaluation.

Data gathering. A considerable number of complete surveys were made of the mill circuit. These surveys were quite conventional and were aimed at providing data which, either directly or by a validated calculation procedure, could be used to test the fit of available circuit models.

Circuit simulation. Simulation studied were carried out using the ball mill and wedge-wire screen models discussed in Chapters 4 and 6. Good agreement between simulation and experimental results were obtained for a wide range of conditions and the models were then used for a further study of the circuit.

The main conclusions were that the properties of the circuit product were very stable under widely variable circuit conditions and that grinding circuits closed by screens were more stable than those closed by cyclones or hydraulic classifiers. A corollary of this is that ball-mill—screen circuits are less responsive than conventional circuits to the manipulation of control variables.

The simulations showed that, provided the circuit feed rate was controlled and the screen feed conditions were constant, the circuit operations were stabilised. Manipulation of the process variables which were thought to govern the sizing action of the screen indicated no real regulation of sizing could be derived by their control.

Control system design. Following the results of simulation, the objectives of the control system were defined as follows:

(1) to stabilise the screen feed in terms of volumetric flow rate and pulp density, that is, mass flow rate of solids to the screen;

(2) to maintain a steady new feed rate to the circuit;

(3) to stabilise circuit operations with a view to improving plant availability.

The assumption was made that the mill could be run at maximum capacity by raising the new feed tonnage to the physical limit under which the circuit could be stabilised by the screen feed pump. This had been verified by simulation.

A variable-speed motor was installed on the screen feed pump and this improved circuit stability and overall availability.

The control strategy was now fairly closely defined. The variable-speed pump would control the screen volumetric feed rate. Water addition to the pump would control the screen feed density and the new feed would be controlled by existing variable-speed conveyors. Plant experience indicated the desirability of some control of the sump level so this provision was included.

Control strategies based on these concepts were tested on a pilot mill circuit. This investigation was aimed at manipulating control blocks to find how control actions were best related to achieve the aims of the system. The control flow sheet which resulted from this investigation is shown in Fig. 13-3.

Equipment. The control logic shown in Fig. 13-3 formed the basis for equipment selection. Long-term metallurgical planning indicated that several computer-based control or monitoring applications were likely to be justifiable within five years. For this reason an initial decision was made to base the grinding circuit control system around a digital computer. Because it had little experience with digital computer-based systems, Renison chose to utilise the manufacturer-supported Fox 2/10 systems of Foxboro Proprietary Limited. The major measurement devices are: (1) rolls crusher-feed rate: Ramsay Belt weigher; (2) screen-feed density: Ohmart density gauge type ED6; (3) screen-feed volumetric flow: Foxboro 2800 series magnetic flowmeter; (4) pump sump level: Foxboro electronic differential pressure transmitter.

The variable-speed motor chosen for the circuit pump device was an ASEA thyristor-controlled DC motor. Availability of all this equipment has been of the order of 99%. Total installation costs including the pump motor, but neglecting Renison salaries, was just under $A 100,000.

Commissioning. Prior to control system commissioning, the mill circuit pump sump was replaced with a larger unit. This move was indicated by the extreme sensitivity of the original sump level to alterations in the circulating

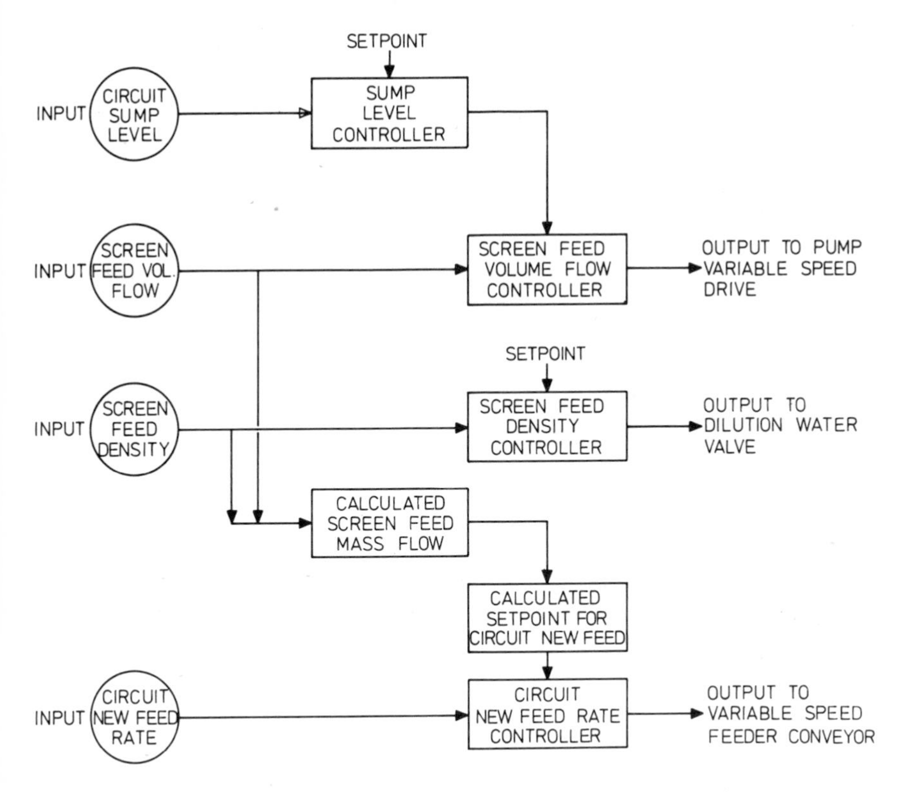

Fig. 13-3. Control logic for Renison Limited primary grinding circuit.

load around the mill. It is a considered view that this type of sensible engineering very largely reduces the problems that can be encountered in commissioning a control circuit.

A modification was also made to the controlled speed range of the variable-speed motor. As received, the motor speed was controllable over a full 0–1400 rpm. At Renison direction, ASEA modified this range so that the motor speed could be varied in a range 1100–1560 rpm. Both these modifications were deemed essential.

Apart from some extended delays caused by inconsistencies in the computer software, all the remaining commissioning, including control loop programming, testing, commissioning and tuning, and support software generation was accomplished by Renison metallurgists without a single disruption to plant operations.

System evaluation. Referring to the above mentioned initial expectations for the control circuit, the following results have been obtained from use of the computer system:

(1) Gains in circuit throughput have been slight due to the installation of polyurethane screens to replace the original stainless steel screens. The

screening section is to be expanded to allow throughput to be maximised and it is expected that a significant increase in throughput will be obtained when this has been completed.

(2) The circuit simulation indicated the stability of the circuit with respect to the final product sizing. Long-term experience indicates the primary circuit product sizing has become marginally more stable on a daily basis and a little coarser than prior to the control circuit commissioning. The installed polyurethane screens have been primary causes of these results. It is thought that the effect of control on circuit sizing has been one of minimising short-term fluctuations in product sizing.

(3) Circuit availability has improved by some 1.5% with the commissioning of the control system.

Basing calculation of the economic pay-back period on this availability increase only, the pay-back period for the Renison system was six months.

13.2.3 *General*

The work outlined in this study was very largely carried out by staff with no great sophistication in the application of computer techniques to mineral processing. It was essential at some stages of the project to seek assistance from experts outside the company. Where suitable staff exist or competent consultants are available, the execution of a project of this type is eminently feasible. The point should be made that careful consideration of the engineering of the circuit to be controlled is vital to the successful commissioning of a well controlled system.

13.3 CASE STUDY 2. ROD-MILL, BALL-MILL—HYDROCYCLONE CIRCUITS AT MOUNT ISA MINES LIMITED*

13.3.1 *Introduction*

Digital computer control of wet mineral grinding circuits is utilized in both the lead—zinc and copper concentrators of Mount Isa Mines Limited. The No.4 copper concentrator has had grinding control since its commissioning in 1973, while control was introduced into the No.2 lead—zinc concentrator in late 1975.

The first digital computer control system was commissioned on one grinding line in the superseded No.1 Concentrator in 1970. This followed successful automatic control studies using an analogue computer in 1968.

Development work with the No.2 Concentrator system is still in progress. The control system software in No.4 Concentrator has undergone an

* By Mount Isa Mines Limited Staff.

extensive revision commencing in late 1975, and is still in progress. This has coincided with the replacement of the previous 16K digital computer by a real-time disc-based computer system which was introduced to allow the development of flotation control. The decision to restructure the grinding control software system takes advantage of the real-time operating system and disc storage. The aim is to develop an overall control strategy which is concerned ultimately with the optimisation of the entire concentrating operation, rather than consideration of grinding operation alone. This work is now in progress.

Features of both control systems will be discussed. General gains from computer applications are discussed and specific comparisons of grinding performance before and after commissioning of the grinding control system in the No.2 Concentrator are presented.

13.3.2. *Grinding control objectives*

The objectives of grinding control have varied with changed production requirements, the introduction of a higher-capacity concentrator, and the availability of more powerful computer systems. This ability to vary control objectives with reasonable ease illustrates one of the major advantages of the digital control approach—its flexibility in terms of changing requirements and policies. There have been two principal objectives: (1) constant product sizing at constant tonnage throughput, (2) constant product sizing at maximum tonnage throughput.

These were the strategies originally developed in No.1 Concentrator, with throughput maximisation achieving gains of 5% in treatment rates (Fewings, 1971). Tonnage maximisation is also the principal reason for introduction of automatic grinding control to No.2 Concentrator.

In No.4 Concentrator plant throughput is determined by an interaction between total blister copper production, mining rates and mine filling requirements (No.4 Concentrator tailings are the major component of underground fill). At the treatment rate required there are no throughput limitations in the concentrator. Development work in No.4 Concentrator is extending grinding strategies to include overall concentrator performance and the philosophy of maximum grinding efficiency to obtain optimum flotation performance at required treatment rates is being developed.

In meeting any of these overall aims, a number of intermediate control objectives must be satisfied. These include control of sump levels and cyclone feed flow rates by variable-speed pumps, control of circulating loads by water additions and ore/water control on new feed (to the rod mills).

13.3.3 *Plant description*

No.2 Concentrator. A general flowsheet of a typical grinding line and its relationship to the rest of the plant is illustrated in Fig. 13-4. The grinding

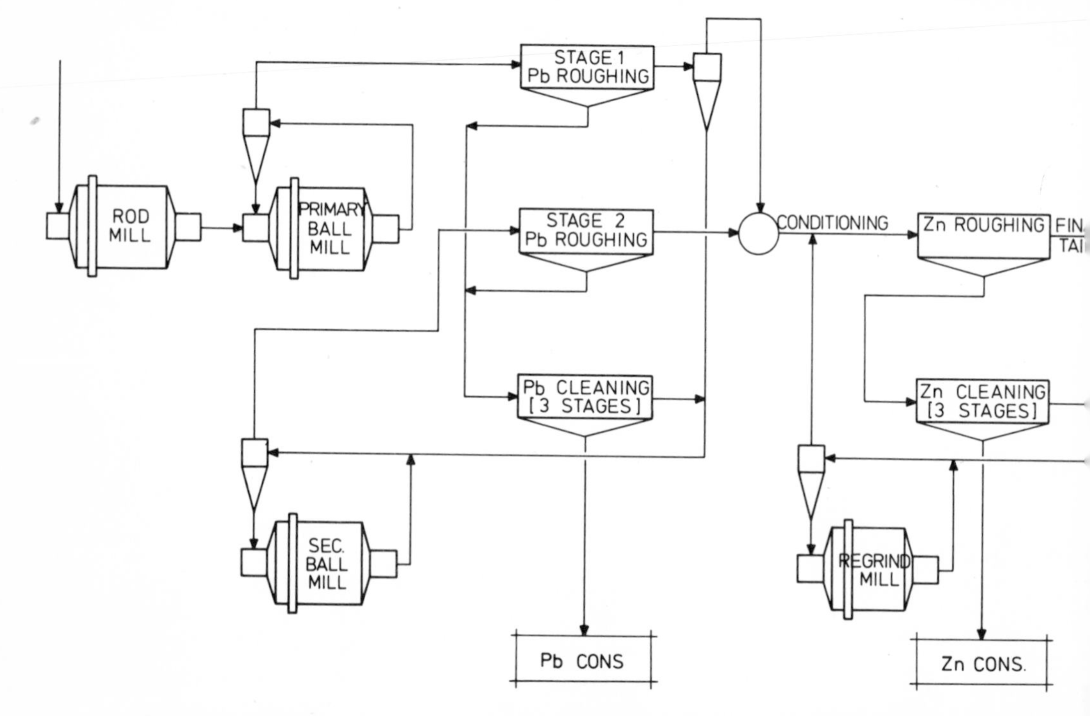

Fig. 13-4. Flowsheet of the lead—zinc ore grinding and flotation circuits, Mount Isa Mines Limited.

section consists of four identical parallel grinding circuits, feeding two parallel flotation circuits. Each section consists of a rod mill, primary ball mill and secondary ball mill. Average treatment rate is 75 t/h.

The rod mills are 2.74 m diameter by 3.66 m and charged with 75-mm diameter rods. Primary ball mills and secondary ball mills are drum-fed 3.20-m diameter by 4.27-m overflow discharge mills. Each ball mill is operated in closed circuit with two Krebs D20B cyclones and the cyclones are fed by a variable-speed motor driving an 8/6 Warman Series A pump. The standby system is a fixed-speed motor driving an 8/6 Warman pump feeding two Krebs D15B cyclones. Pump motor size is 80 kW on both primary and secondary ball mills.

Ore is fed to the rod mill from three of six slot-discharge belt feeders. Two of the three feeders used at any one time are manually adjustable while the speed of the third feeder is remote-controlled.

No. 4 Concentrator. The concentrator has two paralled grinding lines each operating at 360—370 t/h. Fig. 13-5 shows a simplified flowsheet of the grinding and flotation circuit.

Fine ore to the rod-mill feed conveyors is reclaimed by sixteen slot feeders, half of which are variable-speed. The overflow rod mills are 3.81 m

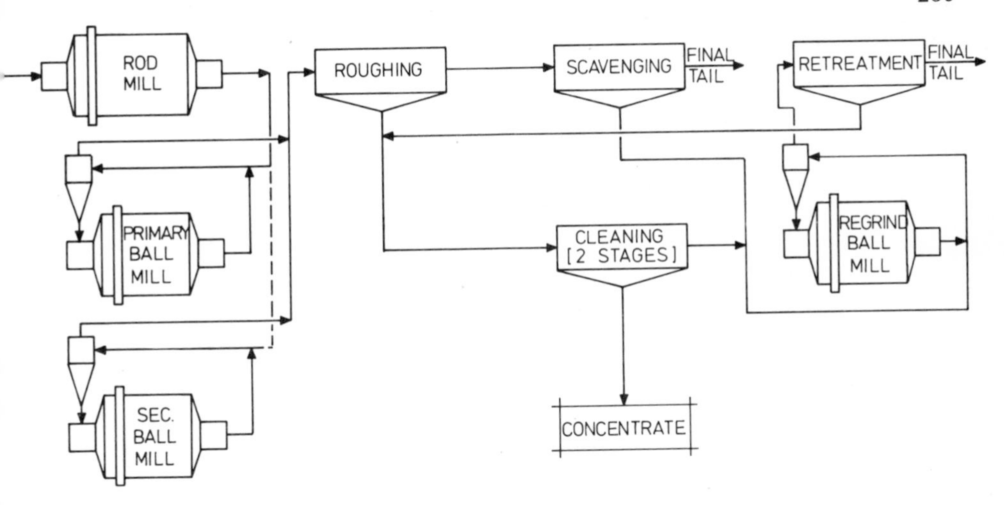

Fig. 13-5. Flowsheet of the copper ore grinding and flotation circuits, Mount Isa Mines Limited.

diameter by 5.55 m. Primary and secondary ball mills are overflow-type with dimensions of 5.03 m diameter by 6.1 m.

Primary ball-mill discharge gravitates to the primary cyclone feed sump where it joins rod-mill discharge. It is pumped to a Krebs D20B cyclopak using a 14/12 Warman Series A pump. Primary cyclone overflow and secondary mill discharge are classified with the overflow product pumped to flotation. Each sump throughout grinding and flotation is equipped with a variable-speed pump and a standby fixed-speed unit. Pump motor size is 300 kW on both primary and secondary ball mills.

13.3.4 *Instrumentation*

As a result of experience, instruments requirements both in the field and in the control rooms have been closely defined to minimise capital expenditure, maintenance problems and maintenance and operating labour establishment. The types of instruments and their location in the circuit are similar for both concentrators as described below.

Plant instruments. The instrument diagram for No. 4 Concentrator is included in Fig. 13-6. Instrumentation in No. 2 Concentrator is similar. Comments on the different units follow.

(1) Solids tonnage measurement is by a mechanical displacement belt weightometer.

(2) Slurry density measurement is by a radiation absorption gauge. Linearisation of the signal is carried out in the secondary unit but a final calibration is necessary in the computer. Smoothing of the signal is also carried out digitally rather than by analogue methods.

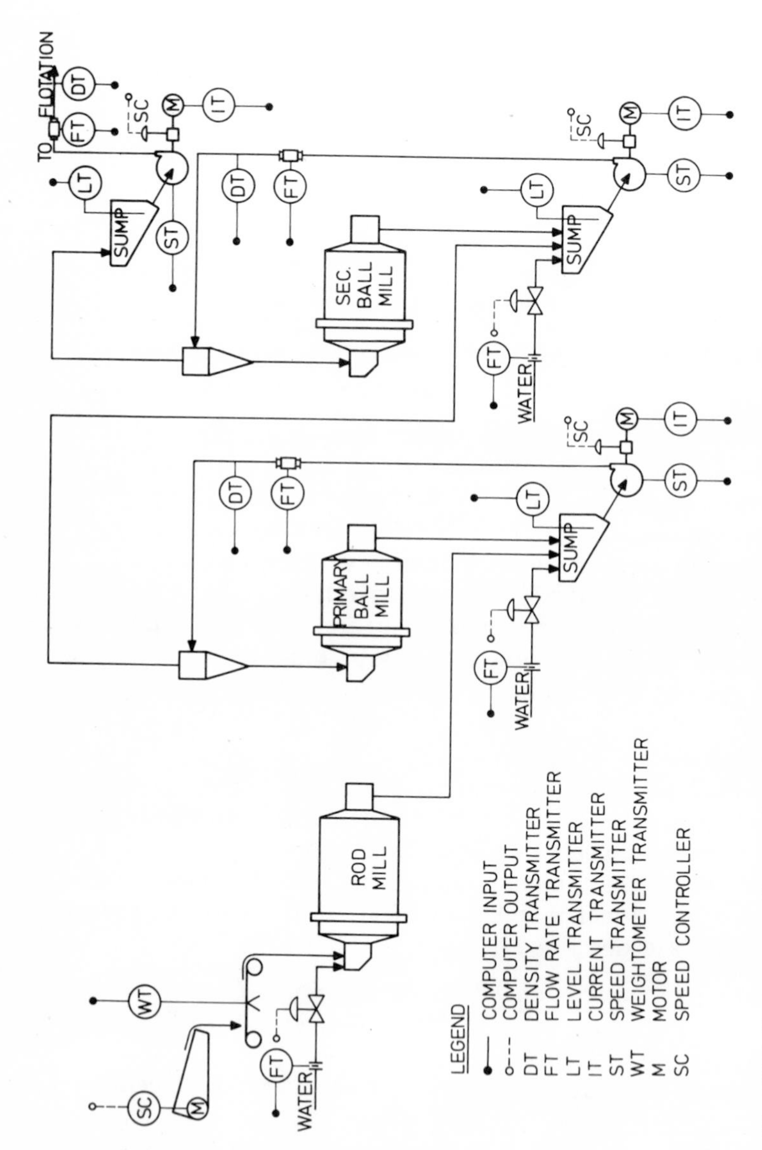

Fig. 13-6. Instrument diagram for copper ore grinding circuit, Mount Isa Mines Limited.

(3) Volume flow measurement is done by magnetic flow-meter primary elements. Rack-mounted secondary units provide a 4—20 mA output.

(4) Sump level measurement is by an air—water purged probe in the sump and differential pressure transmitter.

(5) Water flow measurement is by standard orifice plates with electronic differential pressure transducers.

(6) Water control valves are standard pressure valves after 4—20 mA current to pressure transducers. Control valves on rod-mill feed water are fail open but all other water addition control valves are fail shut. No valve positioners are used.

(7) Variable-speed motors are thyristor-controlled. This mode of speed control was chosen in preference to hydraulic couplings due to higher efficiency in utilisation of power even though capital costs were higher. A potential disadvantage of thyristor-controlled drives is that large, high-frequency voltage spikes may be fed back into the power supply. This requires that extra care be taken in isolating any digital logic circuits from such interference on the power supply. If possible, it is advisable that computer power be drawn off a different source from that of the thyristor-controlled drives.

(8) Cyclone-inlet pressure measurement (No.2 Concentrator only) is obtained using a water-filled diaphragm primary element inserted in the cyclone-inlet feed line.

Control room instruments. Both installations have fully centralised control with an arrangement of one room containing marshalling racks and the computer adjacent to the actual control room. In the case of No.4 Concentrator control room instrumentation is by Honeywell, while Foxboro SPEC-200 instruments are used in No.2 Concentrator.

Backup analogue control is provided on some grinding loops in No.4 Concentrator. However, with computer availability in excess of 99%, the additional cost of providing the facility is not considered justified. There is no analogue backup in No.2 Concentrator. Remote manual control is available from the control room in both concentrators.

13.3.5 *Computer details*

No.4 Concentrator. The original digital computer in No.1 Concentrator was a Hewlett-Packard 2114B, and this was replaced first by a 16K 2116C and in 1975 by the HP21MX and disc. Details of this computer system are:

HP21MX
32K of 16-bit semi-conductor memory
2.5-m word moving head disc
128 analogue inputs
40 analogue outputs

32 digital inputs
24 digital outputs
Teletype system console
Decwriter logging console
Decwriter operator's console
Paper tape reader and punch

Analogue inputs are in the ranges 1—5 V to 4—20 mA, while field outputs are in the range 2—10 V. Plant inputs are scanned every second. All software, with the exception of some device drivers, has been developed locally. The languages used are FORTRAN IV and HP ASSEMBLER. The presence of the disc and real-time operating system allows on-line program development. Over a third of disc space is allocated to a plant data file which maintains the most recent 25-h data of 380 variables stored on a 1-min frequency.

No. 2 Concentrator. A Foxboro 2/30 system is installed, primarily for compatability with the SPEC-200 instrumentation. Details of the system are:

Digital Equipment Corporation PDP 11/20
28K of 16-bit word memory
Fixed-head disc storage of 496K words
Teletype system console
Decwriter logging console
VDU operator's console
High-speed paper tape reader
High-speed paper tape punch
Programmable real-time clock
Power outage timer
144 analogue inputs
48 analogue outputs
24 digital inputs
24 digital outputs

The programming languages used are Foxboro IMPAC and FORTRAN IV. The Foxboro IMPAC software simplified implementation of standard control strategies and was utilised in bringing control loops on line as soon as the instrumentation was complete.

13.3.6 *Process control techniques*

The initial requirement in a control system is to develop stabilising control loops. These loops are similar in both concentrators. Before discussing particular control loops, one important comment can be made on grinding control. The process response is usually slow and derivative action is not required. The desired response is usually one approaching critical damping.

Rod-mill feed water. This variable is controlled to provide a constant rod-mill feed and hence discharge density. The percent solids in the rod-mill feed is calculated digitally and a P & I controller used to cascade a set point to another P & I controller setting a valve opening by reference to an orifice-plate flow signal. Advantage again is taken of digital computer capacity to set a delay on the rod-mill feed tonnage signal corresponding to the time lag between weightometer and rod mill.

Cyclone feed pump level. The aim of pump level control is to minimise the flow disturbances in cyclone feed. The pump sump is used to absorb short-term disturbances as level fluctuations rather than passing such disturbances to the cyclone feed flow rate. A P & I controller with a non-linear term to provide faster response at large errors is used.

Circulating load control. Grinding efficiency improves with increasing circulating load until the mill overload condition is reached and mill power draught begins to decrease. The aim with circulating load control at this stage has been to maximise circulating load by controlled water additions to cyclone feed pumps. Output from a P & I controller with cyclone mass flow input is cascaded to a controller which controls pump water addition. Large increases in circulating loads (10—20%) are obtained compared with manual control. This is further discussed in the next section.

Tonnage maximisation. This loop is being developed in the No.2 Concentrator, but has operated successfully at No.1 and No.4 Concentrators where 5% increases in throughput have been achieved at constant product size (Fewings, 1971). The prediction of product sizing is based upon the work of Rao (1965) where the sizing term also was calculated from cyclone parameters and water terms. From this relationship, a % − 200-mesh sizing equation was developed:

$$\log(\% - 200\text{-mesh}) = K1 - K2 \cdot RMF - K3 \cdot MFW - K4 \cdot P$$

where RMF = rod-mill feed, MFW = mass flow water to cyclone, P = cyclone operating pressure, $K1$—$K4$ = constants.

The pressure term was subsequently excluded and an equation of the format below has been found satisfactory:

$$\log(\% - 200\text{-mesh}) = K5 + K6 \cdot MFW - K7 \cdot MFO - K8 \cdot RMF$$

where MFO = mass flow ore to cyclone.

A model of this type is being developed for No.2 Concentrator tonnage maximisation. These equations contain rod-mill feed rate as a variable, and it is this variable which is regulated in the tonnage maximisation loop.

13.3.7 *Gains from grinding control*

Throughput gains of greater than 5% at constant grind have previously been established. Recent comparisons of the performance of No.2

TABLE 13-I

Comparison of flotation feed sizing before and after control

Period	Status of control	Average grind	Standard deviation
10 × 4 weeks	without control	58.2% minus 74 μm	0.8
5 × 4 weeks	with control	61.2% minus 74 μm	1.0

The difference between these means is significant at the 99.6% level.
Note: It is expected that the improved efficiency of grinding will be utilised by obtaining at least a 5% increase in capacity.

Concentrator primary mills before and after control have indicated significant increases in circulating loads (15—20%) and fineness of grind (Table 13-I) under control. Three factors contributing to the finer grind are: improved cyclone classification efficiency, increased circulating loads, and improved smoothness of operation.

Classification efficiency. The improved cyclone classification efficiency is illustrated in Fig. 13-7. At comparable operating conditions, the new system gives better classification of coarse particles to the ball mill and fines to the cyclone overflow. The gains can be attributed to: a control strategy which maximises water addition to cyclone feed, smooth cyclone feed conditions with variable-speed pumps, and installation of Krebs D20B cyclones.

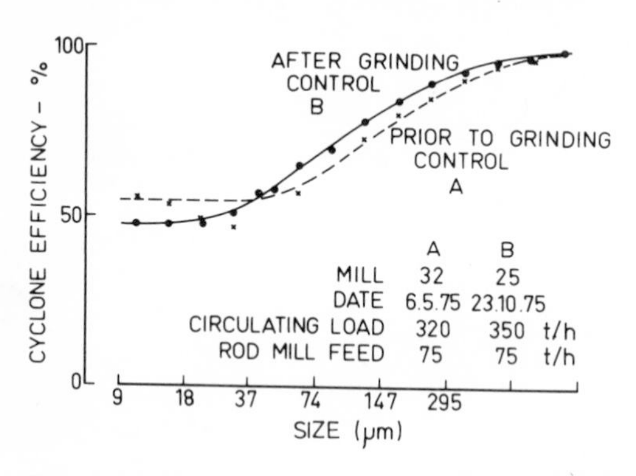

Fig. 13-7. Improved cyclone classification efficiency following installation of control system on lead—zinc grinding circuit, Mount Isa Mines Limited.

Circulating load control. Prior to computer control, circulating loads in the primary mills were in the range 200—300 t/h. The highest recorded was 300 t/h. Currently the circulating loads are controlled to 400 t/h.

13.3.8 *Maintenance support*

The high degree of automation in both concentrators makes it essential to maintain effective support for both instruments and computing equipment. A group of tradesmen are attached to each concentrator for instrument installation, repair, maintenance and calibration. A separate group provides similar support for the computer equipment.

13.3.9 *Conclusions*

Several years of digital computer control of grinding circuits have established the system as almost indispensable to grinding operation. Gains in grinding circuit stability have resulted, and increase in circuit throughput of at least 5% has been proven. Full centralisation of control has also contributed to improved overall performance and allowed a reduction in operating staff. Finally, the use of powerful disc-based computer systems has provided enormous advantages in software flexibility and control system development and testing. The quality and availability of data to the operator mean that operating decisions are taken on the basis of fact rather than supposition.

13.4 CASE STUDY 3. LARGE BALL-MILL—HYDROCYCLONE CIRCUIT AT BOUGAINVILLE COPPER LIMITED*

13.4.1 *Introduction*

The open-pit mining and processing operations of Bougainville Copper Limited are located at Panguna on Bougainville Island. Politically part of Papua New Guinea, Bougainville is geographically the northernmost island of the Solomon Islands group. The island is 192 km long by 48 km wide with Panguna located on the western slope of the Crown Prince Range in south central Bougainville at an elevation of 670 m. The terrain is precipitous and jungle-covered and the annual rainfall exceeds 5000 mm.

The Panguna deposit is a porphyry-copper type occupying an area of approximately 1600 m by 2300 m. Ore reserves are estimated to be 900,000,000 tonnes averaging 0.48% copper and 0.55 g/t gold. The mineralization is predominantly chalcopyrite, associated with gold and silver with some bornite, pyrite, magnetite and molybdenite also present.

Production commenced in April, 1972, and averages over 30,000,000 t of ore per year yielding approximately 650,000 t of copper concentrate assaying 29% copper, 30 g/t gold and 60 g/t silver.

---

* By Bougainville Copper Limited Staff.

13.4.2 *The BCL concentrator*

The run-of-the-mine ore is delivered by 100 or 170 t rear-dump haul trucks to a primary crusher comprising two 1.37 by 1.88 m gyratory crushers. The − 200-mm product is conveyed to a coarse ore stockpile with capacity of 140,000 t.

Ore reclaimed from the stockpile is fed onto seven 2.44 by 6.10 m double-deck vibrating screens. The −13-mm screen undersize is conveyed to a covered fine-ore stockpile while the screen oversize is crushed in seven 2.13-m secondary cone crushers. The secondary crusher product is conveyed to a separate tertiary screening facility containing twenty-two 2.44 by 6.10 m single-deck vibrating screens. The −13-mm screen undersize is conveyed to the 225,000 t capacity fine-ore stockpile. Screen oversize is returned to fourteen 2.13-m tertiary cone crushers, the product from which recirculates to the tertiary screening plant.

Variable-speed fine-ore reclaim conveyors located in separate tunnels beneath the fine-ore stockpile feed the nine primary ball mills each in closed circuit with a cluster of five 762-mm hydrocyclones.

The product slurry containing 40—50% solids is pumped to one hundred and eight 17-m^3 rougher-scavenger flotation cells. Rougher and scavenger concentrates are reground separately in three 3.0 by 6.1 m overflow ball-mills in closed circuit with eight 381-mm cyclones. The rougher concentrate then undergoes three stages of cleaning and the scavenger concentrate one stage of scavenging. The final concentrate is cycloned to remove oversize in two 507-mm cyclones prior to thickening to 65% solids in a 64-m diameter gravity thickener. A simplified flowsheet of the grinding and flotation circuits is given in Fig. 13-8.

The thickened slurry is pumped via a 26.6-km pipeline to the port of Anewa Bay where it is filtered in eight 3.1-m diameter rotary disc filters, dried in two 2.7 by 15.2 m rotary kiln dryers and stockpiled in a 60,000 t capacity storage-shed prior to shipment.

13.4.3 *Primary grinding circuit*

The nine primary ball mills are single-stage overflow mills, 5.5 m in diameter and 6.4 m long, driven by a 3,170-kW, 3.3-kV synchronous motor through an air clutch, pinion and ring gear. The mill operates at 12.45 rpm or 68% of critical speed.

The primary grinding circuit, as commissioned, was comprised of only eight mills, each in closed circuit with a cluster of eight 508-mm hydrocyclones fed by a fixed-speed 406 by 406 mm centrifugal slurry pump operating at 317 rpm. Cyclone overflow was pumped to the rougher flotation distributors by 305 by 254 mm slurry pumps while cyclone underflow was returned with new feed to the mill feed spout. Mill water addition was

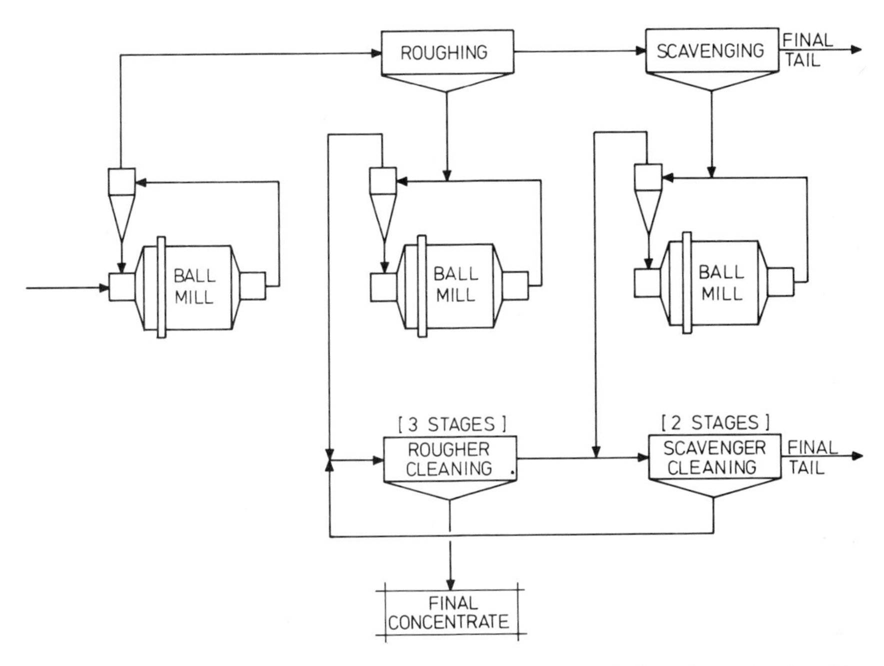

Fig. 13.8. Flowsheet of the copper ore grinding and flotation circuits, Bougainville Copper Limited.

generally to the mill discharge sump although facilities were available for addition to the head of the mill.

Following commissioning, it was found that the designed mill efficiency could not be achieved. This appears to be due to lower than expected breakage rates in the coarse fractions (+ 3360 μm) of the ore. The resulting high circulating loads (400—500%) exceeded the capacity of the cyclones and, unless the feed rate was restricted, caused frequent spigot blockages. Apart from the feed rate restrictions, the spigot blockages caused further production losses as the coarse material reporting to the overflow following a blockage frequently resulted in blocked overflow lines and bogged pumps and sumps, which took several hours to clear.

The coarse grind was also responsible for the very high wear rate of the rougher flotation mechanisms. For these reasons, the clusters of eight 508-mm cyclones were replaced by clusters of five 762-mm cyclones by early 1973. The spigot blockages still occurred, particularly on hard ore, but the frequency was significantly reduced. Although the grind was coarser and the flotation recovery lower than design, the installation of the larger cyclones resulted in the achievement of the copper production targets.

In June 1973, the ninth ball mill was commissioned. The major differences between this and the other eight circuits were the use of a split feed conveyor consisting of a fixed-speed weigher conveyor and a variable-speed reclaim conveyor and the provision of a variable-speed eddy-current coupling drive for the cyclone feed pump. The pump drive motor was also increased from 261 to 373 kW while the pump speed ranged from 210 rpm. The changes in the grinding circuit control system will be discussed in a later section.

Apart from the general improvement to the control of the grinding circuit, the variable-speed pump allowed maximum production to be maintained over the full 750-h life of the pump impeller where previously the drop in pump efficiency as the impeller wore, had restricted production. Because the spigot blockages continued and, in fact, increased with the higher efficiency of the variable-speed pump, the cyclone spigots were enlarged from 178 to 203 mm. Although the product sizing was not significantly affected when only four of the five cyclones were used, the increased circulating load with five cyclones exceeded the capacity and available power of the pump.

Based upon the findings from operation of the ninth mill, major modifications to all nine grinding circuits were carried out in early 1975. The 406 by 406 mm cyclone feed pumps were replaced by 508 by 457 mm pumps. The fixed-speed direct v-belt pump drives were replaced by variable-speed eddy-current coupling drives with the pump being driven through a right-angled gear-box. The drive motors were increased in power from 373 to 560 kW. At the same time, the cyclone spigots were changed from white iron to a bonded silicon carbide ceramic. The speed of the larger pumps was ranged from 180 to 280 rpm.

Since the installation of the larger pumps, no problems with pump capacity and pump power have occurred. The slower speed and larger size of the pump have resulted in a significant reduction in pump wear with services now being scheduled every 1100 h instead of 375 h with the smaller pumps. This increase in plant availability was only possible, however, because of the increased life of the ceramic cyclone spigots as the frequent service schedule was due in part to the 450 h maximum life of the white iron spigots.

While the pump capacity is no longer a circuit limitation, the cyclone cluster still causes problems. Experiments with 203-mm spigots indicated serious problems would occur as the spigots wore out to 215—230 mm. The problems are associated with the restricted outlet from the cyclone underflow collection box to the ball-mill feed spout. With spigots of 215 mm or greater, the high circulating load resulted in the underflow box filling up and eventually causing all five spigots to block simultaneously. Therefore, until the underflow box and feed spout are redesigned to eliminate the flow restriction, 190-mm spigots will be used and replaced on wearing to 215 mm.

Mill operating conditions are quite variable with feed rates ranging from 340 to 650 t/h per mill. Overflow density is normally kept within the range 40—50% solids and is adjusted to achieve a particular recovery target.

TABLE 13-II

Typical mill-circuit sizings and flows, Bougainville Copper Limited.

	Ball-mill feed	Ball-mill discharge	Cyclone (U/Flow)	Cyclone (O/Flow)	Sump water
Solids (t/h)	2301	2301	1910	391	
Water (t/h)	670	670	654	394	378
Sizings, micrometres (weight on, %)					
9510	11.6	3.0	10.4	—	
4760	22.2	11.4	19.5	—	
2380	11.7	7.4	9.6	—	
1190	9.0	8.2	8.8	—	
595	10.4	12.2	12.2	2.6	
297	11.4	15.9	14.2	12.5	
149	5.8	9.7	6.6	18.7	
74	4.3	7.5	4.3	17.7	
−74	13.6	24.7	14.4	48.5	

Overflow sizings vary from 45 to 55% − 74 μm and 15 to 30% + 297 μm. Typical circuit sizings and flows are given in Table 13-II.

13.4.4 *Grinding circuit control*

The control instrumentation associated with the grinding circuit is as follows:

(1) A strain-gauge weightometer to measure the new feed rate and a variable speed conveyor for controlling feed rate.

(2) A differential pressure transducer to measure sump level.

(3) Gamma density gauges to measure cyclone feed and overflow density.

(4) Orifice plates for water flow rate measurement and pneumatically operated butterfly valves for flow control.

(5) With the variable-speed pumps only, a magnetic flow meter to measure the cyclone-feed volumetric flow rate.

With the fixed-speed cyclone feed pump in the grinding circuit, the control loops were as follows:

(1) Mill discharge sump level was controlled by sump water addition.

(2) New feed rate was cascade-controlled by the cyclone overflow density.

(3) Overflow density was controlled by the new feed rate ratioed to the sump water addition.

(4) Water to the head of the mill was controlled from a manual set point.

All controllers are standard proportional plus integral analog controllers with manual, automatic and cascade facilities.

This system suffered from a number of problems. On restarting the mill after a shutdown or spigot blockage, the control loops had to be run on manual because restoration of both sump level and overflow density were incompatible objectives with a low circulating load. A more serious problem was that with changes of ore hardness, the control loops became unstable unless the controller gains were adjusted. Because of the nature of the mining operations at Panguna, frequent changes in ore hardness occurred. Control during water shortages was also a problem as any reduction in volume of the sump water addition had to be compensated for by a corresponding increase in the solids volume in order to maintain sump level with the fixed-speed pump. As sump level had to be maintained to prevent the feed pump from cavitating and to ensure reliable cyclone performance, water shortages resulted in either a shutdown of one or more mills or, at best, a very coarse grind and consequent low recovery.

With the installation of the variable-speed drive pump on the ninth mill, the control loops were as follows:

(1) Mill-discharge sump level was controlled by the cyclone feed pump speed.

(2) Overflow density was controlled by sump water addition ratioed to the new feed rate.

(3) New feed rate was controlled by the mass flow rate of solids to the cyclone cluster.

(4) Water to the head of the mill was still controlled from a manual set point.

The above control system overcame all of the problems associated with the previous system without incurring any further significant problems, the main problems now being caused by out-of-calibration instruments.

The instrumentation and control loops for the larger cyclone feed pumps are essentially identical to those of the ninth mill.

13.4.5 *Digital computer control*

An 8K Digital Equipment Corporation PDP-11 minicomputer was installed in May, 1974, to carry out research on digital control of the grinding circuit. The installation was designed to control a single mill with any future expansion dependent upon the results obtained during the course of the research project. The computer was also required to provide complete data-logging facilities on one mill to acquire reliable data on mill operation. The ninth ball mill was chosen for computer control because of its variable-speed cyclone feed pump.

The initial control scheme essentially duplicated the analog control scheme already installed on the mill although a number of modifications were incorporated.

Because the circuit limitation was normally the cyclone spigot capacity, the new feed rate was controlled by the cyclone underflow mass flow which was inferred from the cyclone feed mass flow and new feed rate. This had the effect of maximising the feed rate for any ore hardness without requiring the frequent adjustments of the cyclone feed mass flow set point associated with the analog control system. The cyclone underflow mass flow set point was also continuously updated to allow for spigot wear which had a significant effect on the circulating load over the 375 h life of the spigot.

A routine was incorporated to detect the occurrence of a spigot blockage based upon changes in overflow density. This routine cut off the feed until the blockage clears, restored the feed and controlled the circulating load back to the required value. Additional loops controlling circulating load included to prevent the cyclone feed pump reaching maximum speed and losing control of the sump level as the pump impeller wore, and to prevent the pump motor from drawing maximum power and tripping which occurred with new pump impellers and casings. With the larger pumps, neither of these routines appear necessary.

A variety of data-logging and alarm-monitoring routines were written to keep the control room operator completely aware of the mill operating conditions.

While the above program has been operating satisfactorily for a considerable period, this control system does result in an extremely variable product sizing whenever any changes in ore hardness occur. There is currently under development a sizing control program which has the aim of maximising the mill throughput at a particular product sizing, the sizing being selected to achieve a target recovery required for a particular ore type.

This program is based upon a mathematical model of the grinding circuit which infers the product sizing from readily measureable circuit variables. Currently this model infers the percentage of $+297$-μm product using as parameters new feed rate, cyclone-feed volumetric flow rate and cyclone feed density. Variations in cyclone overflow density and underflow mass flow are used to control the actual product sizing.

Development work on this control system is expected to be complete before 1977.

13.5 CASE STUDY 4. ROD-MILL, BALL-MILL—RAKE CLASSIFIER CIRCUIT AT NEW BROKEN HILL CONSOLIDATED LIMITED*

13.5.1 *Introduction*

The New Broken Hill Consolidated Limited (NBHC) concentrator treats high-grade lead—zinc sulphide ore from the southern end of the line of lode at Broken Hill, New South Wales, Australia. Approximately 1.5 million

* By New Broken Hill Consolidated Limited Staff.

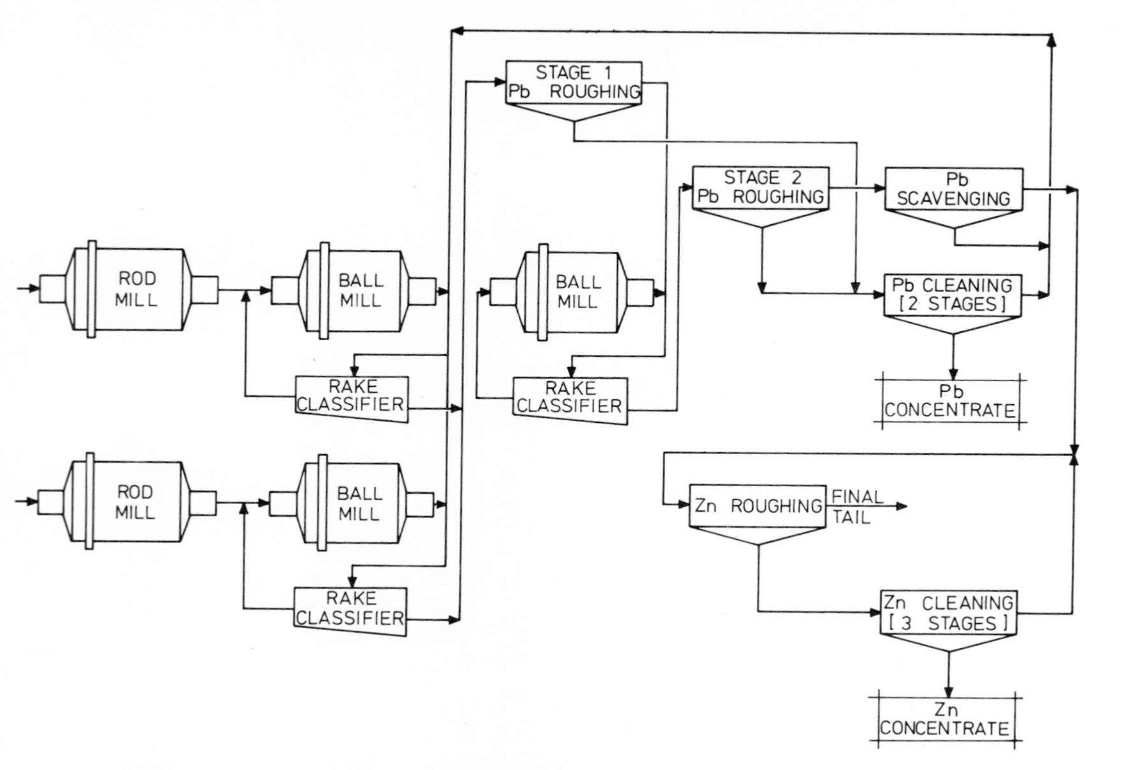

Fig. 13-9. Flowsheet of the lead—zinc ore grinding and flotation circuits, New Broken Hill Consolidated Limited.

tonnes of ore are mined and milled annually at an average feed grade of 8.8% lead and 13.9% zinc. The metals occur in galena and marmatite, respectively. The ore is treated using selective flotation to produce separate lead and zinc concentrates.

The concentration process at NBHC is inherently efficient but major variations in feed properties, particularly ore hardness and grade, require alterations in plant conditions which are best achieved through on-line sensing and control. At the planned treatment rate of 240 t/h, the grinding capacity is barely adequate. Under manual operation, poorly controlled water addition to the rake classifiers often resulted in the overloading of the classifiers and ball mills. To avoid such overloads, operators tended to run the circuit at elevated densities with a consequent decrease in fineness of grind. Flotation performance suffered accordingly.

The ability to operate safely at the process limits, that is, to make maximum use of grinding capacity while eliminating ball-mill overloads, and to stabilise the sizing distribution and flow rate of feed to flotation, was the main incentive for the development of a computer control scheme for NBHC.

Circuit arrangement. The NBHC grinding plant consists of two parallel primary circuits and one secondary circuit. The flowsheet of the grinding and flotation circuits is given in Fig. 13-9. Each primary section consists of a rod mill discharging into a rake classifier in closed circuit with a ball mill. The combined classifier overflows are fed to a bank of lead-rougher flotation cells in which 65—75% of the galena is recovered. This is a recent circuit modification which has reduced the amount of galena ground into the slow floating ultra-fine fraction. Less flotation capacity has been required to almost maintain the previous level of lead recovery. This has enabled an increase in zinc recovery because of the availability of the spare capacity for zinc-roughing duty — an overall net gain. The primary lead-rougher tailing is fed to two parallel rake classifiers in closed circuit with a single secondary ball mill. The secondary circuit limits the capacity of the grinding circuit. The combined secondary classifier overflows provide the feed to the secondary lead-rougher circuit and, subsequently, the zinc circuit.

Instrumentation. The instrument diagram for the grinding circuit is given in Fig. 13-10. In 1971, a Digital Equipment Corporation PDP 11/20 was installed at NBHC to act as a dedicated processor for on-stream analysis and computer control of plant operations. The proprietory high-level language, FOCAL, is used. The computer is equipped with 12K of core, 32 analogue to digital channels to monitor instruments in the plant, 12 digital to analogue channels to initiate control action and 48 digital input/output channels which are used to monitor on-stream analysis (OSA) signals and alarms. As a first step towards more comprehensive control in the concentrator, priority

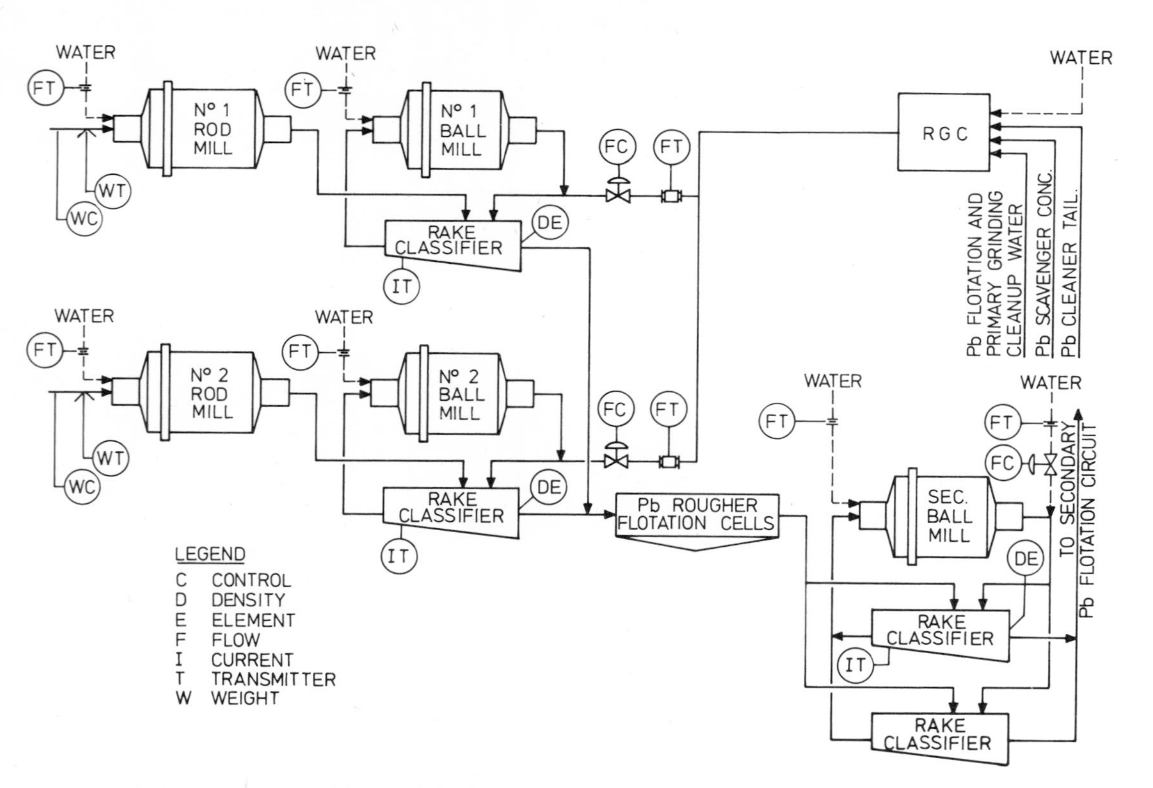

Fig. 13.10. Instrument diagram for lead—zinc ore grinding circuit, New Broken Hill Consolidated Limited.

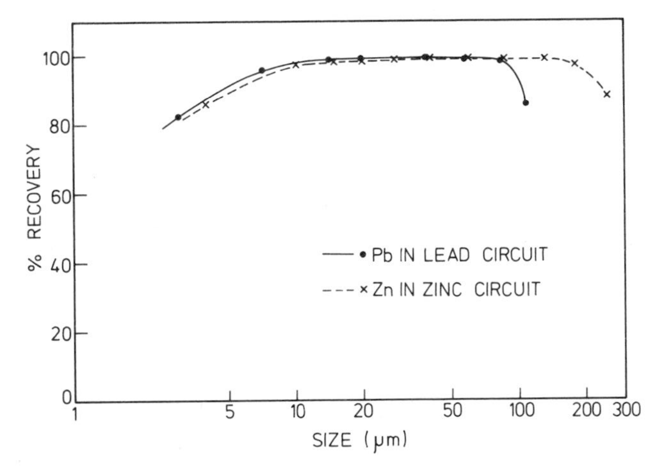

Fig. 13-11. Particle-size—mineral-recovery relationship for major constituents in the NBHC ore.

was given to the control of the grinding section. Existing analogue controllers were incorporated in the control strategies developed, and the digital channels to monitor instruments in the plant, 12 digital to analogue additional control loops. The inherent flexibility of the computer assisted development.

Control schemes in the NBHC grinding section have relied on two measurements; classifier overflow density, and rake classifier power draft in each ball-mill—rake classifier circuit. Classifier overflow density gives a good indication of classifier product size and rake classifier power draft provides a qualitative measure of mill circulating load. Moreover, both measurements have proved extremely reliable, requiring very little attention or maintenance.

A typical size-recovery plot for major constituents at NBHC is shown in Fig. 13-11. It can be seen that a very high plateau exists for both lead (10—100 μm) and zinc (10—150 μm). Because of this feature, it was considered unnecessary to have very accurate measurement of grinding section product size which would necessitate high-maintenance equipment. Instead, particle size of the secondary section product is inferred from an empirical equation involving classifier density and ore feed rate. The equation is:

$$\% +63\,\mu\text{m} = 0.0667 \cdot (\text{ore feed rate in t/h}) + 26.7 \cdot (\text{classifier overflow density in t/m}^3)$$

The use of classifier density alone is considered sufficiently accurate for primary classifier control. Overflow densities are measured by Z-CRA probes (Hinckfuss and Stump, 1971), using the absorption of 660 keV gamma radiation from a Caesium-137 source.

13.5.2 *Grinding section control schemes*

The performance of the grinding circuit is controlled using variations in the water addition to the rake classifiers and ball mills and variation in ore feed rate. Control schemes have been used which:

(1) maximised the fineness of grind at a manually-set feed rate by maximising the circulating loads in the ball-mill—rake classifier circuits;

(2) maintained constant density in the rake classifier overflows at a manually-set feed rate and only reverted to the circulating load criterion when overload conditions of the ball mills were approached; and

(3) maximised the rate of throughput at a constant product size by having both water addition and ore feed rate under computer control.

The development of these control systems and the disadvantages associated with them are discussed.

Control to maximise grind at a manually-set ore feed rate. Initial control of the grinding section aimed at eliminating the classifier and ball-mill overloads. The current drawn by the rake classifier motor when the ball mill was on the point of spilling was determined by observation. The average current was used as the measured variable in a digital proportional plus integral cascade control loop which controlled water addition to the classifier baths.

Varying the water addition to the classifier altered the quantity of sands returned to the ball-mill. The more water added, the lower the density in the classifier bath, the finer the average particle size in the overflow and the greater the weight of sands returned to the mill. Water addition was controlled to maintain the maximum circulating load consistent with the capacity of the mill as defined by the classifier current set point. Therefore, the maximum fineness of grind was obtained at any set ore feed rate.

Water was added to the three ball-mill circuits via three independent control loops. The flow rate of water to the rod-mill feed was set manually as it has been in all subsequent control schemes. The flow of returns to the primary grinding circuits (RGC) was also set manually to a fixed level. The returns consisted of lead cleaner tailing and lead scavenger concentrate streams, and cleanup water from the primary grinding and lead flotation sections. Water only was added to the secondary circuit.

This control system, used between October, 1971, and July, 1972, worked reasonably well; the number of ball-mill overloads was reduced considerably while metallurgy improved. However, a disadvantage of the system was that, because the volume of water to each classifier was constantly changing, the flow rate to the flotation circuit fluctuated. Also, because a high circulating load involved a high water addition, excessive coarse material was swept to the rake classifier overflow, thus reducing the advantage of a high circulating load. The practice of using water control over and above a fixed RGC addition to the primary circuits was often undesirable. If the use of cleanup water was considerable, a large surplus of RGC formed which often became unmanageable.

To eliminate the disadvantages of this control system, a new scheme was introduced in July, 1972.

Control to maintain constant product size at a manually set ore feed rate. The buildup of RGC was avoided by controlling the addition of RGC, rather than water, to the primary rake classifiers. Make-up water is added to the RGC sump to maintain constant level in the sump. When cleanup water and middling returns from flotation are plentiful, minimal make-up water is required. The only disadvantage of this arrangement has been an increase in the wear rate of rubber sleeves in the automatic valves controlling RGC flow.

The new control scheme involved the operation of the rake classifier overflows at constant density when circulating load conditions permitted. At constant classifier density and ore feed rate, the volume flow rate and product sizing to flotation can be assumed to be steady. The density, measured by Z-CRA probe, was the measured variable in a cascaded control system with water or RGC flow as the controlled variable. Control reverted to the circulating load criterion if the critical current draft in the classifiers was reached, that is, current replaced density as the measured variable. Control returned to density control when conditions eased.

This control scheme produced a steady feed more conducive to flotation.

Control to maximise throughput at constant product size. In the previous control scheme, ore feed rate was set manually. A further development, and the scheme currently used at NBHC, uses computer control of feed rate to maximise throughput at a constant product size (Stump and Roberts, 1974; Whiten and Roberts, 1974). As described previously, approximate product size is inferred directly from the classifier overflow density in the primary circuits, but, in the secondary circuit, size is calculated from an empirical equation involving both density and ore feed rate.

The control scheme is presented graphically in Fig. 13-12. Line *ABC* is the set point for water control. While the circulating load is below the safe maximum, water control produces a constant product size *AB* in each of the classifier overflows. If, however, the safe load is exceeded, the water loop uses classifier current as the measured variable to maintain the circulating load at *BC*.

The computer calculates the two errors: E_A = set point classifier amps − actual classifier amps; and E_D = actual classifier density − set point classifier density.

The errors, E_A and E_D are weighted so that one unit of classifier amps error has the same effect as one unit of classifier density error. (At NBHC, the weighting is $1E_A + 5E_D = E_{DW}$). The expression $E_{DW} - E_A$ is evaluated and if it is positive, E_A is used as the input to the proportional plus integral water controller, otherwise E_{DW} is used as the controller input. A change in water addition rate always moves the status of the particular circuit in a direction parallel to *DE*.

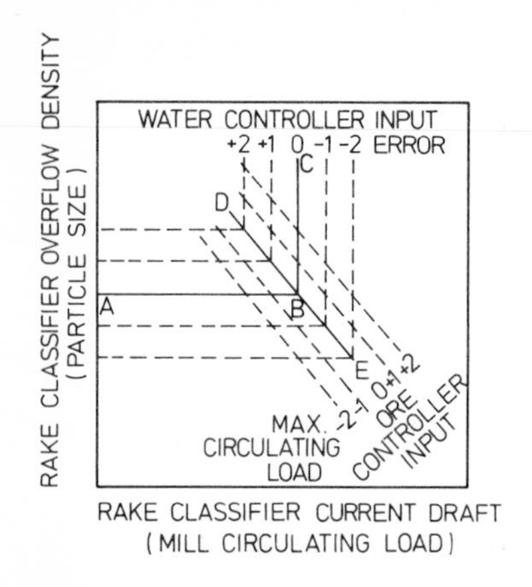

Fig. 13-12. Graphical representation of the NBHC control scheme. Objective is to maximise throughput at constant product size.

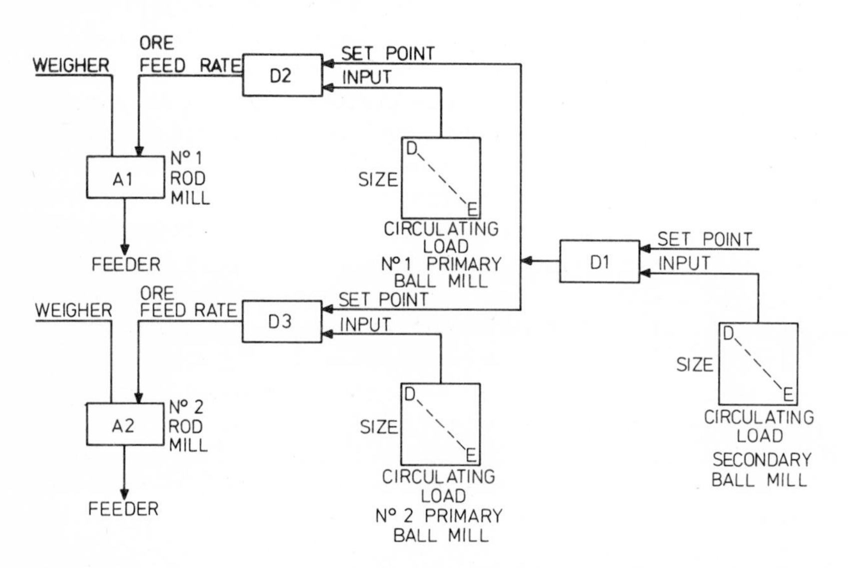

Fig. 13-13. The system of cascade controllers for regulation of ore feed rate at NBHC.

At NBHC, no facility exists for the surface stockpiling of crude-ore and fine-ore bin capacity is sufficient for only four milling shifts. Thus, a close dependence exists between the mining and milling sections of the operation. Periodically, the concentrator is under pressure to maintain maximum possible throughputs at the expense of optimum product size to avoid causing haulage delays. At all other times, however, tonnage is controlled by computer to maximise throughput at a constant product size.

Tonnage is controlled by a cascade system of five controllers as shown in Fig. 13-13. For each ball-mill—rake-classifier circuit the value of error

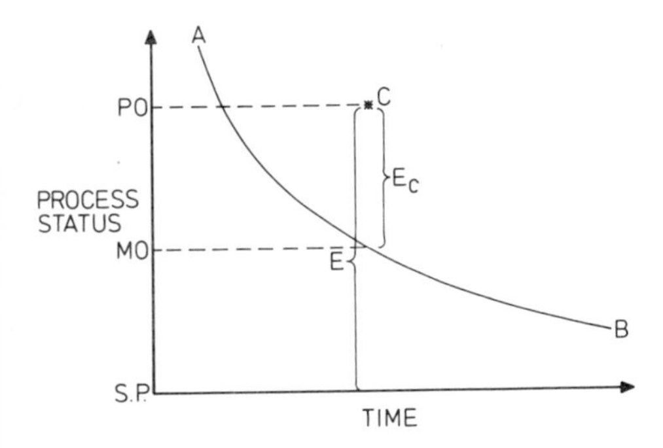

Fig. 13-14. Dynamic compensation of controller action. ***AB*** **is the predicted process response,** ***C*** **is the actual response,** ***E*** **is the normal controller action and** ***EC*** **is the compensated controller action.**

$E_{DW} - E_A$ is used directly in one of the digital controllers. Line *DE* is the set point for the controller, $E_{DW} - E_A$ being zero. A positive error in $E_{DW} - E_A$ will decrease, and a negative error increase, the feed rate. The output of $D1$ indicates the capacity available in the secondary circuit and provides corrected set point for $D2$ and $D3$. The outputs of $D2$ and $D3$ indicate the capacity in each of the primary circuits and provide the set points for analogue controllers $A1$ and $A2$. These controllers, which are proportional plus integral plus derivative, control the ore feed rate to each rod mill. Controllers $D1$, $D2$ and $D3$, which are proportional plus integral, are dynamically compensated to allow for the process lag.

The dynamic compensation uses a simple model to predict the response to a controller action. Let the process response and the model prediction be as in Fig. 13-14.

Line *AB* is the model output which predicts the process response to the controller action and *C* is the actual process output. Normal controller action would include error *E* such that: $E = PO - SP$. The compensated controller action uses the error E_c which is: $E_c = PO - MO$. The compensated controller therefore corrects for offset E_c rather than *E*.

The resultant action of water and tonnage control is to move the status of the secondary grinding circuit to point *B* (Fig. 13-12), that is, maximum product at constant product size. Water requirements are reviewed every minute and compensate for short-term local fluctuations while tonnage control, reviewed every five minutes, accounts for long-term changes in ore characteristics.

13.5.2 *Discussion*

The computer installation at NBHC represented a comparatively low capital investment. The system involved the minimum computer configuration, included existing analogue controllers and relied upon on-line sensing

TABLE 13-III

Metallurgical results at NBHC before and after computer control of the grinding circuit.

	% Pb	% Zn	Pb recovery	Zn recovery
1970:				
Feed	7.98	14.54		
Pb concentrate	75.72	3.61	93.5	2.4
Zn concentrate	1.07	52.11	3.4	92.3
Residue	0.38	1.20	3.1	5.3
1973:				
Feed	8.7	13.8		
Pb concentrate	76.5	3.3	95.3	2.6
Zn concentrate	0.87	53.1	2.4	93.8
Residue	0.31	0.77	2.3	3.6

units which required little attention and maintenance. One metallurgist supervises the control computer while maintenance of field equipment is included in the duties of the mill instrument fitter.

The system is unusual in that it is housed remote from the operations rather than in a control room, and can be left unattended provided there are no hardware or instrument failures.

Operator acceptance has been extremely good; the introduction of video displays at the instrument consoles has considerably improved the operators' confidence in the system.

Significant improvements in plant operation have resulted from the application of computer control to the grinding circuit. Ball mill overloads and downtime associated with them have been eliminated. The closer control of feed sizing and flow rate to flotation has assisted in improving metallurgy. The metallurgical results for the years 1970 (prior to computer control) and 1973 are shown in Table 13-III.

Improvements in lead metallurgy can be largely attributed to grinding control although zinc metallurgy was also improved by circuit modifications.

13.6 CASE STUDY 5. GRATE BALL-MILL—HYDROCYCLONE CIRCUIT AT THE SILVER BELL UNIT OF ASARCO INCORPORATED*

13.6.1 *Introduction*

The Silver Bell Mine is a porphyry-copper—molybdenum operation of 11,000 short tons per day. The sulphide ore body consists of a silicate matrix containing 0.5% copper as chalcocite, chalcopyrite and some oxides; 2—5%

*By ASARCO Inc., Silver Bell Unit Staff.

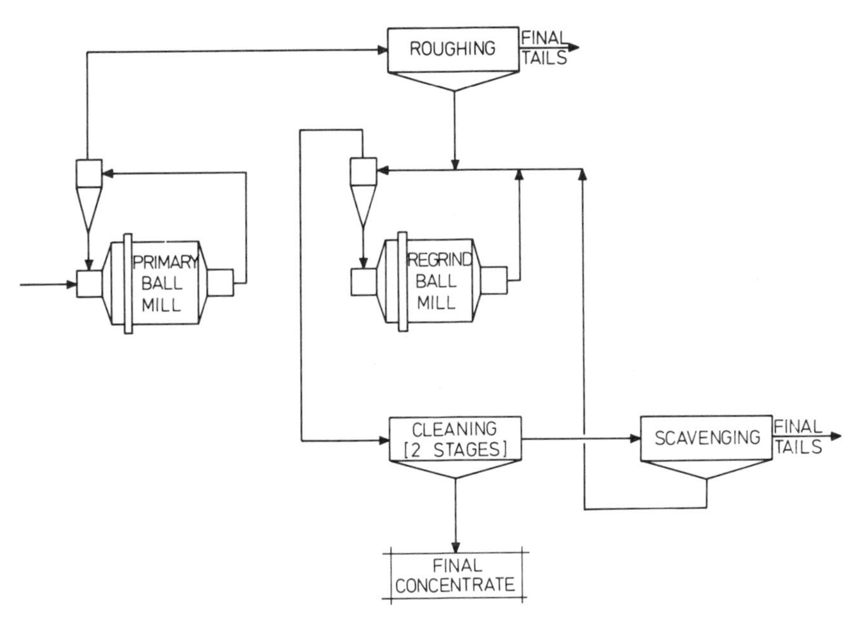

Fig. 13-15. Flowsheet of the copper ore grinding and flotation circuits, ASARCO Inc., Silver Bell Unit.

pyrite and small amounts of galena, sphalerite, magnetite and molybdenite are included. Molybdenite is recovered as grade and market conditions dictate. The average Bond work index of the ore is 13—15. Oxide ore consisting of copper carbonates and chrysocolla is processed via dump leaching and cementation.

A general flowsheet and equipment data for the Silver Bell concentrator is included as Fig. 13-15. Sulphide copper recovery averages 90—95% and molybdenum recovery, when sought, is typically 60—70%.

The Silver Bell mill feed often operates from single drawpoints in two open-pit mines, and the ore hardness can vary widely two to three times per shift. A typical operating rate change is an increase or reduction in individual mill throughput of 20 t/h on an 80 t/h average in a period of 0.5—1.0 h.

The coarse-ore crushing facility is also operating at or exceeding design capacity for most ore types. The resulting necessity that classification screen size must often be changed imposes another major disturbance on grinding circuit operation. In the semi-automatic system, sustained cycling of the cyclone overflow sizing and mill overloads occur when hardness or feed size changes are frequent; this disrupts the performance of unit operations downstream of the grinding circuit.

Grinding circuit. The grinding section consists of six identical lines feeding two copper flotation sections. Each line consists of a 3.20 by 3.66 m Allis Chalmers grate discharge, scoop-fed ball mill in closed circuit with four

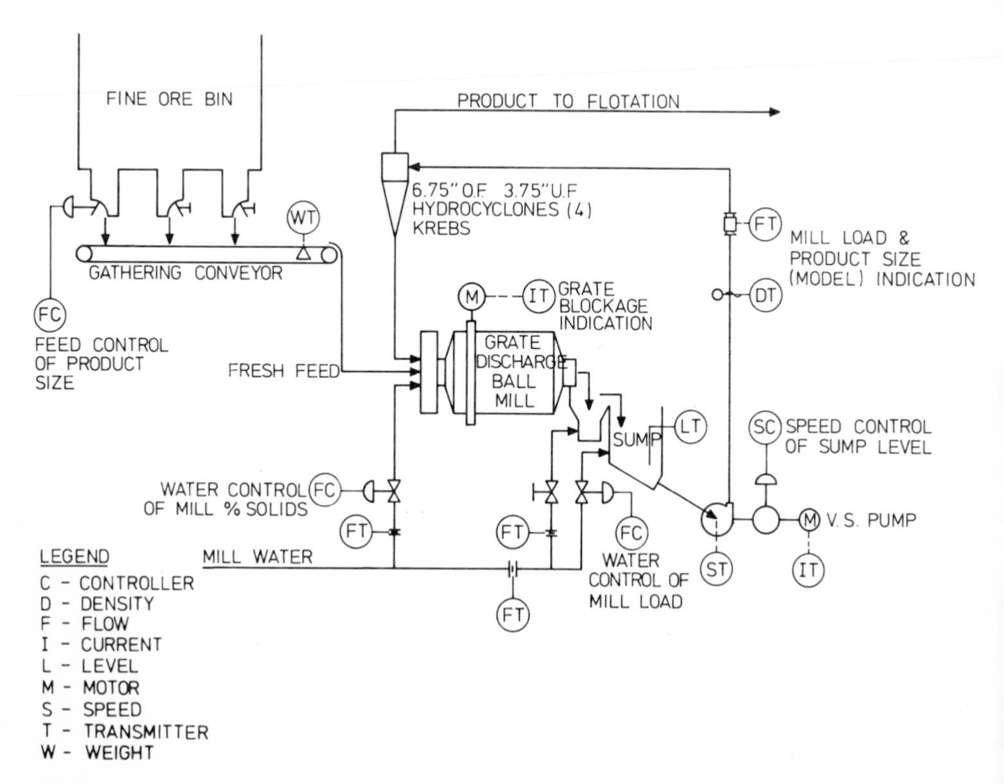

Fig. 13-16. Instrument diagram for copper ore grinding circuit, ASARCO Inc., Silver Bell Unit.

Krebs D20 cyclones in parallel. Fresh feed is added to the feed end of the ball mill with the cyclone underflow and about 20% of the total water addition; the remaining water is added to the ball-mill discharge.

Instrumentation. Fig. 13-16. is a flow diagram of the grinding circuit showing all installed field instruments and their locations.

The existing Merrick weightometer and Honeywell actuator/positioner at the fresh feed control point were satisfactory for conversion to the new control strategy. An I.P. transducer was installed at the actuator to convert the current output to the control point to the required 3—15 psi. A MV/I transmitter was added to the belt scale to convert the noise-susceptible 0—50 MV output to current.

Standard flange-type orifice flow meters were installed on the water addition points at the feed and discharge end of the mill, and at the ball trap. Electronic differential pressure transmitters provide current output proportional to the water flow at each of these points. Pneumatically actuated and positioned diaphragm control valves were placed on the feed and sump water addition points. Diaphragm valves were specified to provide the maximum range of controllability of the flow, a requirement of the control system. An I.P transducer was mounted on the valve actuator to convert current input to the required 3—15 psi for valve drive. The actuators are of the diaphragm-operated, pressure-closing variety to provide fail-open capability.

Cyclone feed sump level is measured by a back pressure bubble tube with air and water purge; the water was provided to minimise plugging. Back pressure is converted to current output by an electronic differential pressure transmitter.

The fixed-speed drive train on the cyclone feed pump was converted to variable-speed by installation of a hydrodynamic drive between the motor and pump. Specification of this type of unit was based on its durability and availability to standard mill maintenance personnel. The slow response and limited control range typical of this type of drive are of no consequence in this application. The drive is controlled by a pneumatically actuated scoop level.

Computer system. A Hewlett-Packard 9600A computer and peripheral system was installed to accept analogue inputs from field sensors and transmit calculated analogue outputs based upon control strategy programming. The system consisted of the following basic components:

(1) 2100A computer mainframe with 16K core memory and time base generator for real-time control programming on a core-resident, single-purpose system.

(2) ASR-33 teleprinter for hard copy print and on-line program access.

(3) High-speed tape input/output for off-line programming.

(4) High-level input subsystem with 32/16 channel single-ended/differential analogue to digital input capability.

(5) Output subsystem with twelve component relay output and four channel digital to analogue output capability.

(6) All required software for Fortran and Assembler programming, system diagnostics, control-system drivers and program editing.

13.6.2 *Automatic control*

The purpose or objective of the control scheme is to maximise the fresh-ore feed throughput while maintaining a specified grind in the milling product. In this control system the grind is predicted from a size prediction equation relating certain process parameters to the percentage of +65-mesh material in the cyclone overflow. The form of the equation is as follows:

$$\ln \% + 65 = K_0 + K_1 \cdot MFO + K_2 \cdot MFW + K_3 \cdot FOF + K_4 \cdot TW \qquad (13\text{-}1)$$

where K_0 through K_4 are constants, MFO is the mass flow of ore to the cyclones, MFW is the mass flow of water to the cyclones, FOF is fresh ore feed into the mill and TW is the total mass of water being added to the circuit. The predicted grind is used to determine the incoming feed rate and as such gives a long-term control of the particle size. The circuit "load" is the mass flow rate of ore to the cylclones and is calculated from signals generated by a magnetic flow meter and a nuclear density gauge on the cyclone feed line. The ore density, which must be given for the mass flow calculations, is preset at a value of 2.68 which is the most common plant feed ore density.

The computer alters four final control elements in four separate control loops in the control system. The four control loops determine the fresh-ore feed rate, feed end water addition, discharge sump water addition and the cyclone feed pump speed. The main function of each control loop is as follows:

(1) the fresh-ore feed loop is used to control the percentage of +65-mesh material in the cyclone overflow;

(2) the feed end water addition or primary water loop is used to control the ball-mill discharge percent solids;

(3) the ball-mill discharge sump water addition or secondary water is used to control the amount of material reporting back to the mill from the cyclone underflow;

(4) the sump level is the circuit stabilising loop with a main function of smoothing the cyclone feed flow thus giving stable particle-size prediction.

Pulp solids content in mill. The first major problem encountered when the digital control system was put on-line was that the mill load fluctuated to such an extent that computer control could not be maintained. This was due to variations in the pulp density inside the grade mill (W.E. Horst, private communication), which were caused by the primary water loop being operated to ratio the primary water to the fresh feed only. The first attempt to solve the problem involved calculating the apparent ball-mill discharge percent solids from the measured mass flow of ore and water to the cyclone minus the water addition to the mill discharge sump.

Initially the control of the primary water and secondary water was done in two completely separate loops but when the primary water loop was changed from being a ratio controller to a ball-mill discharge solids controller, a detrimental interaction between the primary and secondary water additions was observed. If the ore became harder, more material was returned to the mill increasing the discharge solids and the mass flow of material to the cyclones. The increased discharge percent solids caused more primary water to be added and ultimately caused both water additions to oscillate resulting in a loss of system control. This problem was solved by calculating the total water (TW) required to control the mill load and then subtracting the necessary primary water (PW) addition for maintenance of mill discharge percent solids. The mill discharge water addition was then set at the difference ($TW - PW$).

Variations in specific gravity of ore. An unusual control problem resulted from a particular ore type which occasionally entered the plant. This ore was of a higher s.g. (2.8—3.0) than the more normal ores and it produced a finer than normal product size distribution under manual control.

Because of the higher s.g. the calculated ore and water mass flow rates were inaccurate in opposite directions. That is, the mass flow of ore was actually lower and the mass flow of water was actually higher than the calculated values. With computer control the result of this error was to predict a coarser than actual grind, thus lowering the fresh-ore feed rate. In addition the high mass flow of ore term caused the control system to try and bring the load down by decreasing the total water. This drop in total water caused the grind prediction to be coarser thus reducing the fresh-ore feed rate further.

Initially the appropriate control set points were manually adjusted to make the necessary corrections for the high-density ore. That is, the mass flow of ore set point and the ball-mill discharge solids set point were raised. This effectively kept the mill from unloading but increases in the discharge solids set point did not prevent the secondary water (load control) from closing. This meant that the primary water was controlling the mill load. In order to prevent this from taking place, a maximum value was placed on the primary water. This was possible only because of the fluidity of the pulp of high specific gravity ore at a higher than normal discharge solids value.

After making the necessary mass flow and ball-mill discharge set point changes, cyclone overflow samples were taken and determinations of the percent +65-mesh material were made. The samples showed that the separation size was finer than normal, indicating an increase in grind set point was needed.

In order to utilise the changes in set points for the high specific gravity ores, a method for detecting these ores had to be found. There were several possibilities but, by trial and error, the most consistent indicator was a ratio of the mass flow of ore to the cyclone feed volume flow. As the calculated mass flow of ore increases and, if the specific gravity of the ore is higher than usual, the volume flow will decrease from a normal value. If the ore is of normal density and the calculated mass flow is high, then the volume flow will also be high. By checking this ratio (cyclone feed volume flow/ calculated mass flow of ore) a consistent indicator of the presence of high density ore can be obtained.

In order to make the corrections in the set points more flexible, and to accomodate mixtures of high-density ores and normal ore types, three different set points (see Table 13-IV) are used depending upon the indicator ratio value. The computer can be used to detect ore changes and to increment set point changes in such a manner that the operation of the mill is not seriously disturbed.

TABLE 13-IV

Correction for change in specific gravity of ore

Probable average ore s.g.	MFO S.P. (t/h)	Ratio cyclone feed volume flow/MFO	+65 S.P.
Normal 2.68	680	$\geqslant 3.4$	20
2.75	700	$3.4 > r_1 \geqslant 3.14$	21
2.90	720	$3.4 > r_2 \geqslant 2.92$	22
3.00	740	$2.92 \geqslant r_3$	23

Refinement of model for grind control. The first model used for grind control was:

$$\ln \% +65 = K_0 + K_1 \cdot MFO + K_2 \cdot MFW + K_3 \cdot FOF \qquad (13\text{-}2)$$

This model was suitable for most operating conditions but at times the size prediction was up to 4% in error.

It was observed that when the total water addition (TW) to fresh-ore feed (FOF) ratio deviated from a norm, the percent +65 also deviated. After establishing the TW to FOF norm, an equation relating increment in the intercept constant, K_0, in the model to this norm was developed:

$$Q = TW - (x + y \cdot FOF) \cdot \text{Constant}$$

where Q is the amount by which K_0 is incremented and x and y are the slope and intercept for the TW—FOF normal relationship. This change in the model intercept was made at equilibrium only and helped in correcting the final equilibrium prediction, but it still had limitations in accuracy for some ores. Eq. 13-2, however, demonstrated the importance of the total water addition term in the size prediction model and a regression equation of the form given in eq. 13-1 was developed. This equation has proved to be suitable for size prediction at all operation conditions.

13.7 CASE STUDY 6. SEMIAUTOGENOUS-MILL, BALL-MILL—HYDROCYCLONE CIRCUIT AT CYPRUS PIMA MINING COMPANY*

13.7.1 *Introduction*

Cyprus Pima Mining Company currently treats about 58,000 t/day of ore containing 0.48% copper and recovers about 85% of the copper in a concentrate assaying 27% copper which is shipped to smelters at Douglas and San Manuel, Arizona. Slightly higher recoveries are possible but the relatively

* By J.A. Bassarear, Mill Manager, Cyprus Pima Mining Company, E.J. Freeh and W.E. Horst, Consultants Industrial Nucleonics Corporation.

coarse grind of 20% +65 mesh (and about 50% −200 mesh) has been proven best for optimum economics. The ore body was discovered in 1950 and operations began in late 1956 treating 3,000 t/day of 1.85% copper ore. As additional low-grade reserves were proven, expansions were made in 1963, 1965 and 1967 (Komadina 1965, 1967; Martin et al., 1966; Martin, 1969) to increase tonnage to 40,000 t/day. The original mill and the first three expansions used conventional crushing and grinding but the fourth expansion (Bassarear and Sorstokke, 1973), in 1971, incorporated semiautogenous grinding.

The mine and mill are located 22 miles southwest of Tucson, Arizona. Cyprus Pima Mining Company is owned 50.01% by Cyprus Mines Corporation, 25% by Union Oil Company of California, and 24.99% by Utah International, Inc., with Cyprus having management responsibility. Cyprus Pima's ore grindability ranges from a Bond work index of 12, for the relatively soft high-grade ores, to 25 kWh/t for the very hard arkose ore. There are three broad groups of ore minerals including carbonates, hornfels, and arkose or igneous rocks such as quartz monozite porphyry. The bulk of the copper occurs in the arkose group as widely disseminated chalcopyrite.

13.7.2 *Cyprus Pima's Mill II*

During the late 1960's a feasibility study was initiated and pilot plant work begun to evaluate various flowsheets including autogenous and semiautogenous grinding circuits. The two-stage semiautogenous ball-mill grinding circuit was chosen and the decision was made to proceed with Mill II composed of two sections designed to process 14,500 t/day. Subsequent improvements have increased the capacity to 19,000 t/day.

Mill II includes a new primary crusher and a radial stacker that feeds stockpiles for both concentrators, Mill I and Mill II. A detailed account of the overall operation is available in the literature (Ramsey, 1976). The − 8 inch primary crusher product is fed from the reclaim stockpile to Mill II by two pairs of pan feeders, each pair feeding a separate conveyor. The grinding circuit consists of two identical sections that can be used as separate circuits or combined. Pan-feeder discharge is conveyed directly to two 28 ft diameter by 12 ft long Koppers semiautogenous mills. Each mill is driven by dual 3,000-hp wound-rotor motors through Falk reducers resulting in a total connected horsepower of 6,000. Semiautogenous-mill speed is 10.95 rpm or 75% of critical. The mills are loaded with 8% by volume of 4-inch forged steel grinding balls. However, recent development work indicates that larger balls are more effective. The semiautogenous mill discharges to a 6 ft by 14 ft Tyler double-deck vibrating screen equipped with rubber covered punch plate with 7/8 inch holes. Oversize from both decks is conveyed to a common recycle conveyor that receives screen oversize from both semiautogenous mills. The recycle conveyor product is proportionally split to two additional

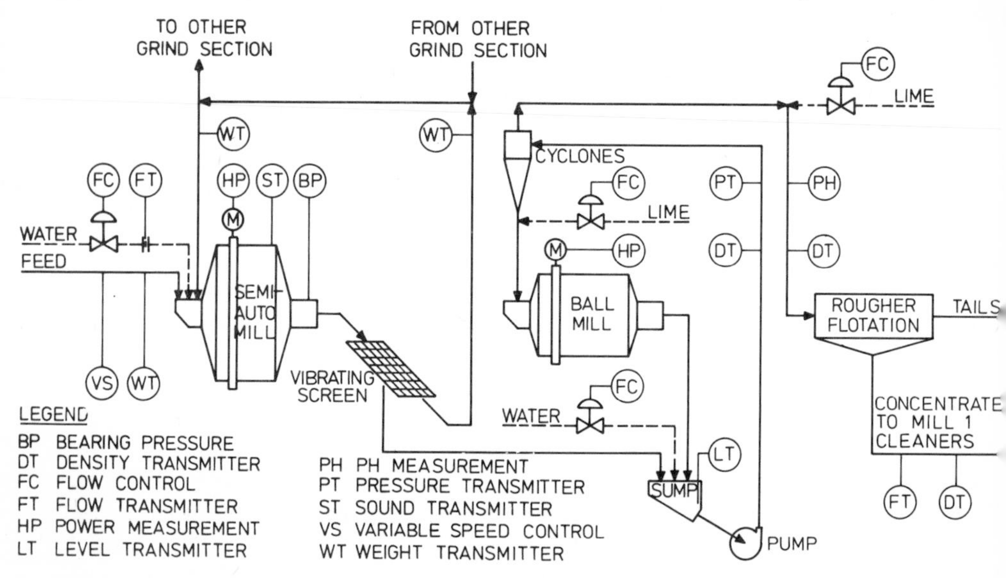

Fig. 13-17. Flowsheet for Mill II, Cyprus Pima Mining Company.

conveyors that feed the original reclaim conveyors. The screen oversize thus recycles back to the semiautogenous mills.

Screen undersize is pumped to a cyclone feed sump by Georgia Iron Works 10 by 12 pumps. Similar pumps feed a nest of four radially mounted 26-inch diameter Krebs hydrocyclones per ball-mill section. Hydrocyclone underflow provides feed to the $16\frac{1}{2}$ ft diameter by 19 ft long Allis Chalmers overflow-type ball mills that operate at 13.09 rpm or 67% of critical. The ball mills each use one Falk reducer and a 3,000-hp, 720-rpm synchronous motor. Hydrocyclone overflow from each section is combined and flows by gravity to a feed distributor and then to flotation. A flowsheet for Mill II, including instrumentation, is shown in Fig. 13-17. Rougher flotation concentrate is pumped to Mill I for further processing.

Instrumentation as originally designed included a horsepower control loop for the semiautogenous-mill circuit which utilised sensors measuring the mill-discharge flow rate (mass flow) and percent solids (density gauge). The control loop depended on accurate measurements for proper operation, but the errors introduced by these two sensors were greater than the variable being sensed for control purposes. Tonnage variations of 200 to over 1200 t/h per section, because of changes in ore hardness, compounded the problems of satisfactory operation of the original analogue control system.

It was recognised that a more sophisticated control system would be required to provide reliable and acceptable control of the Mill II grinding sections. Consequently, a plant-scale development project was begun in late 1973 conducted jointly by Cyprus Pima Mining Company and Industrial Nucleonics Corporation to analyse the process behaviour and design a viable control system for the semiautogenous-mill and ball-mill circuits.

13.7.3 *Process analysis and control system development*

The digital computer system employed for investigation of process behaviour and evaluation of control was capable of capturing information on magnetic tapes which could be either removed from the system for analysis in a computer laboratory, or limited data reduction could be performed on site. In some instances, only a preliminary analysis was made on site to verify that the desired information had been obtained and the data tapes were then forwarded for processing.

The process interface was bi-directional, thus permitting the use of the computer for regulation as well as monitoring of process variables. In the development phase, this capability was used solely for adjustment of set points in existing analogue loops, the early objective of which was dynamic stabilisation of the process to permit quantification of important process variable inter-relationships.

The video output was used by both operating and technical personnel to monitor current process status displayed both as derived variables, such as a measure or ore grindability, and as the interpretation of process measurements in terms of customary engineering units.

As noted above, the grinding circuit was subjected to large and frequent dynamic forcing as a consequence of the ore grindability variations normally encountered. The grindability variations required considerable adjustment of the new feed to the semiautogenous mill to maintain satisfactory grinding conditions in the mill. Operators were reluctant to load the mill to the higher levels of power draft, because an abrupt increase in ore hardness could cause a mill overload condition. The grinding efficiency of a mill is a unimodal function of the inventory of ore for a given steel loading. That is, as inventory is increased grinding efficiency is increased to a maximum, and then decreases as the inventory continues to increase. As long as the operator maintains the mill at an inventory which is below the inventory corresponding to optimum efficiency, he obtains the benefit of self-regulation. That is, if ore hardness were to increase, there would be a gradual increase in inventory, which in turn would increase the grinding efficiency of the mill, thus reducing the rate of inventory build-up that would have occurred without the compensating action. On the other hand, if the operator is at an inventory beyond the optimum, then instead of realising the benefit of self-regulation, he finds he is faced with a positive feed-back situation where there tends to be an acceleration in the inventory build-up as the grinding efficiency continues to decrease with increasing inventory.

The desirable situation is, of course, to maintain the mill operating at as close to optimum grinding conditions as possible and be able to maintain control in spite of strong disturbances. Consequently, a major objective of the study phase of this program was the exploration of the process response surface. The principal variables overtly adjusted were semiautogenous-mill horsepower and slurry density. The effect of other variables, such as cyclone feed density and flow rate and other parameters associated with the ball mill portion of the circuit, were studied, but minimal effort was made to force these variables beyond that obtained consequential to the forcing mentioned above and that introduced naturally.

Effective process analysis and control development required adequate process visibility, which is obtained through a combination of sensors and an appropriate means to capture, correlate and display this information in forms which exhibit the salient dynamic and steady-state process characteristics. A heavily used technique was the employment of time—history plots of selected variables that were line-printer generated, often with time-shifting of individual variables to account for known system (process or sampling) lags. This technique is especially useful in the development of an inferential quantity, such as particle size, when an actual measurement of the inferred variable is temporarily available. Thus, it is possible to plot both the inferred and the measured variable as a function of time, along with the other variables, and observe under what conditions the evolving correlation is found wanting.

A considerable amount of detail concerning the development activities, which led to the design of the grinding circuit computer control system, has been presented by Bailey and Carson (1975a, b). The control philosophy which was finally implemented was appreciably more complex than the stabilisation loops implemented during development.

Extended evaluation of mill performance at different levels of inventory allowed the development of a relationship between optimum horsepower draft and ore grindability. It was found that as the throughput increased, because of softer ore, the horsepower required to obtain maximum tonnage through the semiautogenous mills decreased. The process analysis also provided a more quantitative understanding of the ball-mill limiting condition which occurs at higher tonnage rates. From this information, it was apparent that at least two different general operating regimes must be recognised and appropriate control strategy developed to optimise performance in each regime.

The two operating conditions were identified as: (1) semiautogenous-mill limiting, and (2) ball-mill limiting conditions. With harder ore and the ball mill capable of grinding finer than required, it is desirable to adjust the semiautogenous-mill inventory and solids so as to maximise the new feed rate to the semiautogenous mills. When the ore becomes sufficiently soft to supply feed to the ball mills at a rate greater than they can accept and still meet grind specifications, it is necessary to shift from a throughput maximisation philosophy to one where the semiautogenous mills tend to deliver a finer product at a rate which does not exceed that which the ball mills are capable of accepting.

In spite of the advantages which accrue to utilisation of an optimising control system, there are two major reasons why one should consider designing a control system which can also be operated in an alternate mode, wherein the direction of the process is more closely dependent upon operator decisions. Firstly, it is only prudent to recognise that conditions will occur, both foreseen and unforeseen, which will either interfere with the proper functioning of the system (for example, loss of a process measurement) or an unanticipated combination of circumstances may arise which could be handled more appropriately by an operator making use of the computer-provided process visibility. Secondly, operation of the control system will result in a greater understanding of process behaviour as time goes on, and this will inevitably lead to the desire to evaluate modifications of the operating strategy or at least use the system to assist in performing some additional controlling tests. With the above in mind, it was decided to implement two levels of automatic control. The principal difference between the two is that in the lower level of automatic control, the operator, through his control of horsepower set points, can implicitly control the inventory levels. Mill solids can also be modified by the operator.

Thus, three levels of control are distinguished:

(1) local, or non-computer, where the control of the process is not directly influenced by the computer;

(2) operator-directed automatic control, which allows operator determination of parameters, such as semiautogenous-mill horsepower and solids targets;

(3) optimising automatic control, where the determination of operating levels of selected key variables are dynamically determined by the computer.

For the sake of brevity, the control approach will be illustrated by describing only the control loop employed for regulation of new feed.

In local or non-computer control, new feed to the mills is regulated by employing a feedback control system which involved a weight transmitter to measure present feed rate and a proportional-error-integrating-controller (PEIC) to drive a pair of pan feeders at the speed required to deliver new feed at the rate determined by an operator set point entry.

The automatic modes of control are best explained with the aid of the control loop schematic presented in Fig. 13-18. The righthand side of the diagram depicts the analogue portion of the loop and includes the three primary measurements made on each semiautogenous mill. The bearing pressure sensor was used to provide an inferential measure of mill loading. The innermost loop is seen to be a direct digital control (*DDC*) loop which controls new feed to the mill by adjustment of pan feeder speed. The *DDC* algorithm is a combined feedforward, feedback loop with capability for adaptively adjusting gain. The nonlinearity of feed rate versus pan-feeder speed prompted the inclusion in the software of a table-look-up to provide a speed-dependent gain. Also, at low feed rates and for periodic maintenance, the loop is operated on only one feeder. The number of operating feeders is automatically sensed and the gain doubled whenever single-feeder operation is detected.

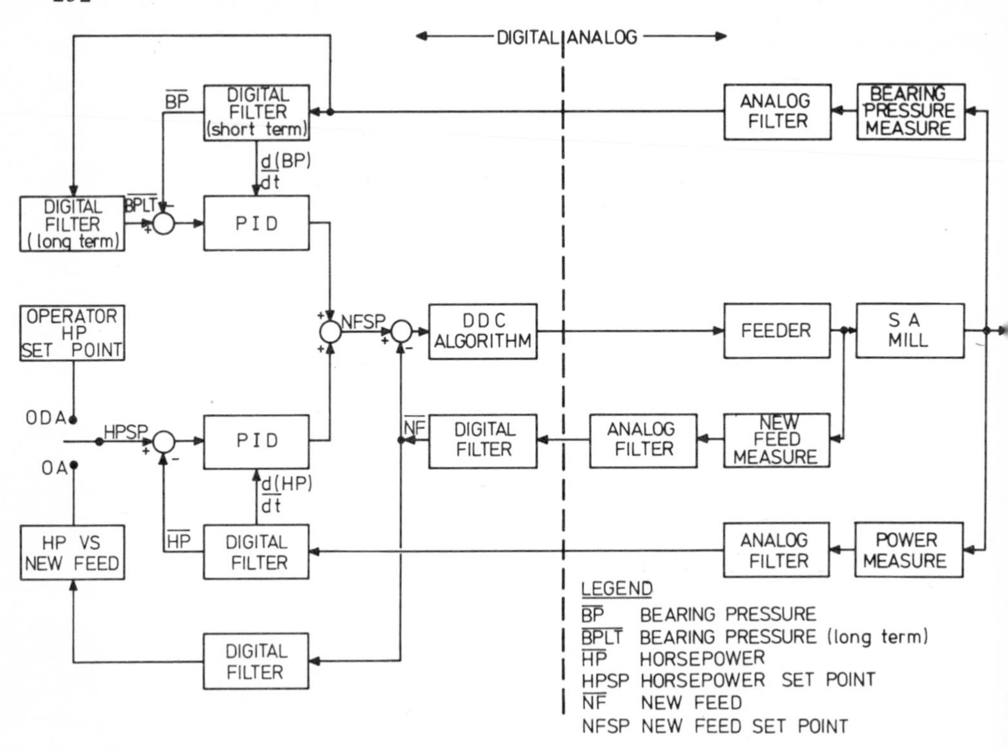

Fig. 13-18. Optimising-automatic (*OA*) and operator-directed (*ODA*) automatic control of new feed to semiautogenous (*SA*) mills.

The new feed set point is formed by summing the output from two three-mode, proportional integral derivative (*PID*), controllers: one employing bearing pressure as the feedback signal and the other mill horsepower. The bearing pressure loop is used purely for stabilisation and can exert no effect on the average (long-term) operating level of the mill, since the reference signal is merely a well filtered value of the feedback signal.

For the control loop under discussion, the difference between optimising control and operator-directed automatic control is the origin of the set point to the mill horsepower loop. In optimising control, instead of using an operator provided set point, the control algorithm provides its own set point by using a well filtered value of new feed to define an average production rate. From this inferred value of ore hardness, the computer calculates the target set point using the previously defined relationship between ore hardness and mill inventory or mill loading as gauged from horsepower draft.

13.7.4 *Computer-based control system*

The grinding circuit control system, designed by Industrial Nucleonics Corporation, consists of a dedicated digital computer, operator station, and all required signal input and output processing. The custom designed operator station effectively provides communication between the computer, the process, and the operator. It contains process displays, interactive operator controls, and includes the operator's video monitor which displays selected process reports upon operator request. The operator panel provides all monitoring, alarms, and control devices necessary for safe and efficient operator interface to the process.

A magnetic tape unit is used to load programs and data into memory. It also provides a recording medium on which diagnostics, special data analysis, and system performance data may be stored for future evaluation. The local teletype is used primarily to print shift and daily summary reports. It is also used for printing data-logging results recorded on the magnetic tape unit. A remote video monitor and teletype unit are located in supervisor's offices providing current process information for milling department management.

The Mill II grinding-circuit control system was installed in October, 1974. Several modifications in hardware were made regarding plant instrumentation concurrent with the installation of the computer-based control system. Implementation of direct digital control of feed rate resulted in by-passing the previously employed PEIC controller. Capacitance-probe sensors installed in the cyclone feed sumps provided accurate and reliable measurements of sump level. Also larger water valves were provided for the water additions to cyclone feed sumps. Discrete signals were provided to the computer to monitor the number of hydrocyclones active in each grinding section.

Control of Mill II is a function of the operating configuration present at any given time. Software is included to facilitate identification of the various operating modes depending on horsepower levels and operator inputs. Implementation of the control philosophy depends on the operating configuration, and it is allowed to change for each of the various operational modes regarding equipment arrangements. The normal method of operating the four grinding units in Mill II is with section 101 semiautogenous mill feeding section 101 ball mill, and section 102 semiautogenous mill feeding section 102 ball mill with the sump splitter active. Seven other operating configurations for a total of eight are defined in Table 13-V. A ninth condition, when the operating configuration flag (OCF) equals zero, indicates all mills are down. The semiautogenous-mill horsepower flags, SAHP1 and SAHP2, are set to 1 if the semiautogenous mills are operating and zero if they are down. The ball-mill horsepower flags, BMHP1 and BMHP2, are set similarly. The split-sump flag, SS is set to zero if the sump is not split and to unity if split.

TABLE 13-V

Mill II operating configurations

Flag	Value of OCF flag 0	1	2	3	4	5	6	7	8
SAHP1	0	1	0	1	0	1	0	1	1
SAHP2	0	0	1	0	1	0	1	1	1
BMHP1	0	1	0	1	1	0	1	1	1
BMHP2	0	0	1	1	1	1	0	1	1
SS	—	—	—	—	—	—	—	0	1

The symbol (—) denotes condition not sensed.

13.7.5 *Control system performance*

Benefits derived from implementation of the computer-based control system include: process stability, increased process visibility, increased throughput at the desired grind size, and improved power utilisation in Mill II. Operator participation during the development phase and acceptance of the final control system was outstanding. By providing increased process visibility for the operator and easy communication with the control system through the customised operator station, the operator's command of the grinding sections was enhanced. Furthermore, through his use of the control system, overloading of the semiautogenous mills has been eliminated. Typical monthly performance shows the grinding circuits under computer control in excess of 98% of the time operated.

An operating baseline for Mill II was developed from plant data collected prior to installation of the control system and details regarding this development have been described in the literature (Pena, 1974; Horst and Bassarear, 1975). With an operating baseline established for Mill II, it was possible to quantify improvements in plant performance following installation of the computer-based control system despite the fact that day-to-day and long range changes in ore grindability occur.

The improvement in throughput rate varied depending on ore type. The relationship between throughput increase per grinding section and ore type is illustrated in Fig. 13-19. A larger increase in throughput was experienced for the softer ores, at throughput rates per section of 500—600 t/h, than for the harder ores at 250—300 t/h per section. The computer control system can respond quickly to changes in ore conditions and process variables thereby minimising the possibility of overloading the semiautogenous mills. This feature is especially important at the higher throughput rates where mass flow through the grate-discharge section can be the limiting condition.

Operating data from Mill II are summarised in Table 13-VI showing results

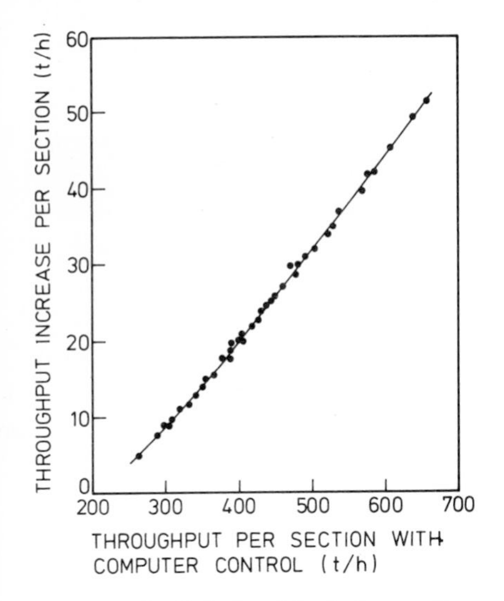

Fig. 13-19. Relationship between throughput increase per section and ore type, Cyprus Pima Mining Company.

TABLE 13-VI

Mill II operating performance

	1972	1973	1974	1975*
Tons milled	4,882,630	6,122,914	6,015,386	6,353,014
Milling rate (t/day)	13,377	16,822	16,526	18,432
Mill II (kWh/t)	16.56	17.11	17.19	16.12
ΔkWh/t, (Mill II — Mill I)	4.46	3.60	3.98	2.38
Grinding media consumption (lb/t)				
Mill II				
Balls, SA mills	1.35	1.19	0.83	0.99
Balls, ball mills	0.53	0.32	0.72	0.55
Total	1.88	1.51	1.55	1.54
Mill I				
Rods, rod mills	0.61	0.58	0.61	0.63
Balls, ball mills	1.12	1.15	1.17	1.13
Total	1.73	1.73	1.78	1.76

* With computer-based control system.

obtained over the four-year period that the plant has operated. These data include one year's operation under computer control. The comparative grinding media consumption data for Mills I and II show that the ore grindability was relatively constant over the four-year period thereby making production rate comparisons straightforward.

Although improvement in plant performance had been achieved since start-up of Mill II, the production rate had levelled off during 1973 and 1974. Throughput was appreciably increased during 1975 at comparable grind size and metallurgical performance. The major portion of this improved capacity can be attributed to the computer-based control system. Also, improved grinding efficiency was achieved through computer control which resulted in reduced energy consumption, as compared with Mill I.

Based on increased productivity, and neglecting the significant energy savings, the payout time before taxes for the capital expenditure including the development program and computer-based control system was about one year.

13.7.6 *Current and future programme*

Control of the grinding circuit has been viewed from the beginning as only the first phase of applying computer technology to the various mineral processing steps involved in the overall Mill II operations. It is anticipated that further modifications in the existing control system for grinding will be made because the improved process knowledge will suggest both process flowsheet changes and equipment modifications which will impact the operating policy.

To ensure that such modifications are forthcoming, management has continued to utilise the process analysis capabilities of the computer system to support an active research and development programme. An example of the type of work being done is the study to determine the optimum ball charge and ball size for the semiautogenous mills. The economic consequences of optimising these two variables ranks with that of the original computerisation of the mill. With ball size, for example, the penalty for using too small a ball is reduced throughput and increased steel cost. Too large a ball size reduces productivity because of insufficient grinding of material that is in the size range being recycled. Somewhat similar arguments can be advanced to show the existence of an optimum steel inventory.

In evaluating the effect of ball size, it was necessary to collect a rather large amount of data over an extended period. After completing a reference period of operation, one semiautogenous mill was charged with a different size ball and the performance monitored as the steel charge equilibrated to a new size distribution, but with steel addition adjusted so as to maintain an equal inventory in both semiautogenous mills. A steel consumption model was developed and fitted to the data, and used to determine charging frequency for each ball size. The computer maintained a record of cumulative hours operated since the last ball charge and signalled the operator when the next charge was to be made.

The time required to reach an equilibrium distribution for the larger balls being tested is approximately 1,400 operating hours. Also, the optimum throughput—horsepower relationship is influenced by the level of steel loading employed which shifts toward higher horsepower levels as the percentage of steel in the charge increases. Hence, the optimisation study in progress is regarded as a relatively long term study. The test is far from consummated, but, thus far a move to a larger ball size and higher steel loading has resulted in significant economic gain.

The process computer and peripherals can be expanded to accommodate the anticipated support required to place the flotation circuit under computer control. The current system is capable of supporting a process analysis study of flotation. Tentative plans are to initiate such a study in the future.

13.7.7 *Summary*

The design and implementation of a computer-based control system for Mill II was based on the results obtained from a detailed plant-scale development project conducted jointly by Cyprus Pima Mining Company and Industrial Nucleonics Corporation. During the initial phase of the development programme, computer-based data-logging equipment was employed to develop a thorough analysis of the process behaviour under a broad spectrum of process conditions.

Control of the grinding sections was predicated on maximum throughput limited by the necessity to meet a particle-size criterion for flotation feed. Horsepower, bearing pressure, and slurry solids were considered as the key variables in semiautogenous-mill control. Slurry-solids measurements of the hydrocyclone feed and overflow streams formed the basis for control of the ball-mill circuit.

Process stability has been achieved along with increased productivity and more efficient energy usage. The process analysis capabilities of the computer system have been useful in supporting continuing research and development activity. Improvements in productivity and efficiency in Mill II have resulted in significant economic gains that clearly demonstrate the efficacy of the process analysis feature incorporated in the control system.

13.7.8 *Acknowledgements*

The authors wish to thank the Cyprus Pima Mining Company of Tucson, Arizona, and the Industrial Nucleonics Corporation of Columbus, Ohio, for permission to publish this case study.

13.8 CASE STUDY 7. CRUSHING CONTROL SYSTEMS AT MOUNT ISA MINES LIMITED*

13.8.1 *Introduction*

A crushing control system was first commissioned in the lead/zinc concentrator of Mount Isa Mines Limited in 1970, following developmental work which commenced in 1968. Subsequently, a similar control system was installed in the new copper concentrator in 1973. The control system installed in the lead/zinc concentrator is described. Details of development are given by Whiten (1971a, b; 1972a, b, c) and Fewings and Whiten (1973).

13.8.2 *Plant description*

The crushing section of Mount Isa Mines Limited No.2 Concentrator (Fig. 13-20) consists of one primary and two secondary units. Each of the primary and secondary crushers are 2.13-m Symons shorthead gyratory cone crushers with extra coarse cavity bowls on the primaries and extra fine cavity bowls on the secondaries. Normal gap settings are 19 mm and 9.5 mm on primary and secondary, respectively, and normal operation utilises one primary and two secondary crushers with one primary crusher on standby.

Two-stage jaw crushing underground reduces the run-of-mill lead—zinc ore to nominally 100% passing 16 cm. Crude ore from underground is stored in three 4,000-t capacity surface storage bins from where it is conveyed to either of the 150-t capacity surge bins ahead of primary crushing.

Crude ore feed rate into the primary surge bins is remote manually controlled from the control room through vibratory feeders under the crude ore storage bins. Screening before primary crushing is effected by a 1.52 m by 3.65 m rod deck screen with four 1.83 m by 4.86 m woven wire screens before secondary crushing. Surge capacity ahead of secondary crushing is 300 t. Variable-speed flat belt feeders provide control of the feed rate to secondary crushing.

Crushing was originally carried out on a 2-shift per day, five days per week basis at a nominal rate of 550—600 t/h. Labour establishment was two men per shift.

13.8.3 *Implementation of a crusher control system*

The development of the crushing control system was undertaken in two distinct phases. The first involved the development of a simulation model of the crushing plant so that control strategies could be proposed and tested. The second phase involved the actual installation and commissioning

* By Mount Isa Mines Limited Staff.

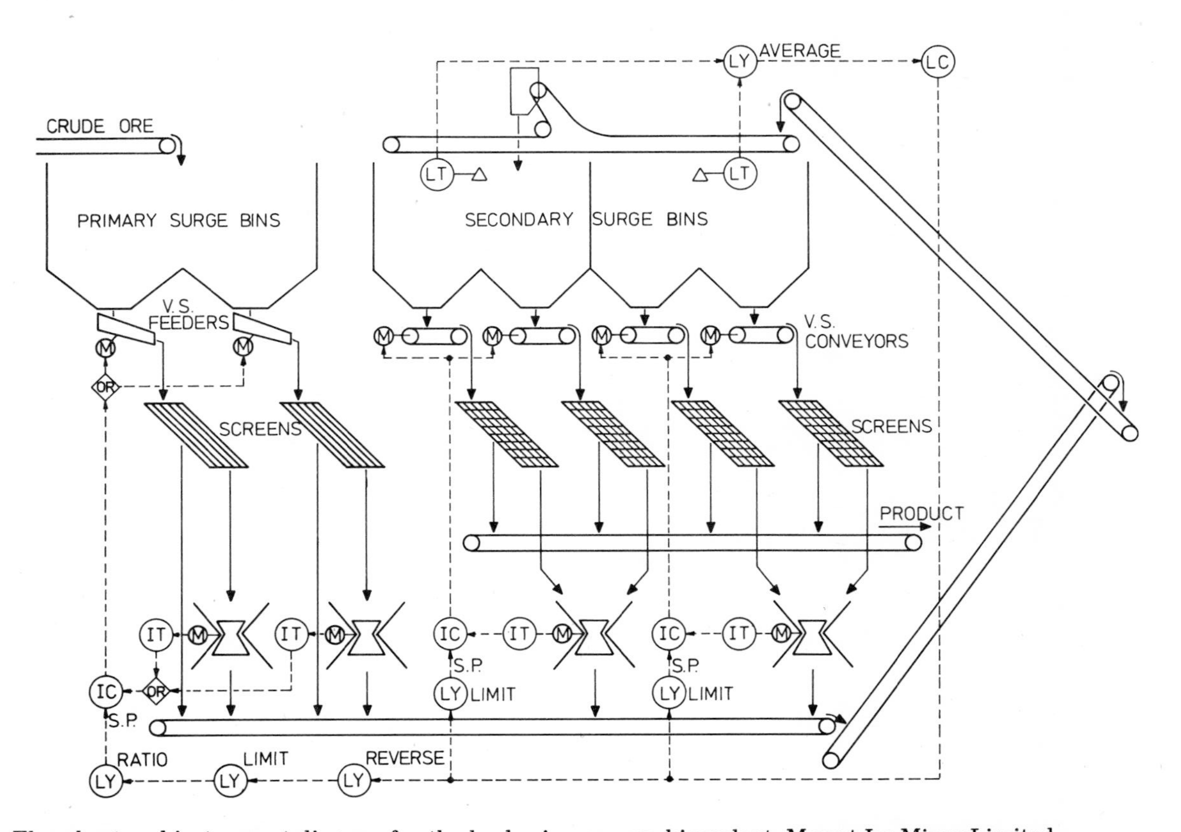

Fig. 13-20. Flowsheet and instrument diagram for the lead—zinc ore crushing plant, Mount Isa Mines Limited.

of a system based upon the simulation studies. Each aspect is discussed briefly.

Simulation model development and control system design. Mathematical models of cone crushers and vibrating screens were developed and combined to provide a crushing plant simulation (Whiten, 1972c).

The simulation model of the plant was used to investigate the effect of the major variables on the operation of the plant. This study showed that the plant throughput could be increased if the current drawn by the crushers was allowed to increase. An increase of 50 t/h in plant feed required the primary crusher current to increase by 1.5 amps and the secondary crushers current by 2.5 amps. This increase in throughput was found to make the plant product slightly finer.

A safety trip-out prevents the crusher being operated at currents above 50 amps and as these currents fluctuate very considerably a wide safety margin is required. The first part of the crusher control system uses controllers to hold each crusher current at a given set point by adjusting the feed rate to the vibrating screen. The aim of these controllers is to reduce the fluctuation in the crusher current so that the crushers can be run at higher currents without increasing the probability of an overload trip-out.

The second part of the control system is to maintain a balance between the primary and secondary sections and at the same time to ensure that at least one section is operating at its maximum capacity. To do this, the secondary bin level is used as the input to a controller whose output after being limited sets the remote set points of the current controllers. These limiters and a reversing unit are arranged so that for low output signals from the bin level controller, the primary crusher current set point is at the limit value while the secondary current set points vary with the signal. A high output signal varies the primary crusher current while the secondary crusher set point is held at its limit. Gain units are provided on the limiter outputs to allow the maximum currents to be set independently on the limit points. The effect of the second part of the control system is to ensure that whichever of the primary or secondary crushers reaches its limit first is held at the limit while the other is used to ensure the surge bin neither overflows nor empties. Fig. 13-20 shows the major components of this control system.

After the control system had been proposed from the steady-state model of the plant, the dynamic behaviour of the proposed system was examined. The bin level controller was not critical provided its action was not too rapid. As this controller is only required to ensure the feed bins remain between empty and full, a low-gain proportional control with possibly some very low-gain integral action was expected to be quite sufficient.

The current controllers are different. The input signal is very noisy and there is an almost pure delay in the process. A study was made, (Whiten

1972c), of a pure delay being controlled by a simulated controller and subject to a given series of disturbances. The set point and controller constants were optimised by non-linear programming to obtain maximum throughput without exceeding an overload value. The results of this study indicated that a two-term controller would provide as good control as could be expected, the proportional gain should be about one third, and two resets per delay time of the process were required.

Installation and commissioning of the control system. The decision to proceed with crushing plant control followed from an assessment of the predicted gains from simulation considered against the costs of system installation. Testwork at Mount Isa was commenced to investigate critical process measurements, and in particular ore bin level measurement.

The secondary surge bin level measurement is essential as it determines whether the primary or secondary crushers are constraining overall operation. If, for example, the ore level in the secondary surge bins is rising with the secondary crushers drawing maximum current, then the primary crushers are not limiting overall throughput. Conversely the primaries are constraining throughput if the level in the secondary surge bins cannot be maintained. The availability of a primary sensing element to provide a continuous, reliable and accurate measure of ore bin level was essential to balance the crusher operation.

Plant testing of various commercial ultrasonic level detectors was carried out to ensure reliability in operation, lack of instrument drift and to observe the degree of sensitivity to the adverse operating environment. The comprehensive test program undertaken has subsequently been fully vindicated by the performance of the installed units. Calibration of the untrasonic level detectors was carried out in position, with the only precaution being to ensure falling ore did not intercept the emitted ore reflected signal.

Smoothing of both primary and secondary crusher motor current measurements was necessary to ensure best system performance. In particular, the primary crusher motor current signal is an extremely variable measurement with frequent spiking due to variable ore size and hardness. Simple resistance-capacitance filtering with a one-second time constant for primary current measurement and one-half second for secondary crusher current, provided input signals with the desired noise characteristics without distortion.

The secondary crusher current loops were commissioned first on local set point and then with the set points determined externally. Commissioning of the primary section and integration of the total system followed. No difficulties of any significance were encountered during the commissioning phase or in subsequent operation. All installed instrumentation has performed reliably, consistently and accurately. Operator acceptance of the control system has been excellent, and has contributed to its very satisfactory performance.

TABLE 13-VII

Crushing plant product sizings

Tyler mesh (% retained)	Crushing plant feed	Treatment rate (580—600 t/h) manual control	Treatment rate (660—680 t/h) automatic control
+ 152 mm	6.4	—	—
76/152	19.6	—	—
38/76	19.5	—	—
19/38	17.2	—	—
13.3/19	11.0	3.3	5.0
9.5/13.3	5.2	28.7	12.4
4 mesh/9.5 mm	7.3	28.7	43.5
4/8 mesh	2.0	14.7	14.1
8/14	3.7	7.7	8.1
14/28	1.7	4.4	4.4
28/48	1.2	2.9	2.9
48/65	1.0	0.9	1.1
−65	4.2	8.7	8.5

13.8.4 *Assessment of system performance*

With the installation of automatic control the crushing rate over the first six months of operation increased to 650—700 t/h at an average of 675 t/h. The average crushing plant product sizings, before automatic control on 580—600 t/h, and after, on 660—680 t/h are shown in Table 13-VII.

The cumulative fractions above 4 mesh and the size distribution below 4 mesh and above 65 mesh are similar for both manual and automatic crusher control. Immediately after commissioning the control system, the size distribution above 4 mesh changed as indicated in Table 13-VII. That is, a 50% reduction in the proportion reporting above 9.5 mm occurred with a corresponding increase in the fraction retained on 4 mesh. Thus, by continually ensuring balanced, near-choke-fed crusher operation, some 10—20% of the total ore previously reporting above 9.5 mm now reports below 9.5 mm but predominantly above 4.7 mm with automatic control.

In conjunction with an automatic grinding control system, the reduction in 9.5 mm material is of fundamental significance in maximising grinding throughput without loss of flotation feed sizing. The simulation model predicted an increased (50 t/h) treatment rate at a slightly finer product sizing, and the actual plant performance after control system installation validated the accuracy of the gyratory cone crusher and vibrating screen models. Plant simulation must use sufficiently sensitive models to allow accurate determination of the effects of changes in process variables on critical operating parameters. Only then can control systems design be confidently approached from simulation.

The system has now been operational for over four years, and is unchanged from its original configuration. Operating requirements for the concentrator have changed, and the plant now operates on a 3-shift basis with one operator. The nominal operating time under automatic control has remained at around 95% of available time since commissioning. The long run increase in treatment rate with automatic control is in the range 5–10%.

13.8.5 *Conclusions*

The crushing control system described here has provided significant gains in plant throughput while improving product quality and enabled a reduction in operating staff. The control system is simple enough to be readily understood by operating and maintenance staff and this has contributed to its success.

Appendix 1. The general curve-fitting problem

The problem is to find the parameters β_l (for example, flow rates or model parameters) and the residuals Δa_{ji} which minimise the weighted sum of squares S:

$$S = \sum_j \sum_i \Delta^2_{ji} / V_{ji}$$

subject to:

$$C^k(a_{ji}, \beta_l) = 0$$

where a_{ji} is the ith experimental measurement of type j (for example, stream j size component i or metallic component i), V_{ji} is the estimated variance of a_{ji} and Δa_{ji} is the adjustment required to make a_{ji} satisfy the constraint functions C^k.

residual = observed − calculated

$$\Delta a_{ji} \quad = \quad a_{ji} \quad - \quad \bar{a}_{ji}$$

The constraint functions can be linearised to $\bar{C}^k$ where:

$$\bar{C}^k = \Delta^k_0 + \sum_i \sum_j \frac{\partial C^k}{\partial a_{ji}} \cdot \Delta a_{ji} + \sum_l \frac{\partial C^k}{\partial \beta} \cdot \Delta \beta_l$$

$\Delta^k_0 = C^k(a_{ji}, \beta_l)$ and $\Delta\beta_l$ is an adjustment to our best estimate so far of β_l, that is:

parameter residual = this estimate − next estimate

$$\Delta\beta_l \quad = \quad \beta_l \quad - \quad \bar{\beta}_l$$

The minimisation sum of squares can be constrained by using a Lagrange multiplier λ^k for each constraint:

$$S_m = \sum_j \sum_i \frac{\Delta a^2_{ji}}{V_{ji}} + \sum_k \lambda^k \cdot \bar{C}^k$$

The modified sum of squares S_m is differentiated with respect to each unknown (Δa_{ji}, $\Delta\beta_l$, λ^k) and set to zero. This gives a large set of linear simultaneous equations. However, all the residuals can be expressed in terms of λ^k

to give a banded symmetric matrix in λ^k and $\Delta\beta_l$. This matrix can be reduced by eliminating the λ^k to give a symmetric matrix in terms of $\Delta\beta_l$ and the derivatives. The sum of squares can be evaluated without calculating the residuals. This general solution is due to Lees (1973). A simple case (with the notation expanded for clarity) is set out below.

Four variables (or streams) are given in four data sets (or components) with a constraint containing three parameters (or flow rates):

Variables a_i, b_i, c_i, d_i $\quad i = 1, 4$
Variances Va_i, Vb_i, Vc_i, Vd_i
Parameters α, β, γ
Constraints C^k or C^i as in this case the constraint applies to each component set.

Zero points of the constraints at the experimental variable levels and the estimated parameters:

$$\Delta_0^i = C^i(a_i, b_i, c_i, d_i, \alpha, \beta, \gamma)$$

The derivatives of the constraint functions are written more compactly:

$$C_a^i = \frac{\partial C^i}{\partial a_i} \quad \text{and} \quad C_\alpha^i = \frac{\partial C^i}{\partial \alpha}$$

$$S = \sum_i \left[\frac{\Delta a_i^2}{Va_i} + \frac{\Delta b_i^2}{Vb_i} + \frac{\Delta c_i^2}{Vc_i} + \frac{\Delta d_i^2}{Vd_i}\right]$$

$$S_m = S - 2 \cdot \sum_i \lambda^i \cdot \bar{C}^i$$

where

$$\bar{C}^i = \Delta_0^i + C_a^i \cdot \Delta a_i + C_b^i \cdot \Delta b_i + C_c^i \cdot \Delta c_i + C_d^i \cdot \Delta d_i + C_\alpha^i \cdot \Delta\alpha + C_\beta^i \cdot \Delta\beta + C_\gamma^i \cdot \Delta\gamma$$

$$\frac{\partial S_m}{\partial \Delta a_i} = \frac{2 \cdot \Delta a_i}{Va_i} - 2 \cdot \lambda^i \cdot C_a^i$$

Therefore:

$$\Delta a_i = \lambda^i \cdot C_a^i \cdot Va_i; \quad \Delta b_i = \lambda^i \cdot C_b^i \cdot Vb_i, \quad \text{etc.} \tag{A1-1}$$

$$\frac{\partial S_m}{\partial \lambda_i} = -2 \cdot \bar{C}^i = 0 \tag{A1-2}$$

$$\frac{\partial S_m}{\partial \alpha} = -2 \cdot \sum_i \lambda^i \cdot C_\alpha^i = 0 \tag{A1-3}$$

Substituting eqs. A1-1 into A1-2 where:

$$s_i = (C_a^i)^2 \cdot Va_i + (C_b^i)^2 \cdot Vb_i + (C_c^i)^2 \cdot Vc_i + (C_d^i)^2 \cdot Vd_i$$

$$-\Delta_0^i = \lambda^i \cdot s_i + C_\alpha^i \cdot \Delta\alpha + C_\beta^i \cdot \Delta\beta + C_\gamma^i \cdot \Delta\gamma \qquad \text{(A1-4)}$$

Writing out eqs. A1-3 and A1-4 in full and adding the sum of squares expression:

$$\begin{aligned}
-\Delta_0^1 &= s_1 \cdot \lambda^1 + C_\alpha^1 \cdot \Delta\alpha + C_\beta^1 \cdot \Delta\beta + C_\gamma^1 \cdot \Delta\gamma \\
-\Delta_0^2 &= s_2 \cdot \lambda^2 + C_\alpha^2 \cdot \Delta\alpha + C_\beta^2 \cdot \Delta\beta + C_\gamma^2 \cdot \Delta\gamma \\
-\Delta_0^3 &= s_3 \cdot \lambda^3 + C_\alpha^3 \cdot \Delta\alpha + C_\beta^3 \cdot \Delta\beta + C_\gamma^3 \cdot \Delta\gamma \\
-\Delta_0^4 &= s_4 \cdot \lambda^4 + C_\alpha^4 \cdot \Delta\alpha + C_\beta^4 \cdot \Delta\beta + C_\gamma^4 \cdot \Delta\gamma \\
0 &= C_\alpha^1 \cdot \lambda^1 + C_\alpha^2 \cdot \lambda^2 + C_\alpha^3 \cdot \lambda^3 + C_\alpha^4 \cdot \lambda^4 \\
0 &= C_\beta^1 \cdot \lambda^1 + C_\beta^2 \cdot \lambda^2 + C_\beta^3 \cdot \lambda^3 + C_\beta^4 \cdot \lambda^4 \\
0 &= C_\gamma^1 \cdot \lambda^1 + C_\gamma^2 \cdot \lambda^2 + C_\gamma^3 \cdot \lambda^3 + C_\gamma^4 \cdot \lambda^4 \\
-S &= -\Delta_0^1 \cdot \lambda^1 - \Delta_0^2 \cdot \lambda^2 - \Delta_0^3 \cdot \lambda^3 - \Delta_0^4 \cdot \lambda^4
\end{aligned} \qquad \text{(A1-5)}$$

Eqs. A1-1 and A1-5 provide the complete solution. However, eq. A1-5 can be further reduced by substituting for λ^i from the first four equations into the last four. This yields three linear simultaneous equations for $\Delta\alpha$, $\Delta\beta$, and $\Delta\gamma$ which contain only the derivatives, the variances and the experimental residuals. The equations are symmetric. The last equation provides the sum of squares directly.

This can be written as: $\mathbf{D} \cdot \mathbf{p} = \mathbf{b}$ where **D** is a matrix containing the combinations of the derivatives, **p** is a vector of parameter adjustments and **b** is a vector of parameter derivatives and constraint residuals.

A1.1 ACCURACY CONSIDERATIONS

A pooled estimated variance for all the measurements can be obtained by calculating the variance of the adjustment residuals.

$$\bar{V}a_{ji} = \sum_j \sum_i \frac{\Delta a_{ji}^2}{Va_{ji}} \qquad \text{No. of degrees of freedom (or no. of data points} - \text{constraints} - \text{parameters} - 1)$$

The accuracy of the parameters can be estimated by comparing the sum of squares with and without that parameter (see Li, 1964, for more detail). The reduction in the sum of squares due to each parameter turns out to depend on the diagonal term of the inverse of matrix **D** corresponding to that parameter, that is:

$$V\alpha = d_{11}^{-1} \cdot V\bar{a}_{ji}; \quad V\beta = d_{22}^{-1} \cdot V\bar{a}_{ji}$$

and so on.

If the pooled estimated variance is 1 or less, the model is as good as, or better than the data and the adjusted data should be better than experimental data in the latter case.

Appendix 2. Theory of comminution machines

A2.1 INTRODUCTION

It is shown how the matrix models of comminution machines can be developed into a rigorous theory. The relation of the matrix models to continuous size-distribution models (for example, Austin et al., 1966) is shown to be essentially one of notation. The discussion in this Appendix is abbreviated and a complete exposition of the theory is given by Whiten (1974).

A2.2 PERFECT MIXING MILLS

The perfect mixing mill can be used as the basis for the construction of general models for comminution machines. The matrix equations for a perfectly mixed mill may be derived as follows.

The contents of the mill are described by a vector s each element of which gives the amount in the mill of one component of the mill contents. Normally the elements of s contain the mass or volume in each size fraction and the water content of the mill; however, if required, s may contain the size distribution of several types of particles possibly even including particles of composite composition. The rate of change of the contents of the mill is $\partial s/\partial t$. This can be equated to the sum of the effects which are causing the contents of the mill to change. These effects are:

(1) Removal of material for breakage. Some of this material may reappear in the same size fraction after breakage. The rate of removal for breakage is $\mathbf{R} \cdot s$ where $\mathbf{R}$ is a diagonal matrix giving the breakage rate of each component of s.

(2) Increase in the components of the mill contents due to receiving the products of breakage. The rate at which material appears from breakage is $\mathbf{A} \cdot \mathbf{R} \cdot s$. Each column of the matrix $\mathbf{A}$ gives the relative distribution after the breakage of the corresponding component of s.

(3) Flow of material into the mill. The rate of new material entering the mill is $\mathbf{f}$.

(4) Flow of material out of the mill. This material is obtained from the contents of the mill; however, a classification of the mill contents can occur. Hence the product from the mill can be written as $\mathbf{p} = \mathbf{D} \cdot s$ where $\mathbf{D}$ is the

diagonal discharge matrix which gives the fractional rates at which the components in the mill are discharged.

These effects are added to obtain the equations of the perfect mixing mill:

$$\frac{\partial \mathrm{s}}{\partial t} = (\mathbf{A} \cdot \mathbf{R} - \mathbf{R} - \mathbf{D}) \cdot \mathrm{s} + \mathrm{f} \tag{A2-1}$$

and:

$$\mathrm{p} = \mathbf{D} \cdot \mathrm{s} \tag{A2-2}$$

In obtaining these equations no assumptions have been made about the quantities **A**, **R**, **D** and **f**. They may in fact be functions of time, the mill feed rate or the mill contents as appropriate.

An integral equivalent of eq. A2-1 can be obtained when s is a single distribution by changing s to a continuous distribution function of the particle size x, that is, $s(x)$. The summations in eq. A2-1 become integrals using corresponding continuous functions for **A**, **R**, **D** and **f**.

The integral form is:

$$\frac{\partial s(x)}{\partial t} = \int_{x_i}^{\infty} a(x, y) \cdot r(y) \cdot s(y)\, \mathrm{d}y - r(x) \cdot s(x) - d(x) \cdot s(x) + f(x) \tag{A2-3}$$

The relation between the matrix elements is as follows: Let x_{i+1}, x_i be the upper and lower size limits of the ith element of s. Then:

$$s_i = \int_{x_j}^{x_{i+1}} s(x)\, \mathrm{d}x$$

$$a_{ij} = \frac{1}{x_{i+1} - x_i} \cdot \int_{x_i}^{x_{i+1}} \int_{x_j}^{x_{j+1}} a(x, y)\, \mathrm{d}y\, \mathrm{d}x$$

$$r_{ii} = \frac{1}{x_{i+1} - x_i} \cdot \int_{x_i}^{x_{i+1}} r(x)\, \mathrm{d}x$$

$$d_{ii} = \frac{1}{x_{i+1} - x_i} \cdot \int_{x_i}^{x_{i+1}} d(x)\, \mathrm{d}x$$

where the left sides are the elements of s, **A**, **R** and **D**. The functions $a(x, y)$, $r(x)$ and $d(x)$ can be regarded as the more basic values describing the mill; however, the matrix description is more convenient with regard to notation and, usually, computation. An integral form exists either explicitly or implicitly for all of the equations derived from eq. A2-1.

In the case of constant coefficients, eq. A2-1 can be solved to give:

$$\mathrm{s}(t) = \mathrm{e}^{(\mathbf{A} \cdot \mathbf{R} - \mathbf{R} - \mathbf{D}) \cdot t} \cdot \mathrm{s}_0 + \int_0^t \mathrm{e}^{-(\mathbf{A} \cdot \mathbf{R} - \mathbf{R} - \mathbf{D}) \cdot \tau} \cdot f(\tau)\, \mathrm{d}\tau \tag{A2-4}$$

where s_0 is the contents of the mill at time zero. If **A**, **R** and **D** are functions of time the expression $(\mathbf{A} \cdot \mathbf{R} - \mathbf{R} - \mathbf{D}) \cdot t$ in eq. A2-4 should be replaced by:

$$\int_0^t (\mathbf{A} \cdot \mathbf{R} - \mathbf{R} - \mathbf{D}) \mathrm{d}t$$

One method of finding the product from a continuous mill is to take the solution of the batch-mill equation and a residence time function $\sigma(t)$ (Reid, 1965). This gives a product of the form:

$$\mathbf{p}(t) = \int_0^t \sigma(\tau) \cdot \mathbf{q}(t, \tau) \, \mathrm{d}\tau \qquad \text{(A2-5)}$$

where $\mathbf{q}(t, \tau)$ is the description of the contents of a corresponding batch mill which has been run for time τ after being filled with material described by $\mathbf{f}(t-\tau)$. The solution of the batch mill equation is eq. A2-4 with $\mathbf{f} = 0$ and $\mathbf{D} = 0$. Hence eq. A2-5 becomes:

$$\mathbf{p}(t) = \int_0^t \sigma(\tau) \cdot \mathrm{e}^{(\mathbf{A} \cdot \mathbf{R} - \mathbf{R})} \cdot \mathbf{f}(t-\tau) \, \mathrm{d}\tau \qquad \text{(A2-6)}$$

Eqs. A2-2 and A2-4 give, when the initial contents can be neglected:

$$\mathbf{p}(t) = \mathbf{D} \cdot \int_0^t \mathrm{e}^{(\mathbf{A} \cdot \mathbf{R} - \mathbf{R} - \mathbf{D})\tau} \cdot \mathbf{f}(t-\tau) \cdot \mathrm{d}\tau \qquad \text{(A2-7)}$$

Eqs. A2-6 and A2-7 are the same provided:

$$\begin{aligned} \sigma(\tau) \cdot \mathbf{I} &= \mathbf{D} \cdot \mathrm{e}^{(\mathbf{A} \cdot \mathbf{R} - \mathbf{R} - \mathbf{D}) \cdot \tau} \cdot \mathrm{e}^{-(\mathbf{A} \cdot \mathbf{R} - \mathbf{R}) \cdot \tau} \\ &= \mathbf{D} \cdot [\mathbf{I} - \mathbf{D} \cdot \tau + \tfrac{1}{2} \cdot \mathbf{D}^2 \cdot \tau + \tfrac{1}{2} \cdot (\mathbf{D} \cdot \mathbf{A} \cdot \mathbf{R} - \mathbf{A} \cdot \mathbf{R} \cdot \mathbf{D}) \cdot \tau^2] + 0(\tau^2) \end{aligned} \qquad \text{(A2-8)}$$

As $\mathbf{A} \cdot \mathbf{R}$ is not a diagonal matrix $\mathbf{D} \cdot \mathbf{A} \cdot \mathbf{R} - \mathbf{A} \cdot \mathbf{R} \cdot \mathbf{D}$ is not a diagonal matrix unless **D** is a multiple of the unit matrix.

When $\mathbf{D} = k \cdot \mathbf{I}$, eq. A2-8 becomes: $\sigma(\tau) = k \cdot \mathrm{e}^{-k\tau}$.

Hence a solution of the form A2-5 holds when the initial contents of the mill can be neglected and there is not classification of the mill product.

For steady-state conditions eq. A2-1 becomes:

$$(\mathbf{D} + \mathbf{R} - \mathbf{A} \cdot \mathbf{R}) \cdot \mathbf{s} = \mathbf{f} \qquad \text{(A2-9)}$$

In the case of constant **A**, **R** and **D** this is a set of triangular simultaneous equations for **s** and hence **s** is easily calculated. Austin and Klimpel (1966) have published an integral equivalent of eq. A2-9 when $d(x)$ is constant. Eqs. A2-2 and A2-9 may be combined to give:

$$\mathbf{p} = \mathbf{D} \cdot (\mathbf{D} + \mathbf{R} - \mathbf{A} \cdot \mathbf{R})^{-1} \cdot \mathbf{f} \qquad \text{(A2-10)}$$

which may be written:

$$\mathbf{p} = \mathbf{D} \cdot \mathbf{R}^{-1} \cdot (\mathbf{D} \cdot \mathbf{R}^{-1} + \mathbf{I} - \mathbf{A})^{-1} \cdot \mathbf{f} \tag{A2-11}$$

and if **D** is non-singular:

$$\mathbf{p} = (\mathbf{I} + \mathbf{R} \cdot \mathbf{D}^{-1} - \mathbf{A} \cdot \mathbf{R} \cdot \mathbf{D}^{-1})^{-1} \cdot \mathbf{f} \tag{A2-12}$$

Eqs. A2-11 and A2-12 show the interdependence of **D** and **R** for feed product relations. $\mathbf{D} \cdot \mathbf{R}^{-1}$ may be calculated from a mill feed and product from the relation:

$$\mathbf{p} = \mathbf{D} \cdot \mathbf{R}^{-1} \cdot (\mathbf{I} - \mathbf{A})^{-1} \cdot (\mathbf{f} - \mathbf{p}) \tag{A2-13}$$

if the value of **A** is assumed to be known. Additional information is required to separate the values of **D** and **R**. If the mill contents **s** are also measured, **D** can be calculated from eq. A2-2 and then **R** and $\mathbf{D} \cdot \mathbf{R}^{-1}$ from the following equation:

$$\mathbf{R} \cdot \mathbf{s} = (\mathbf{I} - \mathbf{A})^{-1} \cdot (\mathbf{f} - \mathbf{p}) \tag{A2-14}$$

In the case of a constant volume mill, **D** varies with the feed rate. If $\mathbf{D} = \beta \cdot \mathbf{I}$, β may be calculated as: $\beta = f(\mathbf{p})/v(\mathbf{s})$ where $f(\mathbf{p})$ is the volume flow rate of **p** (which is the same as that of **f**) and $v(\mathbf{s})$ is the volume of **s**. This relation comes from eq. A2-2. In other cases an iterative procedure can be used to calculate $\mathbf{D} = \mathbf{D} \cdot (\beta)$ so that the calculated $v(\mathbf{s})$ is the same as the volume of material that the mill is required to contain, that is, find β such that $v(\mathbf{s})$ = actual mill volume. Numerical techniques for finding the zero of an arbitrary equation may be applied to solve this problem. The assumption $\mathbf{D} = \beta \cdot \mathbf{D}_1$ where $\mathbf{D}_1$ is fixed would appear appropriate for mills with a physical screen over the discharge end.

A2.3 MULTIPLE-SEGMENT COMMINUTION MACHINES

Many comminution machines may be assumed to be divided into segments each of which is perfectly mixed. Using this assumption a model of the comminution machine may be constructed using the perfect mixing mill model of the previous section.

If the comminution machine has n segments and the content of the ith segment is $\mathbf{s}_i$ then the transfer of material from the ith segment to the jth segment is $\mathbf{T}_{ij} \cdot \mathbf{s}_i$. The $\mathbf{T}_{ij}$ may of course be a function of several variables, for example, $\mathbf{s}_i$ or t. Two additional segments are introduced for the feed and the product to provide a uniform notation. The contents of these segments are calculated from:

$$\mathbf{f} = \mathbf{I} \cdot \mathbf{s}_0, \quad \text{and} \quad \mathbf{p} = \mathbf{I} \cdot \mathbf{s}_{n+1}$$

The equation for the ith segment can be derived in the same manner as the perfect mixing mill equation was derived. The equation of the ith segment is:

$$\frac{\partial \mathbf{s}_i}{\partial t} = (\mathbf{A}_i \cdot \mathbf{R}_i - \mathbf{R}_i) \cdot \mathbf{s}_i - \sum_j \mathbf{T}_{ij} \cdot \mathbf{s}_i + \sum_k \mathbf{T}_{ki} \cdot \mathbf{s}_k \qquad \text{(A2-15)}$$

for the feed:

$$\mathbf{f} = \sum_j \mathbf{T}_{0j} \cdot \mathbf{s}_0 \qquad \text{(A2-16)}$$

and for the discharge:

$$\mathbf{p} = \sum_k \mathbf{T}_{k,n+1} \cdot \mathbf{s} \qquad \text{(A2-17)}$$

A2.4 THE GENERAL CASE OF STEADY-STATE CONDITIONS

For the steady state we have: $\partial \mathbf{s}_i/\partial t = 0$ and **A**, **R** and **T** are not functions of time. Now summing eq. A2-15 for $i = 1$ to n gives:

$$\sum_i (\mathbf{I} - \mathbf{A}_i) \cdot \mathbf{R}_i \cdot \mathbf{s}_i = \mathbf{f} - \mathbf{p} \qquad \text{(A2-18)}$$

and if $\mathbf{A}_i$ and $\mathbf{R}_i$ are constant over i then:

$$\mathbf{R} \cdot \sum \mathbf{s}_i = (\mathbf{I} - \mathbf{A})^{-1} \cdot (\mathbf{f} - \mathbf{p}) \qquad \text{(A2-19)}$$

Eq. A2-19 relates the contents of the comminution machine ($\Sigma \mathbf{s}_j$) to the feed and product. It assumes only steady state and constant **A** and **R** and hence applies to many comminution machines.

A2.5 BATCH MILLS

For a batch mill $\mathbf{f} = \mathbf{p} = 0$; hence, summing eq. A2-15 for $i = 1$ to n gives:

$$\frac{\partial \Sigma \mathbf{s}_i}{\partial t} = \sum (\mathbf{A}_i \cdot \mathbf{R}_i - \mathbf{R}_i) \cdot \mathbf{s}_i \qquad \text{(A2-20)}$$

If the mill contents are homogeneous, we can write $\mathbf{s} = \Sigma \mathbf{s}_i$ and obtain the equation:

$$\frac{\partial \mathbf{s}}{\partial t} = \sum (\mathbf{A}_i \cdot \mathbf{R}_i - \mathbf{R}_i) \cdot \mathbf{s} \qquad \text{(A2-21)}$$

Alternatively, if $\mathbf{A}_i = \mathbf{A}$ and $\mathbf{R}_i = \mathbf{R}$, eq. A2-20 becomes:

$$\frac{\partial \mathbf{s}}{\partial t} = (\mathbf{A} \cdot \mathbf{R} - \mathbf{R}) \cdot \mathbf{s} \qquad \text{(A2-22)}$$

where $s = \Sigma s_i$ again. Note, however, that eq. A2-22 does not require homogeneous contents of the mill. The solution to eq. A2-22 for A and R functions of time only can be written:

$$s(t) = e^{\int_0^t (A \cdot R - R) dt} \cdot s(0) \qquad (A2\text{-}23)$$

where s(0) is the distribution of the total mill contents at time 0. If A and R are constant eq. A2-23 becomes:

$$s(t) = e^{(A \cdot R - R)t} \cdot s(0) \qquad (A2\text{-}24)$$

Several authors have considered these or equivalent equations for the particular case when s consists of a single size distribution. Reid (1965) has considered eq. A2-22 with constant A and R and its solution, namely eq. A2-24, for the case when A has zero diagonal elements. Callcott (1963) has considered A2-23 and some special solutions derived from it. Gardner (1961), Meloy (1966) and Austin (1966) have all considered the integral form of eq. A2-22 with constant A and R.

If the mill contains material with a range of breakage properties it will be shown that average breakage properties do not always adequately describe the material's behaviour. For simplicity a material with two components will be considered. If:

$$s = \begin{bmatrix} q \\ u \end{bmatrix}$$

and A and R are partitioned to match the components of s, that is:

$$A = \begin{bmatrix} A^{11} & A^{12} \\ A^{21} & A^{22} \end{bmatrix}$$

and:

$$R = \begin{bmatrix} R^1 & 0 \\ 0 & R^2 \end{bmatrix}$$

then eq. A2-22 becomes after summing the two components of s;

$$\frac{\partial(q+u)}{\partial t} = (A^{11} + A^{21} - I) \cdot R^1 \cdot q + (A^{22} + A^{12} - I) \cdot R^2 \cdot u \qquad (A2\text{-}25)$$

The right side of eq. A2-25 cannot be put into the form:

$$(A^* \cdot R^* - R^*) \cdot (q + u)$$

unless A^* or R^* is allowed to vary with **q** and **u**. Since, in general, the ratios of the elements of **q**, **u** and **q** + **u** change with the amount of grinding, no mean matrices can be found to describe the breakage properties of the combined material for both large and small amounts of grinding. However, a mean matrix can give a good fit over a limited range.

Since the amount of material in a size fraction during breakage is typically described by a sum of negative exponentials the fitting of which is an ill conditioned problem, the use of two components should give a close fit in almost all cases.

REFERENCES

Allen, T., 1968. *Particle Size Measurement.* Chapman and Hall, London, 454 pp.

Andrews, J.R.G. and Mika, T.S., 1975. Comminution of a heterogeneous material; development of a model for liberation phenomena. *Int. Min. Proc. Congr., Cagliari, 1975:* Pap. 3.

Arbiter, N. and Harris, C.C., 1965. Particle-size distribution—time relationships in comminution. *Br. Chem. Eng.*, **10**: 240—247.

Austin, L.G., 1973. Understanding ball mill sizing. *Ind. Eng. Chem. Process. Des. Dev.*, **12**: 121—129.

Austin, L.G. and Klimpel, R.R., 1964. Theory of grinding operations. *Ind. Eng. Chem.*, **56**: 18—29.

Austin, L.G., Klimpel, R.R. and Beattie, A.N., 1966. Solutions of equations of grinding. *2nd Eur. Symp. Comminution, Amsterdam,* pp. 217—248.

Bailey, J.E. and Carson, H.B., 1975a. Cyprus Pima Mining Company's control system for a copper concentrator. *A.I.M.E. Annu. Meet., N.Y., Pap.*, 75-B-72.

Bailey, J.E. and Carson, H.B., 1975b. Development of a grinding control system for a copper concentrator. *Proc. ISA Symp. Instr. Min. Metall. Ind.*, **3**.

Barlin, B. and Keys, N.J., 1963. Concentration at Bancroft. *Min. Eng.*, **15**: 47—52.

Bass, L., 1954. Contribution to the theory of grinding processes. *Z. Angew. Math. Phys.*, **5**: 283—292.

Bassarear, J.H. and Sorstokke, H.W., 1973. Pima expansion IV uses semiautogenous grind. *Trans. A.I.M.E.*, **19**: 297—300.

Beke, B., 1964. *Principles of Comminution.* Maclaren, London.

Bennett, J.G., 1936. Broken coal. *J. Inst. Fuel,* **10**: 22—39.

Bickle, W.H., 1958. *Crushing and Grinding—A Bibliography.* H.M.S.O., London, 425 pp.

Bodziony, J., 1965. On the possibility of application of integral geometry methods in certain problems of liberation of mineral grains. *Bull. Acad. Polon. Sci.*, **13**: 459—467.

Bond, F.C., 1952. The third theory of comminution. *Trans. A.I.M.E.*, **193**: 484—494.

Borisson, U. and Syding, R., 1976. Self-tuning control of an ore crusher. *Automatica*, **12**: 1—8.

Bradley, D., 1958. A theoretical study of the hydraulic cyclone. *Ind. Chem.*, **34**: 473—480.

Bradley, D., 1965. *The Hydrocyclone.* Pergamon Press, London, 330 pp.

Bradley, D. and Pulling, D.J., 1959. Flow patterns in the hydraulic cyclone and their interpretation in terms of performance. *Trans. Ind. Chem. Eng.*, **37**: 34—45.

Broadbent, S.R. and Callcott, T.G., 1956. A matrix analysis of processes involving particle assemblies. *Phil. Trans. R. Soc. Lond., Ser., A*, **249**: 99—123.

Broadbent, S.R. and Callcott, T.G., 1956/57. Coal breakage processes. *J. Inst. Fuel*, **29**: 524—539; **30**: 13—25.

Buckley, P.S., 1964. *Techniques of Process Control.* Wiley, New York, N.Y., 303 pp.

Bush, P.D., 1967. *The Breakage of Mineral Particles in Ball Mills.* Ph.D. thesis, Univ. Queensland.

Callcott. T.G., 1967. Tumbling mills and breakage processes. *Proc. Aust. Inst. Min. Metall.*, **181**: 1—34.

Callcott, T.G., 1960. Study of the size-reduction mechanisms of swing hammer mills. *J. Inst. Fuel*, **33**: 529—539.

Callcott, T.G., 1963. Note—information deduced from closed-circuit grinding. *J. Inst. Fuel*, **36**: 419—421.

Callcott, T.G. and Lynch, A.J., 1964. An analysis of breakage processes within rod mills. *Proc. Aust. Inst. Min. Metall.*, **209**: 109—131.

Canalog, E.M. and Geiger, G.H. 1973. How to optimise crushing and screening through computer-aided design. *Eng. Min. J.*, **174**: 82—87.

Charles, R.J., 1957. Energy-size reduction relationships in comminution. *Trans. A.I.M.E.*, **208**: 80—88.

Chaston, I.R.M., 1958. A simple formula for calculating the approximate capacity of a hydrocyclone. *Trans. Inst. Min. Metall.*, **67**:203.

Dahlstrom, D.A., 1949. Cyclone operating factors and capacities on coal and refuse slurries. *Trans. A.I.M.E.*, **184**: 331—344.

Dahlstrom, D.A., 1954. Fundamentals and applications of the liquid cyclone. *Chem. Eng. Prog. Symp. Ser.*, 15: 41—61.

Danckwerts, P.V., 1953. Continuous flow systems. *Chem. Eng. Sci.*, **2**: 1—13.

Davis, E.W., 1919. Fine crushing in ball mills. *Trans. A.I.M.E.*, **61**: 20.

De Kok, S.K., 1956. Symposium on recent developments in the use of hydrocyclones—a review. *J. Chem. Metall. Min. Soc. S. Afr.*, **56**: 281—294.

Dell, C.C., 1969. An expression for the degree of liberation of an ore. *Trans. Inst. Min. Metall.*, **78**: C152—C153.

Dell, C.C., Bunyard, M.J., Richelton, W.A. and Young, P.A., 1972. Release analysis: a comparison of techniques. *Trans. Inst. Min. Metall.*, **81**: C89—C96.

Deming, W.E., 1964. *Statistical Adjustment of Data.* Dover Publications, New York, N.Y., 261 pp.

Devaney, F.D. and Shelton, S.M., 1940. Properties of suspension media for float and sink separation. *U.S. Bur. Min. Rep. Inv.*, 3469.

Digre, M., 1970. Wet autogenous grinding in tumbling mills. *A.I.M.E. Annu. Meet., Denver. Pap.*, 70-B-10.

Draper, N., Dredge, K.H. and Lynch, A.J., 1969. Operating behaviour of an automatic control system for a mineral grinding circuit. *Comm. Min. Metall. Congr., 9th, Pap.*, 22.

Draper, N. and Lynch, A.J., 1965. An analysis of the performance of multi-stage grinding and cyclone classification circuits. *Proc. Aust. Inst. Min. Metall.*, **213**: 89—128.

Epstein, B., 1948. Logarithmico-normal distributions in breakage of solids. *Ind. Eng. Chem.*, **40**: 2289—2291.

Fagerholt, B., 1945. *Particle-Size Distribution of Products Ground in a Tube Mill.* G.E.C. Gads Forlag, Copenhagen, 227 pp.

Fahlstrom, P.H., 1963. Studies of the hydrocyclone as a classifier. *Proc. 6th Int. Min. Proc. Congr., Cannes*, 87—114.

Fewings, J.H., 1971. Digital computer control of a wet mineral grinding circuit. *Proc. Symp. Automatic Control Systems Miner. Process. Plants—Aust. Inst. Min. Metall, S. Qld. Branch*, pp. 333—357.

Fewings, J.H. and Whiten, W.J., 1974. Crushing control systems development at Mount Isa Mines Limited. *IFAC Symp. Automatic Control Min., Miner. Metal Process.*, pp. 119—124.

Fontein, F.J., 1954. The D.S.M. sieve bend. *2nd Int. Coal Prep. Congr., Essen, 1954*, pp. 1—8.

Fontein, F.J., 1965. Some variables affecting sieve bend performance. *Proc. A.I.Ch.E.—I.Ch.E. Meet. Lond.*, pp. 122—130.

Freeh, E.F., Horst, W.E., Adams, W.C. and Kellner, R.C., 1970. Mathematical modelling applied to analysis and control of grinding circuits. *A.I.M.E. Annu. Meet., Denver, Pap.*, 70-B-28.

Gardner, R.P. and Austin, L.G., 1962. A chemical engineering treatment of batch grinding. *1st Europ. Symp. Comminution, Frankfurt*, pp. 217—248.

Gaudin, A.M., 1939. *Principles of Mineral Dressing.* McGraw-Hill Book, New York, N.Y., 554 pp.

Gaudin, A.M. and Meloy, T.P., 1962. Model and a comminution distribution equation for single fracture. *Trans. A.I.M.E.*, **223**: 40—43.

Gaudin, A.M., Spedden, H.R. and Kaufman, D.F., 1951. Progeny in comminution. *Min. Eng.*, **3**: 969—970.

Gault, G.A., 1973. *Dynamic Behaviour of a Small-scale Grinding Circuit.* M.Sc. thesis, Univ. Queensland.

Gault, G.A., 1975. *Modelling and Control of Autogenous Grinding Circuits.* Ph.D. thesis, Univ. Queensland.

Gilvarry, J.J., 1961. Fracture of brittle solids, 1. Distribution function for fragment size in single fracture. *J. Appl. Phys.*, **32**: 391—399.

Gurun, T., 1973. Design of crushing plant flowsheets by simulation. In: *Application of Computer Methods in the Mineral Industry.* S. Afr. Inst. Min. Metall., pp. 91—98.

Hall, W.H., 1971. The mathematical form of separation curves based on two known ore parameters and a single liberation coefficient. *Trans. Inst. Min. Metall.*, **80**: C213—C221.

Harris, C.C., 1966. On the role of energy in comminution: a review of physical and mathematical principles. *Trans. Inst. Min. Metall.*, **75**: C37—C56.

Harris, C.C., 1968. The application of size distribution equations to multi-event comminution processes. *Trans. A.I.M.E.*, **241**: 343—358.

Haskell, P.E. and Beaven, C.H.J., 1970. Computer simulation of the transient performance of a closed grinding system. *Trans. Inst. Min. Metall.*, **79**, C238—C242.

Hathaway, R.V., 1972. A proven on-stream particle-size monitor system for grinding circuit control. *Min. Metall. Ind. Group Symp., Inst. Soc. Am.*

Hinckfuss, D.A. and Stump. N.W., 1971. System design for on-stream analysis at New Broken Hill Consolidated Limited. *Proc. Symp. Automatic Control Systems Miner. Process. Plants—Aust. Inst. Min. Metall., S. Qld. Branch*, pp. 85—96.

Horst, W.E. and Freeh, E.J., 1970. Mathematical modelling applied to analysis and control of grinding circuits. *A.I.M.E. Annu. Meet., Denver, Pap.*, 70-B-27.

Horst, W.E. and Bassarear, J.H., 1975. Use of simplified ore grindability technique to evaluate plant performance. *A.I.M.E. Annu. Meet., Salt Lake City, Pap.*, 75-B-322.

Hukki, R.T., 1961. Proposal for a Solomonic settlement between the theories of von Rittinger, Kick and Bond. *Trans. A.I.M.E.*, **220**: 403—408.

Kapur, P.C., 1971. The energy size reduction relationships in comminution of solids. *Chem. Eng. Sci.*, **26**: 11—16.

Kellner, R.C. and Edmiston, K.J., 1975. An investigation of crushing parameters at Duval Sierrita Corporation. *A.I.M.E. Annu. Meet. N.Y., Pap.*, 75-B-38.

Kelsall, D.F., 1952. A study of the motion of solid particles in a hydraulic cyclone. *Trans. Ind. Chem. Eng.*, **30**: 87.

Kelsall, D.F., 1953. A further study of the hydraulic cyclone. *Chem. Eng. Sci.*, **2**: 254—273.

Kelsall, D.F., 1964. A study of breakage in a small continuous open circuit wet ball mill. *Proc. 7th Int. Min. Proc. Congr., New York*, pp. 33—42.

Kelsall, D.F. and Reid, K.J., 1965. The derivation of a mathematical model for breakage in a small continuous wet ball mill. *Proc. Am. Inst. Chem. Eng.—Int. Chem. Eng. Joint Meet., Lond., Sect. 4*: 13—19.

Kelsall, D.F., Stewart, P.S.B. and Reid, K.J., 1968. Confirmation of a dynamic model of closed circuit grinding with a wet ball mill. *Trans. Inst. Min. Metall.*, **77**: C120—C127.

Kelly, F.J., 1970. An empirical study of comminution in an open circuit ball mill. *Trans. Can. Inst. Min. Metall.*, **73**: 573—581.

Kick, F., 1883. Contribution to the knowledge of brittle materials. *Dinglers J.*, **247**: 1—5.

Kinasevich, R.S. and Fuerstenau, D.W., 1964. Research on the mechanism of comminution in tumbling mills. *Can. Metall., Q.*, **3**: 1—25.

King, R.P., 1975. A model for the quantitative estimation of mineral liberation by grinding. *Dept. Metall., Univ. Witwatersrand, Res. Rep.*

Klimpel, R.R., 1964. *A Mathematical and Physical Analysis of the Comminution of Brittle Materials.* Ph.D. thesis, Pennsylvania State Univ.

Komadina, G.A., 1965. Pima Mining Company—a progress report. *Min. Eng.*, **17**: 48—53.

Komadina, G.A., 1967. Two-stage program boosts Pima to 30000 tpd. *Min. Eng.*, **19**: 68—72.

Lees. M.J., 1970. *Simulation Techniques Applied to Industrial Comminution Circuits.* M.Sc. thesis, Univ. Queensland.

Lees, M.J., 1973. *Experimental and Computer Studies of the Dynamic Behaviour of an Industrial Grinding Circuit.* Ph.D. thesis, Univ. Queensland.

Lees, M.J. and Lynch, A.J., 1972. Dynamic behaviour of a high capacity, multi-stage grinding circuit. *Trans. Inst. Min. Metall.*, **81**: C227—C235.

Li, J.C.R., 1964. *Statistical Inference.* Edward, Michigan, 1233 pp.

Lilge, E.O., 1972. Hydrocyclone fundamentals. *Trans. Inst. Min. Metall.*, **71**: 285—337.

Linden, A.J. ter, 1949. Investigations into cyclone dust collectors. *Proc. Inst. Mech. Eng.*, **160**: 233—240.

Loveday, B.K., 1967. An analysis of comminution kinetics in terms of size distribution parameters. *J. S. Afr. Inst. Min. Metall.*, **68**: 111—131.

Luckie. P.T. and Austin, L.G., 1972. A review introduction to the solution of the grinding equations by digital computation. *Min. Sci. Eng.*, **4**: 24—51.

Lynch, A.J., 1959. *An Evaluation of Certain Theories of Comminution.* M.Sc. thesis, Univ. New South Wales.

Lynch, A.J., 1964. *Matrix Models of Certain Mineral Dressing Processes.* Ph.D. thesis, University of Queensland.

Lynch, A.J. and Dredge, K.H., 1969. Automatic control system for mineral grinding circuits. *Trans. Inst. Eng. Aust. (Elect. Eng.)*, **EE 5**: 101—108.

Lynch, A.J. and Rao, T.C., 1965. Digital computer simulation of comminution systems. *Proc. 8th Comm. Min. Metall. Congr., Aust., N.Z.*, **6**: 597—606.

Lynch, A.J. and Rao, T.C., 1975. Modelling and scale-up of hydrocyclone classifiers. *Proc. 11th Int. Min. Proc. Congr., Cagliari, Pap.*, 9.

Lynch, A.J. and Wiegel, R.L., 1972. Experiences with a computer-controlled pilot-scale grinding circuit. *Min. Congr. J.*, **58**: 49—56.

Lynch, A.J. Rao, T.C. and Bailey, C.W., 1975. The influence of design and operating variables on the capacities of hydrocyclone classifiers. *Int. J. Min. Proc.*, **2**: 29—38.

Lynch, A.J., Rao, T.C. and Prisbrey, K.A., 1974. The influence of hydrocyclone diameter on reduced efficiency curves. *Int. J. Min. Proc.*, **1**: 173—181.

Lynch, A.J., Rao, T.C. and Whiten, W.J., 1967a. Technical note on on-stream sizing analysis in closed grinding circuits. *Proc. Aust. Inst. Min. Metall.*, **223**: 71—73.

Lynch, A.J., Whiten, W.J. and Draper, N., 1967b. Developing the optimum performance of a multi-stage grinding circuit. *Trans. Inst. Min. Metall.*, **76**: C169—C182.

Lynch, A.J., Rao, T.C., Whiten, W.J. and Kelly, J.R., 1967c. An analysis of the performance of a ball-mill—rake classifier comminution circuit. *Proc. Aust. Inst. Min. Metall.*, **224**: 9—18.

Lynch, A.J., Rao, T.C., Whiten, W.J. and Kelly, J.R., 1968. The behaviour of galena and marmatite in the grinding circuit at New Broken Hill Consolidated Limited. In: *Aust. Inst. Min. Metall., Broken Hill Mines*, pp. 465—473.

Marlow, D., 1973. *A Mathematical Analysis of Hydrocyclone Data.* M.Sc. thesis, Univ. Queensland.

Martin, M.D., Komadina, G.A. and Olk, J.F., 1966. Pima Mining Company's expansion program. *Min. Congr. J.*, **52**: 26—35.
Martin, M.D., 1969. Pima Mining Company—a further major expansion. In: H.L. Hartman (Editor), *Case Studies of Surface Mining.* A.I.M.E., pp. 179—194.
Matschke, D.E. and Dahlstrom, D.A., 1959. Miniature hydrocyclones, 2. Solid elimination efficiency. *Chem. Eng. Progr.*, **55**: 79—82.
Meloy, T.P. and Bergstrom, B.H., 1964. Matrix simulation of ball-mill circuits considering impact and attrition grinding. *Proc. 7th Int. Min. Proc. Congr., New York*, pp. 19—31.
Mitchell, Jr., W., Sollenberger, C.L., Kirkland, T.G. and Bergstrom, B.H., 1954a. Comparison of overflow and end-peripheral discharge mills. *Trans. A.I.M.E.*, **202**: 949—954.
Mitchell, Jr., W., Sollenberger, C.L., Kirkland, T.G. and Bergstrom, B.H., 1954b. Analysis of variables in rod milling. *Trans. A.I.M.E.*, **199**: 1001—1009.
Moder, J.J. and Dahlstrom, D.A., 1952. Fine-size close specification gravity solid separation with the liquid-solid cyclone. *Chem. Eng. Progr.*, **48**: 15—18.
Moore, D.E., 1964. *A Mathematical Analysis of Mineral Breakage*, Ph.D. thesis, Univ. Queensland.
Morrison, R.D., 1976a. A two-stage, least-squares solution to the general material balance problem. *Julius Kruttschnitt Miner. Res. Cent. Internal Rep.*, 61.
Morrison, R.D., 1976b. Statistical analysis of mineral processing plant data. *Julius Kruttschnitt Miner. Res. Cent. Internal Rep.*, 64.
Mular, A.L., 1965. Comminution in tumbling mills—a review. *Can. Metall. Q.*, **4**: 31—74.
Mular, A.L. and Henry, W.G., 1971. Calculation of S matrices for rod and ball-mill matrix models. *Can. Metall. Q.*, **10**: 215—221.
Myers, J.F., 1953. Defining the scope of the open-circuit rod mill in comminution. In: *Recent Developments in Mineral Dressing.* I.M.M., London, pp. 137—150.
Myers, J.F. and Lewis, F.M., 1946. Fine crushing with a rod mill at the Tennessee Copper Company. *Trans. A.I.M.E.*, **169**: 106—118.
Myers, J.F. and Lewis, F.M., 1949. Effects of rod mill speed at Tennessee Copper Company. *Trans. A.I.M.E.*, **184**: 131—132.
Myers, J.F. and Lewis, F.M., 1950. Progress report on grinding at Tennessee Copper Company. *Trans. A.I.M.E.*, **187**: 707—711.
Noble, B., 1969. *Applied Linear Algebra.* Prentice-Hall, Inglewood Cliffs, N.J., 523 pp.
Peachey, C.G., 1960. Distribution of water in large-diameter cyclones under operating conditions. *Proc. Int. Min. Congr., London*, pp 147—156.
Pena, F., 1974. Cyprus-Pima pilot plant predictions vs. actual plant results. *A.I.M.E. Annu. Meet., Acapulco, Pap.*, 74-B-353.
Prentice, T.K., 1943. Ball wear in cylindrical mills. *J. Chem. Metall. Min. Soc. S. Afr.*, **43**: 7, 8.
Ralston, A. and Wilf, H.S., 1960. *Mathematical Methods for Digital Computers.* Wiley, New York, N.Y., 293 pp.
Ramsey, T.C., 1976. The Cyprus Pima concentrator. *Proc. A.M. Gaudin Memorial Symp.—A.I.M.E.* (in press).
Rao, T.C., 1966. *The Characteristics of Hydrocyclones and Their Application as Control Units in Comminution Circuits.* Ph.D. thesis, Univ. Queensland.
Reid, K.J., 1965. A solution to the batch grinding equation. *Chem. Eng. Sci.*, **20**: 953—963.
Rietema, K., 1961. The mechanism of the separation of finely dispersed solids in cyclones. In: *Cyclones in Industry.* Elsevier, Amsterdam, pp. 46—63.
Rittinger, R.P. von, 1867. *Textbook of Mineral Dressing.* Ernst and Korn, Berlin.
Rosin, P. and Rammler, E., 1933. The laws governing the fineness of powdered coal. *J. Inst. Fuel*, **7**: 29—36.

Ruangsak, V., 1963. *The Investigation of Pulp Density and Viscosity Effects in a Wet Cyclone.* M.Eng. Sci. thesis, Univ. Melbourne.

Schubert, H., 1975. Discussion on Modelling and scale-up of hydrocyclone classifiers by A.J. Lynch and T.C. Rao. *Proc. 11th Int. Miner. Process. Congr.*, p. 9.

Schubert, H. and Neesse, T., 1973. The role of turbulence in wet classification. *Proc. 10th Int. Miner. Process. Congr.*, pp. 213—239.

Schuhmann, R., 1960. Energy input and size distribution in comminution, *Trans. A.I.M.E.*, **217**: 22—25.

Stanley, G.G., 1972. Autogenous milling plant design and a mathematical model of an autogenous mill, M.E. thesis, University of Queensland.

Stanley, G.G., 1974. The autogenous mill. A mathematical model derived from pilot- and industrial-scale experiment, Ph.D. thesis, University of Queensland.

Stump, N.W. and Roberts, A.N., 1974. On-stream analysis and computer control at the New Broken Hill Consolidated Limited concentrator. *Trans. A.I.M.E.*, **256**: 143—148.

Tarjan, G., 1962. Discussion on "Hydrocyclone fundamentals" by E.O. Lilge. *Trans. Inst. Min. Metall.*, **71**: 539—546.

Tarr, D.T., 1972. The influence of variables on the separation of solid particles in hydrocyclones. *Proc. 33rd Min. Symp. Univ. Minn., Duluth*, pp. 64—77.

Thornte, W.L., Whaley, H.P. and Morawski, F.R., 1964. Erie Mining Company milling taconite at Hoyt Lakes, Minnesota. In: N. Arbiter (Editor), *Milling Methods in the Americas.* Gordon and Breach, New York, N.Y., pp. 231—256.

Watson, D., Cropton, R.W.G. and Brookes, G.F., 1970. Modelling methods for a grinding/classification circuit and the problem of plant control. *Trans. Inst. Min. Metall.*, **79**: C112—C119.

Weatherburn, C.E., 1961. *A First Course in Mathematical Statistics.* University Press, Cambridge, 271 pp.

Whitby, K.T., 1958. The mechanics of fine sieving. *Symp. Particle Size Measurement—A.S.T.M. Spec. Tech. Publ.*, **234**: 3—24.

Whiten, W.J., 1971a. Model building techniques applied to mineral treatment processes. *Proc. Symp. Automatic Control Systems Miner. Process. Plants—Aust. Inst. Min. Metall., S. Qld. Branch*, pp. 129—148.

Whiten, W.J., 1971b. The use of multi-dimensional cubic spline functions for regression and smoothing. *Aust. Computer J.*, **3**: 81—88.

Whiten, W.J., 1972a. *Simulation and Model Building for Mineral Processing.* Ph.D. thesis, Univ. Queensland.

Whiten, W.J., 1972b. Investigation of optimal controller settings for a delay line. *Aust. J. Inst. Control*, **28**: 33—35.

Whiten, W.J., 1972c. A model for simulating crushing plants. *J. S. Afr. Inst. Min. Metall.*, **72**: 257—264.

Whiten, W.J. and Roberts, A.N., 1974. Control of a multi-stage grinding circuit. *Trans. Inst. Min. Metall.*, **83**: C209—C212.

Wickham, P., 1972. *Comminution of Pebbles and Fine Ore.* M. Eng. Sci. thesis, Univ. Queensland.

Wiegel, R.L., 1964. *A Mathematical Model for Mineral Liberation by Size Reduction.* M.S. thesis, Carnegie Inst. Technology.

Wiegel, R.L., 1965. A quantitative approach to mineral liberation. *Proc. 7th Int. Min. Proc. Congr. New York*, pp. 19—31.

Wiegel, R.L., 1972. Advances in mineral processing material balances. *Can. Metall. Q.*, **11**: 413—424.

Wiegel, R.L., 1975. Liberation in magnetite iron formations. *Trans. A.I.M.E.*, **258**: 247—256.

Wiegel, R.L. and Li, K., 1967. A random model for mineral liberation by size reduction. *Trans. A.I.M.E.*, **238**: 179—189.

Williamson, J.E., 1960. The automatic control of grinding medium in pebble mills. *J. S. Afr. Inst. Min. Metall.*, **60**: 335—345.
Yoshioka, N. and Hotta, Y., 1955. Liquid cyclone as a hydraulic classifier. *Chem. Eng. Jpn.*, **19**: 632—640.

Author Index*

*Author names in text (not including Reference List).

Subject Index